Conference Proceedings and Lecture Notes in Geometry and Topology

Editors
N. Hitchin
R. Kirby
J. Wolf

Volume I	*Lectures on Differential Geometry* R. Schoen and S.-T.. Yau
Volume II	*Lectures on Harmonic Maps* R. Schoen and S.-T.. Yau
Volume III	*Lectures in Low Dimensional Topology* Edited by K. Johannson
Volume IV	*Geometry, Topology, and Physics for Raoul Bott* Edited by S. -T. Yau

International Press Publications

Mathematical Physics

Quantum Groups: From Coalgebras to Drinfeld Algebras
Steven Shnider and Shlomo Sternberg
75 Years of Radon Transform
edited by Simon Gindikin and Peter Michor
Perspectives in Mathematical Physics
edited by Robert Penner and S.-T. Yau
Essays On Mirror Manifolds
edited by S. T. Yau
Mirror Symmetry II
edited by Brian Greene
XIth International Congress on Mathematical Physics
edited by D. Iagolnitzer
Differential Equations and Mathematical Physics
edited by I. Knowles

Number Theory

Elliptic Curves, Modular Forms, and Fermat's Last Theorem
edited by John Coates and Shing Tung Yau

Geometry and Topology

L^2 Moduli Spaces with 4-Manifolds with Cylindrical Ends
by Clifford Henry Taubes
The L^2 Moduli Space and a Vanishing Theorem for Donaldson Polynomial Invariants
by J. Morgan, T. Mrowka, and D. Ruberman
Algebraic Geometry and Related Topics
edited by J.-H. Yang, Y. Namikawa, and K. Ueno
Lectures on Harmonic Maps
by R. Schoen and S.-T. Yau
Lectures on Differential Geometry
by R. Schoen and S.-T. Yau
Geometry, Topology and Physics for Raoul Bott
edited by S.-T. Yau
Lectures on Low-Dimensional Topology
edited by K. Johannson
Chern, A Great Geometer
edited by S.-T. Yau

Surveys in Differential Geometry
edited by C.C. Hsiung and S.-T. Yau

Analysis

Proceedings of the Conference on Complex Analysis
edited by Lo Yang
Integrals of Cauchy Type on the Ball
by S. Gong
Advances in Geometric Analysis and Continuum Mechanics
edited by P. Concus and K. Lancaster
Lectures on Nonlinear Wave Equations
by C. D. Sogge

Physics

Physics of the Electron Solid
edited by S.-T. Chui
Proceedings of the International Conference on Computational Physics
edited by D.H. Feng and T.-Y. Zhang
Chen Ning Yang, A Great Physicist of the Twentieth Century
edited by S.-T. Yau
Yukawa Couplings and the Origins of Mass
edited by Pierre Ramond

Current Developments in Mathematics, *1995*
Collected and Selected Works

The Collected Works of Freeman Dyson
The Collected Works of C. B. Morrey
The Collected Works of P. Griffiths
V. S. Varadarajan

Journals
Communications in Analysis and Geometry
Mathematical Research Letters
Methods and Applications of Analysis

International Press Incorporated, Boston
P.O. Box 2872
Cambridge, MA 02238-2872
USA

Schoen, Richard and Yau, Shing-Tung
Lectures on Harmonic Maps

ISBN 1-57146-002-0

Typeset using the TeX system
Printed on acid-free paper, in the United States of America

Conference Proceedings and Lecture Notes in
Geometry and Topology

Volume II

Lectures on Harmonic Maps

R. Schoen
S. T. Yau

International Press

Preface

There is a natural concept of energy associated to maps between manifolds. Critical points of such an energy functional are called harmonic maps. From the beginning harmonic maps were studied in connection with the theory of minimal surfaces. Böchner was the first one to single out the theory of harmonic maps as generalized minimal surfaces. However, the major existence and regularity theory had to wait until C. Morrey solved the famous Plateau problem in the late forties. Morrey's theory had deep influence on all of the later works on harmonic maps defined on two dimensional surfaces. This includes the fundamental work of Sacks-Uhlenbeck on minimal spheres and the works on incompressible minimal surfaces. In the midseventies the authors realized that much of the theory can be applied to the study of Teichmüller theory and Kähler geometry.

Part I of the book is devoted to harmonic maps defined on Riemann surfaces. While we include topics that we find interesting, we do omit a lot of important developments. Most notable is the subject of harmonic maps as exactly solvable model.

The theory of harmonic maps defined on a higher dimensional manifold was not developed until the major breakthrough of Eells and Sampson in the midsixties. Instead of the variational approach, Eells-Sampson used a heat equation argument which has had deep influence on geometry. Regularity theory was developed much later when the target manifold is not necessary negatively curved. In the first two chapters of Part II, we report on such regularity theory even when the target space need not be a nice manifold. In this setting, it was developed by the first author, and jointly with N. Korevaar later. In the early seventies, the first author realized that the theorem of Eells and Sampson could be used to reprove the famous rigidity theorem of Mostow and superrigidity theory of Margulis. It was not until the late eighties that most of these goals were achieved. This is reported in a joint work by J. Jost and the second author in the last chapter of Part II. In the midseventies the authors had already succeeded in applying the theory of harmonic maps to study topology of manifolds of negative curvature; these works are reported in Part II. We regret that many more applications are omitted because of time constraints.

In 1985, the authors presented a series of lectures at the University of California at San Diego on the subject of harmonic maps. Most of these lectures are collected here as Part I. Part II was added much more recently, and comprises part of the thesis of the first author, the work of N. Korevaar and the second author, and applications to geometry by the authors. The final chapter is due to J. Jost and the second author.

The authors would like to extend their special thanks to Jiagyoung Choe for taking and preparing the copious lecture notes that became Part I of this book.

R. Shoen
Stanford University

S.-T. Yau
Harvard University

Table of Contents

PART I

Chapter I
Harmonic Maps for Surfaces

§1. Energy of a Map

Let (M^n, g), (N^k, h) be Riemannian manifolds and let u be a C^1 map from M^n to N^k. For metrics g and h, we use the following notations:

$$ds_M^2 = \sum g_{\alpha\beta}(x)dx^\alpha dx^\beta \ , \quad ds_N^2 = \sum h_{ij}(u(x))du^i du^j \ .$$

The pull-back of h, $u^*(ds_N^2)$ is a symmetric quadratic form such that

$$u^*(ds_N^2) = \sum_{\alpha,\beta}\Big(\sum_{i,j} h_{ij}(u(x))\frac{\partial u^i}{\partial x^\alpha}\frac{\partial u^j}{\partial x^\beta}\Big)dx^\alpha dx^\beta \ .$$

By finding orthonormal vector fields and their dual 1-forms $\omega^1, \dots, \omega^n$, we can diagonalize $u^*(ds_N^2)$ such that $u^*(ds_N^2) = \sum_{\alpha=1}^{n} \lambda_\alpha(\omega^\alpha)^2$. Then interesting invariants are symmetric functions of $\lambda_1, \dots, \lambda_n$. One of them is the trace:

$$|du|^2 \equiv \mathrm{Tr}_{ds_M^2}(u^* ds_N^2) = \sum_{\alpha=1}^{n} \lambda_\alpha \ .$$

This $|du|^2$ is called the *energy density* of u and its coordinate representation is given by

$$|du|^2 = \sum_{i,j,\alpha,\beta} g^{\alpha\beta}(x)h_{ij}(u(x))\frac{\partial u^i}{\partial x^\alpha}\frac{\partial u^j}{\partial x^\beta} \ .$$

Note that the *energy density* of u is independent of the choice of coordinate system of M^n.

Typeset by $\mathcal{A}\mathcal{M}\mathcal{S}$-TEX

Other symmetric functions of λ_α, e.g. $\prod \lambda_\alpha$, do not seem to lead to geometrically interesting variational problems.

We define the *energy functional* $E(u)$ by

$$E(u) = \int_M |du|^2 dv_M$$

where $dv_M = (\det g)^{\frac{1}{2}} dx$ is the volume element of M^n. Then the critical points of E in the space of maps are called *harmonic maps*.

REMARK.

(1) If $N^k = \mathbb{R}^k$ and $h_{ij} = \delta_{ij}$, then

$$E(u) = \sum_{i=1}^{k} \int | \nabla u^i|^2 dv_M \ .$$

Hence u is harmonic if and only if $\triangle_M u^i = 0$, $i = 1, \dots, k$, that is, component functions are harmonic.

(2) If $M = [0,1]$ (in other words, u is a curve in N), then the energy of u is $E(u) = \int_0^1 \left\| \frac{du}{dt} \right\|^2 dt$ and the critical points of E are geodesics with constant speed parametrization. Note that another energy functional E' defined by $E'(u) = \int_0^1 \left\| \frac{du}{dt} \right\| dt$ gives us the same critical points, geodesics. However, these geodesics need not have constant speed parametrization since E' is independent of parametrization.

§2. Harmonic Map Equation

Suppose N^k is isometrically embedded in $\mathbb{R}^K$. Then we can think of u as $u = (u^1, \dots, u^K)$ with pointwise constraint $u(x) \in N^k$ for any

$x \in M$. Hence

$$E(u) = \sum_{i=1}^{K} \int_M |\nabla u^i|^2 dv_M \ .$$

A nice way of thinking harmonic maps is to look for extremals of E with $u(x) \in N$, $\forall x \in M$. In this way we get the *harmonic map equation*

$$\Delta u^i - g^{\alpha\beta} A^i_{u(x)}\left(\frac{\partial u}{\partial x^\alpha}, \frac{\partial u}{\partial x^\beta}\right) = 0 \ , \quad i = 1, \dots, K \ , \qquad (*)$$

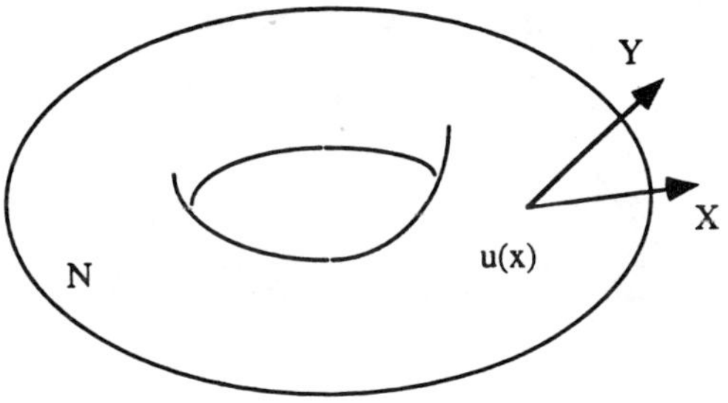

where A is the second fundamental form of N defined by $A_u(X,Y) = (D_X Y)^\perp$. Since the second term of $(*)$ is the normal component, we can simply write $(*)$ as $(\Delta u)^{T_u N} \equiv 0$ with constraint. In local coordinates on N, equation $(*)$ becomes

$$\Delta u^i + g^{\alpha\beta}\Gamma^i_{j\ell}(u(x))\frac{\partial u^j}{\partial x^\alpha}\frac{\partial u^\ell}{\partial x^\beta} \equiv 0 \ , \quad i = 1, \dots, k \ .$$

where Γ^i_{jk} denote the Christoffel symbols of N.

§3. Surface Problems

For the remainder of this chapter we restrict ourselves to the case $n = 2$. The most salient feature in this case is that the energy is conformally invariant: Suppose M is equipped with the metric $g =$

$g_{\alpha\beta}dx^\alpha dx^\beta$ and let $\tilde{g} = e^{2v}g$, $v \in C^\infty(M)$. Then $|\widetilde{du}|^2 = e^{-2v}|du|^2$ and $\sqrt{\det \tilde{g}} = e^{2v}\sqrt{\det g}$, hence $\tilde{E}(u) = E(u)$.

Choose an isothermal coordinate system (x, y) on M such that $g = \lambda(x, y)(dx^2 + dy^2)$, $\lambda > 0$. Let $z = x + iy$. Then the conformal structure obtained from positively oriented isothermal coordinates defines a complex structure so that (M, z) is a Riemann surface with Kähler metric $g = \lambda(z)|dz|^2$. The conformal invariance of energy implies that $E(u)$ depends only on the complex structure of M and not on the metric g of M.

§4. Rado's Theorem

Theorem. *(Rado, 1930's) Assume $\Omega \subset \mathbb{R}^2$ is a convex domain with smooth boundary $\partial\Omega$. Given any homeomorphism $\varphi : S^1 \to \partial\Omega$, there exists a unique harmonic map $u : D \to \Omega$ such that $u = \varphi$ on $\partial D = S^1$ and u is a diffeomorphism.*

Remarks. The map of the Riemann mapping theorem minimizes energy over all orientation preserving homeomorphisms: $D \to \Omega$ which are $1-1$ and continuous on ∂D. If Ω is not convex then the above theorem is not true since folding occurs near the concave part of $\partial\Omega$ if a small neighborhood of a point in ∂D is mapped by φ onto a large concave part of $\partial\Omega$:

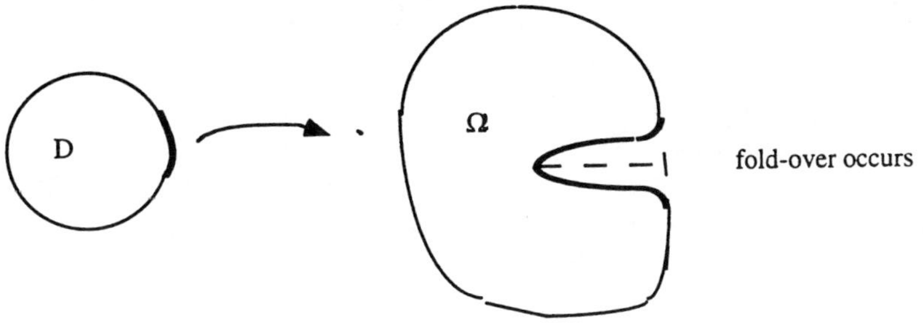

Proof. The existence and uniqueness of u follows from standard PDE theory. Hence it remains to prove that u is a diffeomorphism. (Note that harmonic maps do not necessarily preserve orientation.) It suffices to show $\det J(u) \neq 0$ in D, where $J(u)$ is the Jacobian of u. Suppose

$$\det \begin{pmatrix} \dfrac{\partial u_1}{\partial x} & \dfrac{\partial u_2}{\partial x} \\ \dfrac{\partial u_1}{\partial y} & \dfrac{\partial u_2}{\partial y} \end{pmatrix} = 0$$

at $(x_0, y_0) \in D$. Here we are assuming $u = (u_1, u_2)$, $\varphi = (\varphi_1, \varphi_2)$, and u_1, u_2 are harmonic extensions of φ_1, φ_2 respectively, i.e., $\Delta u_i = 0$ in D and $u_i = \varphi_i$ on S^1. Thus there exists $(a, b) \neq (0, 0)$ such that $\dfrac{\partial}{\partial x}(au_1 + bu_2) = 0$, $\dfrac{\partial}{\partial y}(au_1 + bu_2) = 0$ at (x_0, y_0). Let $h = au_1 + bu_2$. Then h is harmonic and $\nabla h = 0$ at (x_0, y_0). Let $S = \{(x, y) : h(x, y) = h(x_0, y_0)\}$. If $D \sim S$ has a component K which is disjoint from ∂D, then $h = 0$ on ∂K and hence by the maximum principle, $h = 0$ in K, which is a contradiction. Thus $\partial D \cap S$ contains at least 4 points since a neighborhood of (x_0, y_0) consists of at least four arcs emanating from (x_0, y_0), and each of these arcs must extend out to the boundary. Now note that S is the preimage of the line $ax + by = 0$ under u. Since Ω is convex, the line intersects $\partial\Omega$ two points. It follows that u maps at least 4 points on ∂D onto two points in $\partial\Omega$, contradicting the hypothesis (φ being $1-1$). Therefore u must be a diffeomorphism since it is proper and a local diffeomorphism.

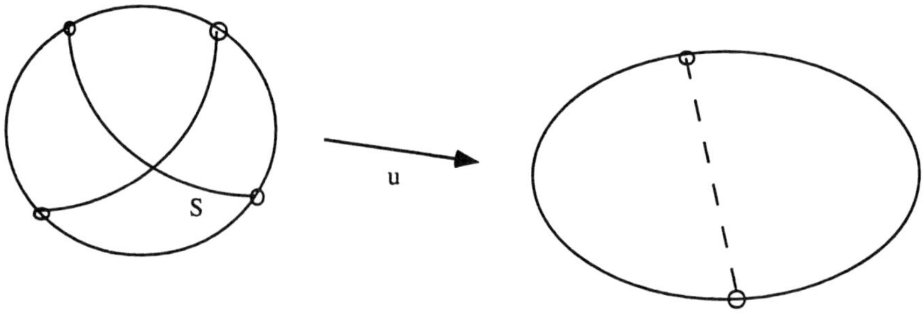

§5. The Hopf Differential

Let $u : M^2 \to N^k$ be harmonic. In normal coordinates centered at $u(x)$, the harmonic map equation reduces to $\triangle u^i(x) = 0$ since $\Gamma^i_{jk}(u(x)) = 0$. On the other hand, $\triangle u^i = 4\dfrac{\partial^2 u^i}{\partial z\,\partial \bar{z}}$. Hence if we define

$$\varphi(z) = \left\langle \frac{\partial u}{\partial z}, \frac{\partial u}{\partial z} \right\rangle ,$$

or equivalently

$$\varphi(z) = \sum_{i,j} h_{ij}(u(z)) \frac{\partial u^i}{\partial z} \frac{\partial u^j}{\partial z}$$

and hence $\dfrac{\partial \varphi}{\partial \bar{z}} = 0$ (i.e. φ is holomorphic) since $\varphi(z)$ is independent of coordinate choice and ∇h_{ij} are zero at $u(z)$ in normal coordinates. Define $\Phi = \varphi(z)dz^2$, which is called the *Hopf differential.* Note that Φ is independent of z whereas φ is not. When u is harmonic, Φ is a *holomorphic quadratic differential.* Since

$$\varphi(z) = \left\langle \frac{\partial u}{\partial z}, \frac{\partial u}{\partial z} \right\rangle = \frac{1}{4}\left(\left|\frac{\partial u}{\partial x}\right|^2 - \left|\frac{\partial u}{\partial y}\right|^2 - 2i\left\langle \frac{\partial u}{\partial x}, \frac{\partial u}{\partial y} \right\rangle \right) ,$$

we see that $\varphi(z) \equiv 0$ if and only if u is conformal. Hence $\Phi \equiv 0$ if and only if $u(M)$ is a *branched minimal surface* in N (because harmonic & conformal $\Leftrightarrow$ minimal).

Proposition. *A harmonic map $u : S^2 \to N^k$ is automatically conformal.*

Proof. We show that there exists no nonzero holomorphic quadratic differential on S^2: Suppose $\Phi = \varphi(z)dz^2$ is a quadratic differential on S^2. Then $\varphi(z)$ is an entire function on $\mathbb{C} = S^2 \sim \{\infty\}$ and φdz^2 is regular at infinity. Let $w = \frac{1}{z}$, and so $\Phi = \varphi(z(w))w^{-4}dw^2$. Hence $\varphi(z(w))w^{-4}$ is regular at $w = 0$ and therefore $\varphi(z(w)) \leq c|w|^4$ near $w = 0$. Then $\varphi(z) \leq c|z|^{-4}$ near $z = \infty$. Thus by the maximum principle $\Phi \equiv 0$ and hence u is conformal.

Corollary.

(1) *A harmonic map $u : S^2 \to S^2$ is holomorphic or antiholomorphic depending on whether u is orientation preserving or reversing.*

(2) *There exists no nontrivial harmonic map from S^2 to a surface N^2 of positive genus, regardless of the metric on N.*

Proof.

(1) By the proposition u is (anti) holomorphic.

(2) Let $\tilde{N}$ be a simply connected covering space of N and $\tilde{u}$ a lifting of u (i.e. $P \circ \tilde{u} = u$). Then $\tilde{N}$ is conformally equivalent to either $\mathbb{C}$ or D, and $\tilde{u}$ is also (anti) holomorphic. Thus in either case $\tilde{u}$ gives rise to a holomorphic function on S^2 which is constant by the maximum principle.

Remark. One can construct a nontrivial harmonic map from T^2 to any surface.

§6. The Complex Form of the Equations

Let M, N be compact oriented surfaces without boundary and u a map from M to N. Let $z = x + iy$, $u = u_1 + iu_2$ be complex coordinates of M, N respectively. Then the metrics of M, N are of the forms $\lambda(z)|dz|^2$, $\rho(u)|du|^2$ respectively. The map u is represented locally by $u = u(z)$. As usual, we denote

$$\frac{\partial}{\partial z} = \frac{1}{2}\Big(\frac{\partial}{\partial x} - i\frac{\partial}{\partial y}\Big), \quad \frac{\partial}{\partial \bar{z}} = \frac{1}{2}\Big(\frac{\partial}{\partial x} + i\frac{\partial}{\partial y}\Big).$$

Then

$$\begin{aligned}
\frac{\partial u}{\partial z} &= \frac{1}{2}\Big[\Big(\frac{\partial u_1}{\partial x} - i\frac{\partial u_1}{\partial y}\Big) + i\Big(\frac{\partial u_2}{\partial x} - i\frac{\partial u_2}{\partial y}\Big)\Big] \\
&= \frac{1}{2}\Big[\Big(\frac{\partial u_1}{\partial x} + \frac{\partial u_2}{\partial y}\Big) + i\Big(\frac{\partial u_2}{\partial x} - \frac{\partial u_1}{\partial y}\Big)\Big], \\
\frac{\partial u}{\partial \bar{z}} &= \frac{1}{2}\Big[\Big(\frac{\partial u_1}{\partial x} - \frac{\partial u_2}{\partial y}\Big) + i\Big(\frac{\partial u_1}{\partial y} + \frac{\partial u_2}{\partial x}\Big)\Big], \\
|du|^2 &= \frac{\rho(u(z))}{\lambda(z)}\Big(\Big|\frac{\partial u}{\partial x}\Big|^2 + \Big|\frac{\partial u}{\partial y}\Big|^2\Big) \\
&= 2\frac{\rho(u(z))}{\lambda(z)}\Big(\Big|\frac{\partial u}{\partial z}\Big|^2 + \Big|\frac{\partial u}{\partial \bar{z}}\Big|^2\Big) = 2(|\partial u|^2 + |\bar{\partial} u|^2), \\
J(u) &= \frac{\rho}{\lambda}\Big(\frac{\partial u_1}{\partial x}\frac{\partial u_2}{\partial y} - \frac{\partial u_2}{\partial x}\frac{\partial u_1}{\partial y}\Big) = |\partial u|^2 - |\bar{\partial} u|^2,
\end{aligned}$$

where $J(u)$ is the Jacobian of u. Hence

$$|du|^2 = 2(2|\partial u|^2 - J(u)) = 4|\partial u|^2 - 2J(u).$$

Now let's derive the *harmonic map equation in complex form.* Let $u_t = u + t\eta$, where η is a C^∞ complex valued function with compact support. Then

$$\begin{aligned}
E(u_t) &= 2\Big(\int_M |\partial u_t|^2 \lambda dx\, dy + \int_M |\bar{\partial} u_t|^2 \lambda dx\, dy\Big) \\
&= 4\int_M |\partial u_t|^2 \lambda dx\, dy - 2\int_M J(u_t)\lambda dx\, dy.
\end{aligned}$$

Note that $\int_M J(u_t)\lambda dx\,dy = \deg(u_t)\cdot \mathrm{vol}(N)$. Hence

$$\begin{aligned}
\frac{d}{dt}E(u_t)\Big|_{t=0} &= \frac{d}{dt}4\int_M |\partial u_t|^2\lambda dx\,dy\Big|_{t=0}\\
&= 4\int (\rho_u\eta + \rho_{\overline{u}}\overline{\eta})|u_z|^2 + \rho(\eta_z\overline{u}_{\overline{z}} + u_z\overline{\eta}_{\overline{z}})dx\,dy\\
&= 8\int \mathrm{Re}(\rho(u)u_z\overline{\eta}_{\overline{z}} + \rho_{\overline{u}}\overline{\eta}|u_z|^2)dx\,dy\\
&= 8\int \mathrm{Re}[\{-(\rho_u u_{\overline{z}} + \rho_{\overline{u}}\overline{u}_{\overline{z}})u_z - \rho u_{z\overline{z}} + \rho_{\overline{u}}|u_z|^2\}\overline{\eta}]dx\,dy\\
&= 8\int \mathrm{Re}[(-\rho u_{z\overline{z}} - \rho_u u_{\overline{z}}u_z)\overline{\eta}]dx\,dy\\
&= 0\ , \quad \text{for all } \eta\ .
\end{aligned}$$

Hence we obtain the *harmonic map equation*

$$u_{z\overline{z}} + (\log\rho)_u u_z u_{\overline{z}} = 0\ , \qquad (*)$$

or

$$\overline{u}_{\overline{z}z} + (\log\rho)_{\overline{u}}\overline{u}_{\overline{z}}\overline{u}_z = 0\ ,$$

which is the complex conjugate of the former.

§7. The Bochner Formulae

Let K_M, K_N be the Gaussian curvatures of M, N respectively. Then

$$K_M = -\frac{1}{2}\Delta_M \log\lambda\ , \quad K_N = -\frac{1}{2}\Delta_N\log\rho\ ,$$

since $\Delta_M = \dfrac{4}{\lambda}\dfrac{\partial^2}{\partial z\,\partial\overline{z}}$. Recall that

$$|\partial u|^2 = \frac{\rho}{\lambda}|u_z|^2\ , \quad |\overline{\partial} u|^2 = \frac{\rho}{\lambda}|u_{\overline{z}}|^2\ .$$

Hence we have

$$(\log|\partial u|^2)_{z\overline{z}} = (\log\rho - \log\lambda + \log u_z + \log\overline{u}_{\overline{z}})_{z\overline{z}} \ .$$

Also we have

$$\begin{aligned}
(\log\rho)_{z\overline{z}} &= ((\log\rho)_u u_z + (\log\rho)_{\overline{u}}\overline{u}_z)_{\overline{z}} \\
&= (\log\rho)_{uu} u_z u_{\overline{z}} + (\log\rho)_{u\overline{u}} u_z \overline{u}_{\overline{z}} + (\log\rho)_u u_{z\overline{z}} \\
&\quad + ((\log\rho)_{\overline{u}}\overline{u}_z)_{\overline{z}} \ , \\
-(\log\lambda)_{z\overline{z}} &= \frac{\lambda}{2}K_M \ , \\
(\log u_z)_{z\overline{z}} &= \left(\frac{u_{z\overline{z}}}{u_z}\right)_z = -((\log\rho)_u u_{\overline{z}})_z \qquad \text{(by (*))} \\
&= -(\log\rho)_{uu} u_z u_{\overline{z}} - (\log\rho)_{u\overline{u}}\overline{u}_z u_{\overline{z}} - (\log\rho)_u u_{z\overline{z}} \ , \\
(\log\overline{u}_{\overline{z}})_{z\overline{z}} &= \left(\frac{\overline{u}_{\overline{z}z}}{\overline{u}_z}\right)_{\overline{z}} = -((\log\rho)_{\overline{u}}\overline{u}_z)_{\overline{z}} \ . \qquad \text{(by (*))}
\end{aligned}$$

Therefore, adding up, we get

$$\begin{aligned}
(\log|\partial u|^2)_{z\overline{z}} &= \frac{\lambda}{2}K_M + (\log\rho)_{u\overline{u}}\overline{u}_{\overline{z}} u_z - (\log\rho)_{u\overline{u}}\overline{u}_z u_{\overline{z}} \\
&= \frac{\lambda}{2}K_M - \frac{\rho}{2}K_N(|u_z|^2 - |\overline{u}_z|^2) \\
&= \frac{\lambda}{2}[K_M - K_N J(u)] \ ,
\end{aligned}$$

and hence

$$\boxed{\begin{aligned}
\Delta_M \log|\partial u| &= -K_N J(u) + K_M \ , \\
\Delta_M \log|\overline{\partial} u| &= K_N J(u) + K_M \ .
\end{aligned}}$$

These are related to the *Bochner formula.*

Proposition. *The functions $|\partial u|$, $|\overline{\partial} u|$ are either identically zero or they have isolated zeroes with well-defined order.*

Proof. Let $h = u_z$. Then $h_{\bar z} = -(\log\rho)_u u_{\bar z} h$. Define $g(z) = -(\log\rho)_u u_{\bar z}$ and let $\zeta(z)$ be a local solution of $\dfrac{\partial\zeta}{\partial\bar z} = -g(z)$; for example

$$\zeta(z) = \int_{\mathbb{C}} \frac{\chi(\zeta)g(\zeta)}{z-\zeta}\, d\zeta d\bar\zeta$$

where χ is a real function with support in the chart which is one in a neibhborhood of the origin. Then

$$(he^\zeta)_{\bar z} = ge^\zeta h - e^\zeta hg = 0 \ ,$$

so he^ζ is holomorphic. Hence $he^\zeta = z^{n_p} f(z)$ where n_p is the order of he^ζ at p and $f(z)$ is nonzero near p. This proves the proposition.

Theorem. *Suppose M, N are closed. If $|\partial u|$ is not identically zero, then*

$$\sum_{\substack{p\in M \\ |\partial u|(p)=0}} n_p = -\deg(u)(2g_N - 2) + (2g_M - 2) \ .$$

If $|\bar\partial u|$ is not identically zero, then

$$\sum_{\substack{p\in M \\ |\bar\partial u|(p)=0}} m_p = \deg(u)(2g_N - 2) + (2g_M - 2) \ ,$$

where n_p, m_p are the orders of $|\partial u|$, $|\bar\partial u|$ at p respectively.

Proof. Since the zeroes of $|\partial u|$ are isolated and M is compact, there exist a finite number of zeroes $p_1, \dots, p_k$ in M. For $\varepsilon > 0$ small enough, let $M_\varepsilon = M \sim \bigcup_{j=1}^{k} D_\varepsilon(p_j)$, where $D_\varepsilon(p_j)$ is the disk of radius ε centered at p_j. From the Bochner formula we have

$$\int_{M_\varepsilon} \triangle \log|\partial u| dv_M = -\int_{M_\varepsilon} K_N J(u) dv_M + \int_{M_\varepsilon} K_M dv_M \ .$$

By the divergence theorem, we have

$$\int_{M_\varepsilon} \triangle \log|\partial u| dv_M = -\sum_{j=1}^{k} \int_{C_\varepsilon(p_j)} \frac{\partial}{\partial r} \log|\partial u| r d\theta \ ,$$

where $\dfrac{\partial}{\partial r}$ is the radial derivative and $C_\varepsilon(p_j) = \partial D_\varepsilon(p_j)$. But $|\partial u| = |z|^{n_{p_j}} g_j(z)$ in $D_s(p_j)$, where $g_j(z)$ is a C^∞ positive function. Hence for each j,

$$\int_{C_\varepsilon(p_j)} \frac{\partial}{\partial r} \log |\partial u| r \, d\theta = 2\pi \cdot n_{p_j} + O(\varepsilon) \ .$$

Therefore, by letting $\varepsilon \to 0$, we get

$$\int_M \triangle \log |\partial u| dv_M = -2\pi \sum_{j=1}^{k} n_{p_j} = -2\pi \sum_{\substack{p \in M \\ |\partial u|(p)=0}} n_p \ .$$

On the other hand,

$$\begin{aligned} &-\int_M K_N J(u) dv_M + \int_M K_M dv_M \\ &= -2\pi \deg(u) \cdot (2 - 2g_N) + 2\pi(2 - 2g_M) \ . \end{aligned}$$

Therefore,

$$\sum_{\substack{p \in M \\ |\partial u|(p)=0}} n_p = -\deg(u)(2g_N - 2) + (2g_M - 2) \ .$$

The second statement can be obtained similarly.

Corollary. *Suppose $N = S^2$, and M, S^2 are equipped with arbitrary metrics. If $u : M \to S^2$ is harmonic and* $\deg(u) > g_M - 1$, *then u is holomorphic.*

Proof. Suppose $|\bar{\partial} u| \not\equiv 0$. Then

$$\begin{aligned} 0 \le \sum_{\substack{p \in M \\ |\bar{\partial} u|(p)=0}} m_p &= \deg(u)(0 - 2) + (2g_M - 2) \\ &< -2(g_M - 1) + (2g_M - 2) = 0 \ , \end{aligned}$$

which is a contradiction. Therefore $|\bar{\partial} u| \equiv 0$ and so u must be holomorphic.

Corollary. *There is no harmonic map from T^2 to S^2 with* $\deg(u) = 1$.

Proof. Suppose u is such a harmonic map. Since $\deg(u) = 1 > genus(T^2) - 1 = 0$, u is holomorphic by the previous corollary, i.e., $|\overline{\partial} u| \equiv 0$. Hence $J(u) \geq 0$. Since u has degree 1, we then see that $\#(u^{-1}(q)) = 1$, for any regular value $q \in S^2$. On the other hand, u is holomorphic, and hence is a branched covering; thus $J(u) > 0$ everywhere and u is a diffeomorphism. This contradiction shows that there is no such u.

The following result shows that harmonic maps which preserve orientation are necessarily branched coverings.

Proposition. *Let $u : \Omega \subset M \to N$ be harmonic on an open connected set Ω of M. If $J(u) \geq 0$ and $J(u)$ is not identically zero in Ω, then the zeros of $J(u)$ are isolated. Moreover if there is a number ℓ such that $\#(u^{-1}(q)) \leq \ell$ for any regular value $q \in N$, then any zero of $J(u)$ is a nontrivial branch point of u.*

Proof. By the Bochner formula,

$$\triangle \log \frac{|\partial u|}{|\overline{\partial} u|} = -2K_N J(u) \;, \quad \text{whenever} \quad |\partial u| \;, \; |\overline{\partial} u| \neq 0 \;.$$

Assume p is a zero of $J(u)$. Since $J(u) = |\partial u|^2 - |\overline{\partial} u|^2$, we get $|\partial u(p)|^2 = |\overline{\partial} u(p)|^2$. We claim that $\partial u(p) = \overline{\partial} u(p) = 0$. To see this, suppose $|\partial u|^2 > 0$, $|\overline{\partial} u|^2 > 0$ in a neighborhood of p. We again use the formula

$$\triangle \log \frac{|\partial u|}{|\overline{\partial} u|} = -2K_N |\overline{\partial} u|^2 \Big(\frac{|\partial u|^2}{|\overline{\partial} u|^2} - 1 \Big) \;.$$

Since $J(u) \geq 0$, $\dfrac{|\partial u|^2}{|\overline{\partial} u|^2} \geq 1$. Hence there exist positive constants $\tilde{c}$, c such that in a neighborhood of p,

$$-2K_N J(u) \leq \tilde{c} \Big(\frac{|\partial u|^2}{|\overline{\partial} u|^2} - 1 \Big) \leq c \Big(\log \frac{|\partial u|^2}{|\overline{\partial} u|^2} \Big) \;.$$

Therefore we have

$$\triangle \log \frac{|\partial u|^2}{|\overline{\partial} u|^2} \leq c\Big(\log \frac{|\partial u|^2}{|\overline{\partial} u|^2}\Big) , \quad c > 0 .$$

Then, by Lemma 6' in [Heinz] (J. D'Analy. Math. 5 (1956/57)), there exist positive constants c', R such that

$$\int_{|z|\leq R} \log \frac{|\partial u|^2}{|\overline{\partial} u|^2} dv \leq c' \log \frac{|\partial u(0)|^2}{|\overline{\partial} u(0)|^2} = 0 .$$

Since $\log \frac{|\partial u|^2}{|\overline{\partial} u|^2} \geq 0$, $\log \frac{|\partial u|^2}{|\overline{\partial} u|^2} \equiv 0$. Hence $J(u) \equiv 0$ in a neighborhood of p. By connectedness of Ω we can conclude that $J(u) \equiv 0$ in Ω. Hence we must have $\partial u(p) = \overline{\partial} u(p) = 0$, proving our claim. This claim implies that the zero set of $J(u)$ and the common zero set of $|\overline{\partial} u|$, $|\partial u|$ are identical. Since the zeros of $|\partial u|$, $|\overline{\partial} u|$ are isolated (Proposition, p.10), the zeros of $J(u)$ are also isolated.

Suppose now that p is an isolated zero of $J(u)$. Then $J(u) > 0$ in a punctured neighborhood of p. By the hypothesis, p is an isolated point of $u^{-1}(u(p))$. Let V be a neighborhood of p such that $J(u) > 0$ in $V \sim \{p\}$ and $V \cap u^{-1}(u(p)) = \{p\}$. By a theorem of Lewy, if $u|_v$ has degree 1, then $J(p) > 0$, which is a contradiction. Therefore $\deg(u) > 1$ in V and so p is a nontrivial branch point of u.

REMARK.

(1) Although there is no harmonic map from T^2 to S^2 with degree 1, we can construct a sequence of maps $u_i : T^2 \to S^2$ whose energy converge to the minimum ($= 4\pi$):

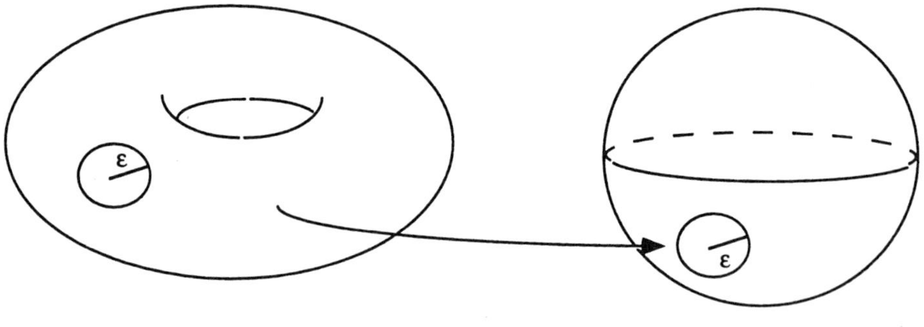

Let $u_i : T^2 \to S^2$ be such that $u_i(D_{\varepsilon_i}) = S^2 \sim D'_{\varepsilon_i}$ and $u_i(T^2 \sim D_{\varepsilon_i}) = D'_{\varepsilon_i}$, where D_{ε_i}, D'_{ε_i} are disks in T^2, S^2 with radius ε_i. As $\varepsilon_i \to 0$, the energy of $u_i \to 4\pi$, but $\{u_i\}$ has no convergent subsequence.

(2) The harmonic maps of degree zero from T^2 to S^2 play an important role in the theory of surfaces of constant mean curvature in R^3 since they are the possible Gauss maps for such surfaces when they are parmetrized conformally. In light of the constructions by Wente and others of constant mean curvature tori in R^3, it would be interesting to obtain a more thorough understanding of the degree zero harmonic maps from a torus to S^2.

§8. When is a Harmonic Map a Diffeomorphism?

The following theorem gives a general criterion under which a harmonic map between surfaces is a diffeomorphism.

Theorem. *Suppose* $g_M = g_N = g$, $\deg(u) = 1$, *u harmonic, and* $K_N < 0$. *Then u is a diffeomorphism.*

Proof. Let $w = \log \dfrac{|\partial u|}{|\bar\partial u|}$. Then

$$\begin{aligned} \triangle w &= -2K_N(|\partial u|^2 - |\bar\partial u|^2) \\ &= -2K_N|\partial u|\cdot|\bar\partial u|\left(\frac{|\partial u|}{|\bar\partial u|} - \frac{|\bar\partial u|}{|\partial u|}\right). \end{aligned}$$

If we define w by

$$e^w = \frac{|\partial u|}{|\bar\partial u|} \quad \text{and note that} \frac{1}{4}|\varphi| = |\partial u|\cdot|\bar\partial u|$$

(recall the Hopf differential $\Phi = \varphi(z)dz^2$). We thus have derived the formula $\triangle w = -K_N|\varphi| \sinh w$.

Now we claim $J(u) \geq 0$, i.e. $w \geq 0$. By the first formula of the Theorem on p.10, we see that $|\partial u|$ has no zeros. It follows that w is bounded below, and achieves its minimum value (unless $\overline{\partial} u \equiv 0$, in which case the theorem is clearly true). Let p_0 be a minimum point of w. Then $\triangle w(p_0) \geq 0$. Hence $-K_N(u(p_0))|\varphi(p_0)| \sinh w(p_0) \geq 0$. Since w is finite at p_0, we must have $|\overline{\partial} u(p_0)| > 0$. It follows that $\varphi(p_0) \neq 0$, and hence $w(p_0) \geq 0$. By the strong maximum principle, we have $w(p_0) > 0$. Hence $J(u) > 0$, and it follows that u is a covering map and therefore a diffeomorphism since it is of degree 1.

§9. Maps Between Hyperbolic Surfaces

The following result of J. Sampson shows that for harmonic diffeomorphisms of hyperbolic surfaces, the Hopf differential determines the image surface uniquely up to an isometry.

Theorem. *Let u_1, u_2 be harmonic maps of degree 1 from (M, z) to (M, ρ_1) and (M, ρ_2) respectively. Suppose $K_{\rho_1} \equiv K_{\rho_2} \equiv -1$. If $\Phi_{u_1} \equiv \Phi_{u_2}$, then $(u_1 \circ u_2^{-1})^* \rho_1 = \rho_2$, i.e., (M, ρ_1) is isometric to (M, ρ_2). (u_1, u_2 are not required to be homotopic to the identity.)*

Proof. Let

$$w_1 = \log \frac{|\partial u_1|}{|\overline{\partial} u_1|}, \quad w_2 = \log \frac{|\partial u_2|}{|\overline{\partial} u_2|}.$$

By hypothesis we have $\varphi(z) = 4\rho_1(u_1(z)) u_{1_z} \overline{u}_{1_z} = 4\rho_2(u_2(z)) u_{2_z} \overline{u}_{2_z}$. Hence $\triangle w_i = |\varphi| \sinh w_i$, $i = 1, 2$. Note that the poles of w_1, w_2

occur at the zeros of φ. Hence $\dfrac{|\partial u_i|}{|\overline{\partial} u_i|} = |a(z)| \cdot |z-p|^{-m_p}$, where $a(z)$ is smooth and $a(p) \neq 0$. Therefore $w_i(z) = m_p \log \dfrac{1}{|z-p|} +$ regular function. Hence $v \equiv w_1 - w_2$ is smooth. We claim $v \equiv 0$. Suppose v is positive somewhere, and let $\Omega_+ = \{x \in M, v(x) > 0\}$. In $\Omega_+, \Delta v = |\varphi| \, (\sinh w_1 - \sinh w_2) > 0$. Moreover Δv is smooth (because $\sinh w_1 - \sinh w_2$ consists of Jacobian and curvature terms). Hence v is subharmonic and $v \equiv 0$ and $\partial\Omega_+$. By the strong maximum principle, $v \equiv 0$ in Ω_+, which is a contradiction. Therefore $v \leq 0$ everywhere. Similarly we can conclude $v \geq 0$ everywhere. Thus $v \equiv 0$, and $w_1 \equiv w_2$; i.e. $\dfrac{|\partial u_1|}{|\overline{\partial} u_1|} \equiv \dfrac{|\partial u_2|}{|\overline{\partial} u_2|}$. On the other hand, $|\partial u_1| \cdot |\overline{\partial} u_1| \equiv |\partial u_2| \cdot |\overline{\partial} u_2|$ since $\varphi(z) \equiv 4\rho_1 u_{1_z} \overline{u}_{1_z} \equiv 4\rho_2 u_{2_z} \overline{u}_{2_z}$. Hence we get

$$|\partial u_1| \equiv |\partial u_2| \quad \text{and} \quad |\overline{\partial} u_1| \equiv |\overline{\partial} u_2| \; . \tag{$*$}$$

Therefore the energy densities of u_1 and u_2 are equal as functions.

Now let $u : (M, z) \to (M, \rho)$, $\rho = \rho(u)|du|^2$. Then

$$\begin{aligned} u^* \rho &= \rho(u(z)) |u_z dz + u_{\overline{z}} d\overline{z}|^2 \\ &= \rho(u(z))(|u_z|^2 + |u_{\overline{z}}|^2)|dz|^2 + \\ &\qquad \rho(u(z))(u_z \overline{u}_z dz^2 + \overline{u}_{\overline{z}} u_{\overline{z}} (d\overline{z})^2) \\ &= \rho(|u_z|^2 + |u_{\overline{z}}|^2)|dz|^2 + \frac{1}{2}\mathrm{Re}(\Phi) \; . \end{aligned}$$

(Note: This is a nice way of introducing the *Hopf differential*. This formula makes it clear that the vanishing of the Hopf differnetial is equivalent to the condition that u is a conformal map.) Hence we have

$$\begin{aligned} u_1^* \rho_1 &= \lambda(z)(|\partial u_1|^2 + |\overline{\partial} u_1|^2)|dz|^2 + \frac{1}{2}\mathrm{Re}(\Phi_{u_1}) \; , \\ u_2^* \rho_2 &= \lambda(z)(|\partial u_2|^2 + |\overline{\partial} u_2|^2)|dz|^2 + \frac{1}{2}\mathrm{Re}(\Phi_{u_2}) \; . \end{aligned}$$

By hypothesis and $(*)$, $u_1^*\rho_1 = u_2^*\rho_2$ on M. Therefore $\rho_2 = (u_2^*)^{-1}(u_1^*)\rho_1$ $= (u_1 \circ u_2^{-1})^*\rho_2$ since u_1, u_2 are diffeomorphisms.

§10. Picard Type Problems

Picard showed that any complex analytic function from $\mathbb{C}$ into $S^2 \sim \{p, q, \infty\}$ is constant. This follows from the fact that any such function can be lifted to a holomorphic map from $\mathbb{C}$ to D, which is a contradiction. More generally, it is natural to ask whether $\mathbb{C}$, D are equivalent under harmonic diffeomorphism.

(1) E. Heinz (1950's) proved that there exists no harmonic diffeomorphism from D onto $\mathbb{C}$ with flat metrics. Notice that this result was a key step in his proof ofthe Bernstein Theorem for minimal surfaces in $\mathbb{R}^3$.

(2) It is conjectured that there exists no harmonic diffeomorphism from $\mathbb{C}$ to (D, ρ) with $K_\rho \equiv -1$.

REMARK. There is another conjecture, a generalization of (1): There exists no proper harmonic map from D to $\mathbb{C}$ with flat metric. Such a map $u = (u_1, u_2)$ will give rise to a map $v = (v_1, v_2)$ with $v : D \to \mathbb{C}^2$ and Re $v_i = u_i$. Actually, analytic functions v_1, v_2 exist such that $(v_1, v_2) : D \to \mathbb{C}^2$ is proper. This conjecture is related to some problems in minimal surface theory. For example, an affirmative answer to this conjecture would imply that hyperbolic minimal surfaces in $\mathbb{R}^3$ cannot properly project onto $\mathbb{R}^2$.

Theorem. *There exists no harmonic diffeomorphism from D to (N, ρ), where $K_\rho \geq 0$, and (N, ρ) is complete.*

Proof. Suppose u is a harmonic diffeomorphism from D_r to (N,ρ) and dist $(u(0),\partial(u(D_r)))\geq R$. Then it suffices to show that

$$|du|^2(0)\geq\frac{CR^2}{r^2}\ ,$$

where C is a universal constant. By hypothesis, we have $|\partial u|^2>|\overline{\partial}u|^2\geq 0$ and

$$\triangle\log|\partial u|=-K_\rho J(u)\leq 0\ . \qquad (*)$$

If we define a Riemannian metric λ on D_r by $\lambda=|\partial u|^2|dz|^2$, then $(*)$ implies that its curvature satiafies $K_\lambda\geq 0$. Let γ be a curve in D_r. The length of γ in (D_r,λ)

$$L_\lambda(\gamma)=\int_\gamma|\partial u|ds\ .$$

On the other hand, the length of $u(\gamma)$ in (N,ρ) is

$$L_\rho(u(\gamma))=\int_\gamma\left|\frac{du}{ds}\right|ds\leq\int_\gamma|du|ds$$
$$=\int_\gamma(2|\partial u|^2+2|\overline{\partial}u|^2)^{\frac{1}{2}}ds\leq 2\int_\gamma|\partial u|ds=2L_\lambda(\gamma)\ .$$

Therefore $\mathrm{dist}_\lambda(0,\partial D_r)\geq\frac{1}{2}\mathrm{dist}_\rho(u(0),\partial(u(D_r)))\geq\frac{1}{2}R$. We now need the following estimate on harmonic functions in terms of curvature [Cheng-Yau, CPAM 28, 333-354 (1975)]:

Gradient Estimate Assume $k\geq 0$ and $\mathrm{Ricci}_M\geq -k$ on B_σ, i.e., the smallest eigenvalue of the Ricci tensor with respect to the metric is at least $-k$. If $h>0$ on B_σ satisfies $\Delta h=0$, then there is a constant $C=C(n)$ such that

$$\frac{|\bigtriangledown h|^2}{h^2}\leq\frac{C(n)(1+k\sigma^2)}{(\sigma-r)^2}\quad\text{at every point of } B_\sigma\ .$$

$$(r=\mathrm{dist}(0,p)<\sigma)$$

We apply this estimate with $h(x,y)=x+r$. By the above estimate,

$$\frac{1}{|\partial u|^2(0)}\leq C\frac{r^2}{R^2}\quad\text{or}\quad|\partial u|^2(0)\geq C^{-1}\frac{R^2}{r^2}\ .$$

This proves the theorem.

Chapter II

Compactifications of Teichmüller Space

§1. Introduction

Teichmüller space $\mathfrak{T}_g$ is the space of conformal (or complex) structures on a topological surface M of genus g where two structures are equivalent if there is a conformal map between them which is homotopic to the identity. Teichmüller proved that when $g \geq 2$, $\mathfrak{T}_g$ is homeomorphic to an open $6g-6$ dimensional ball. Moreover, his proof showed that this homeomorphism could be realized by the radial map along geodesic rays from a fixed base point for a certain metric on $\mathfrak{T}_g$.

In particular, this homeomorphism gives a natural way to compactify $\mathfrak{T}_g$ by adding an endpoint to each ray. We denote the resulting closed $6g-6$ dimensional ball by $\overline{\mathfrak{T}}_g^T$. The natural questions to ask are to what extent $\overline{\mathfrak{T}}_g^T$ depends on the base point from which $\mathfrak{T}_g$ was compactified, and whether or not the action of the mapping class group Γ extends continuously to $\overline{\mathfrak{T}}_g^T$. Kerckhoff [K] showed that the geometry along certain rays depends strongly on their base points and that for $g \geq 2$ there is no continuous extension of Γ to $\overline{\mathfrak{T}}_g^T$.

On the other hand, Thurston (see [T], [FLP]) developed other approaches based on the hyperbolic geometry of the surfaces. He used *measured foliations* which are topological objects on a surface and are obtainable from *holomorphic quadratic differentials* without requiring an underlying conformal structure for their definition. Thurston described a way in which a hyperbolic metric would determine a measured

Typeset by $\mathcal{A}_{\mathcal{M}}\mathcal{S}$-TEX

foliation, and he used this map from metrics to measured foliations to glue the topological sphere of projective measured foliations to the topological ball $\Im_g$. Actually he showed that as one goes to infinity in $\Im_g$ the hyperbolic lengths of curves are approximated well by their intersection numbers with a measured foliation. This allows a compactification, $\overline{\Im}_g^{Th}$, of $\Im_g$ to be defined in terms of ratios of hyperbolic lengths; the boundary of $\overline{\Im}_g^{Th}$ is the space of projective equivalence classes of foliations. Since Thurston's compactification of $\Im_g$ is defined in terms of an intrinsic quantity invariant under isotopy, $\overline{\Im}_g^{Th}$ is both independent of any base conformal structure and is acted upon continuously by the mapping class group Γ. It follows that $\overline{\Im}_g^{T}$ and $\overline{\Im}_g^{Th}$ are distinct even though it is true that a ray of quadratic differentials can determined a projective measured foliation and Teichmüller had described a homeomorphism from $\Im_g$ onto the space of holomorphic quadratic differentials. By examining the fixed points of elements of Γ in $\overline{\Im}^{Th}$ (which exist by Brouwer's theorem), Thurston gave a geometric description of a canonical element of each connected component of $\mathit{Diff}^+(M)$.

Meanwhile, harmonic maps between surfaces have been used to study Teichmüller space: There is a unique harmonic map in the homotopy class of the identity from one hyperbolic closed surface to another, and associated to such a map is a holomorphic quadratic differential on the domain surface. By fixing a base domain metric, and varying the range metric over $\Im_g$, one can get a map $\mathcal{H}$ from $\Im_g$ to $\mathcal{Q}_0$, the space of holomorphic quadratic differentials for the fixed base metric. We will seen that this map is a homeomorphism onto all of the holomorphic quadratic differentials of the base. Thus $\mathcal{Q}_0$ provides a parametrization of $\Im_g$. In his thesis [W], M. Wolf shows that the compactification of $\Im_g$ obtained by adding points at infinity to the rays of that space of differentials agrees with the Thurston compactification of $\Im_g$: to a hyperbolic metric Wolf assigns a measured foliation of the quadratic differential assoicated to the harmonic map, and he shows that the set of projective lengths of isotopy classes of curves agrees asymptotically with the set of projective measures of isotopy classes of curves. In the following sections, we will study Wolf's work in detail.

§2. Teichmüller Space

Let M be a smooth closed oriented surface with genus > 1 and metric ρ. Then we can find a unique function σ on M such that $K_{e^\sigma \rho} \equiv -1$, or $(M, e^\sigma \rho)$ has hyperbolic metric. This implies that a complex structure on M gives rise to a unique hyperbolic metric on M. Motivated by this, we define the following:

$\mathcal{M}$ = the space of Riemannian metrics on M .

$\mathcal{M}_{-1}$ = the space of hyperbolic metrics on M .

$\mathrm{Diff}^+(M)$ = the group of orientation preserving diffeomorphisms of M.

$\mathrm{Diff}_0(M)$ = the group of diffeomorphisms isotopic to the identity on M.

$\Gamma = \mathrm{Diff}^+(M)/\mathrm{Diff}_0(M)$, the discrete quotient group .

Then we have the following facts:

$\mathcal{M}$ is an open subset of $\Gamma(T^*M \odot T^*M)$, the set of symmetric 2-tensors on M.

$\mathcal{M}_{-1}$ is an infinite dimensional hypersurface in $\mathcal{M}$.

$\mathrm{Diff}_0(M)$ is the connected component of $\mathrm{Diff}^+(M)$ containing the identity transformation.

$\text{Diff}^+(M)$ and $\text{Diff}_0(M)$ act on $\mathcal{M}_{-1}$ by

$$\varphi : \mathcal{M}_{-1} \to \mathcal{M}_{-1} , \quad \varphi[\rho] = \varphi^*\rho , \quad K_{\varphi^*\rho} = K_\rho \circ \varphi .$$

$\mathcal{M}_{-1}$ is infinite dimensional because of this action.

Now *Teichmüller space* is defined as the quotient space

$$\Im_g = \mathcal{M}_{-1}/\text{Diff}_0(M) ,$$

and the *moduli space* is defined as the quotient space

$$\mathcal{R}_g = \mathcal{M}_{-1}/\text{Diff}^+(M) .$$

Since Γ acts on $\Im_g$, *Teichmüller space* and the *moduli space* are related by

$$\mathcal{R}_g = \frac{\mathcal{M}_{-1}/\text{Diff}_0(M)}{\text{Diff}^+(M)/\text{Diff}_0(M)} = \Im_g/\Gamma .$$

Our goal is to show that Teichmüller space is diffeomorphic to $\mathbb{Q}_M^{6g-6}$, the *space of holomorphic quadratic differentials* on M. In this section we construct a local diffeomorphism from a neighborhood of $[\rho] \in \Im_g$ to an open set in $\mathbb{R}^{6g-6}$. A slice $S_\rho \subset \mathcal{M}_{-1}$ for the action at each $\rho \in \mathcal{M}_{-1}$ is a C^∞ submanifold of $\mathcal{M}_{-1}$ that is transversal (or orthogonal under appropriate metric on $\mathcal{M}_{-1}$) to the orbit $\mathcal{O}_\rho = \text{Diff}_0(M) \cdot \rho$ of $\text{Diff}_0(M)$ through ρ. $\mathcal{O}_p$ is a C^∞ closed submanifold with tangent space $T_\rho \mathcal{O}_\rho$ given by

$$T_\rho\mathcal{O}_\rho = \{h \in \Gamma\,(T^*M \odot T^*M) : h = L_X\rho \ , \ X \quad \text{is a smooth vector field on } M\} ,$$

that is, the symmetric tensors of type $(0,2)$ which are the Lie derivatives of ρ with respect to a smooth vector field. Note that

$$L_X\rho = \frac{d}{dt}(e^{tX})^*(\rho)\big|_{t=0} = \sum_{i,j}(X_{i;j} + X_{j;i})dx^i\,dx^j .$$

Let $\rho \in \mathcal{M}$, $\rho_t = \sum g_{ij}^t dx^i\, dx^j$. Then $\dot{\rho}_t\big|_{t=0} \in \Gamma(T^*M \odot T^*M)$ and $\dot{\rho}_t\big|_{t=0} := h = \sum h_{ij} dx^i\, dx^j$. We now introduce a natural Riemannian metric on $\mathcal{M}$ and $T_\rho\mathcal{M}$, $\langle h_1, h_2\rangle = \int \langle h_1, h_2\rangle_\rho dv_\rho$. With this metric $\mathcal{M}$ and $T_\rho\mathcal{M}$ become infinite dimensional Riemannian manifolds. Then a slice S_ρ is orthogonal to $T_\rho\mathcal{O}_\rho$ under this metric. Let $h \in S_\rho$ and $X^* = \sum X_i\, dx^i$ be the 1 form associated to a vector field X. Then, if $h \perp L_X\rho$ for any X we have (using orthonormal basis with respect to ρ)

$$0 = \int_M \sum_{i,j} h_{ij}(X_{i;j} + X_{j;i})dv = 2\int_M (\sum h_{ij} X_{i;j})dv \ .$$

By integration by parts, we get

$$\int_M \sum_i (\sum_j h_{ij;j}) X_i dv = 0 \ , \quad \text{for any} \quad X \ .$$

Hence $\sum_{j=1}^{2} h_{ij;j} \equiv 0$ (divergence free condition).

On the other hand,

$$K_\rho = -\frac{1}{2} g^{-1}(g_{11,22} - 2g_{12,12} + g_{22,11}) + \text{Quadratic in } \nabla g_{ij}$$

where $\rho = \sum_{i,j} g_{ij} dx^i\, dx^j \in \mathcal{M}_{-1}$, $g = \det(g_{ij})$. Since $K_\rho \equiv -1$ and quadratic terms in ∇g_{ij} vanish at a point in normal coordinates, we obtain

$$g := \det(g) = \frac{1}{2}(g_{11,22} - 2g_{12,12} + g_{22,11}) \ .$$

Moreover $\dot{g} = g\ \mathrm{Tr}_\rho(h)$, where $h_{ij} = \dot{g}_{ij} = \frac{d}{dt} g_{ij}^t\Big|_{t=0}$. Hence $g\ \mathrm{Tr}_\rho(h) = \frac{1}{2}(h_{11,22} - 2h_{12,12} + h_{22,11})$. Since $h_{12,12} + h_{22,22} = 0$, $h_{12,12} + h_{11,11} = 0$, and $g = 1$ at a point, we have $\mathrm{Tr}_\rho(h) = \frac{1}{2}(h_{11,22} + h_{22,22} + h_{11,11} + h_{22,11})$. Therefore $\Delta_\rho \mathrm{Tr}_\rho(h) = 2\mathrm{Tr}_\rho(h)$. By the maximum principle on a closed surface we conclude $\mathrm{Tr}_\rho(h) \equiv 0$ (Trace free condition). Note that this condition expresses the property that the volume form is preserved pointwise to first order.

Thus

$$T_\rho S_\rho = \{h \in \Gamma(T^*M \odot T^*M) : \sum_{j=1}^{2} h_{ij;j} \equiv 0 \ , \ \mathrm{Tr}_\rho h \equiv 0\} \ .$$

For $h \in T_\rho S_\rho$ with $h = h_{11}dx^2 + 2h_{12}dx\,dy + h_{22}dy^2$, $z = x + iy$, define $\Phi_h = (h_{11} - ih_{12})dz^2$. We can easily show that Φ_h is well defined and holomorphic. Conversely, if Φ_h is holomorphic, then h is divergence free and trace free. Let $\mathbb{Q}_\rho$ be the space of holomorphic quadratic differentials on (M, ρ). We know that

$$\dim_{\mathbb{C}} \mathbb{Q}_\rho = 3g - 3 \ , \quad \dim_{\mathbb{R}} \mathbb{Q}_\rho = 6g - 6 \ , \quad g = \text{genus } (M) \ .$$

Therefore it is not difficult to see that $\Im_g$ has the structure of a smooth manifold of dimension $6g - 6$. Moreover the metric $\langle h_1, h_2 \rangle = \int_M \langle h_1, h_2 \rangle_\rho dv_\rho$ on $\mathcal{M}$ gives a natural Riemannian metric on $\Im_g$ (the Weil-Peterson metric), which is incomplete.

§3. $\Im_g$ is Diffeomorphic to $\mathbb{Q}_0^{6g-6}$

Let φ be a degree 1 map from (M, ρ_0) to (M, ρ). If $K_\rho \leq 0$, then by the theorems of Eells and Sampson and Hartman, which will be proved later, there exists a unique harmonic map u_ρ homotopic to φ. Let $K_\rho \equiv -1$ and let P be the equivalence class of (M, ρ) in $\Im_g$. Let's define a map $\mathcal{H}$ from $\Im_g$ to $\mathbb{Q}_0^{6g-6}$, the space of holomorphic quadratic differentials on (M, ρ_0), by $\mathcal{H}(P) = \Phi_{u_\rho}$. To see that $\mathcal{H}$ is well defined, let $\varphi \in \mathrm{Diff}_0(M)$, then $u_{\varphi^*\rho} = \varphi^{-1} \circ u_\rho$. Since φ^{-1} is an isometry from (M, ρ) to $(M, \varphi^*\rho)$, we see that $\Phi_{u^*_\varphi \rho} = \Phi_{u_\rho}$, so $\mathcal{H}$ is well defined. Moreover, by the theorem in Chapter I, §9, $\mathcal{H}$ is $1 - 1$.

Theorem. *$\mathcal{H}$ is a diffeomorphism.*

Proof. Smoothness of $\mathcal{H}$ is left as an exercise. It suffices to prove $\mathcal{H}$ is proper. (That properness implies $\mathcal{H}$ is onto follows from the fact that $\mathfrak{T}_g$ and $\mathcal{Q}_0$ are C^∞ finite dimensional manifolds of the same dimension.) Let $\{P_j\}$ be a divergent sequence in $\mathfrak{T}_g$. We want to show that $\{\mathcal{H}(P_j)\}$ is diverges to infinity in $\mathcal{Q}_0$. If this were not the case, then we would have the sequence $\{\mathcal{H}(P_j)\}$ bounded. Let u_j be the harmonic map from (M, ρ_0) to (M, ρ_j), where (M, ρ_j) is a representative of P_j. By assumption $\{\Phi_{u_j}\}$ is bounded in $\mathcal{Q}_0$. Hence

$$|\Phi_{u_j}| = |\partial u_j| \cdot |\bar{\partial} u_j| \leq C \ , \quad \text{for any} \quad j \ .$$

We can now use the Bochner formulae to show that the first derivatives of the $\{u_j\}$ are uniformly bounded. First we may apply the minimum principle using the formula $\Delta \log |\partial u| = J(u) - 1 \leq |\partial u|^2 - 1$ to conclude that $|\partial u| \geq 1$ everywhere. Combining this with the bound above we find that $|\bar{\partial} u_j| \leq C$. Evaluating the same formula at a maximum point p_0 of $|\partial u_j|$, we find

$$|\partial u_j|^2(p_0) \leq |\bar{\partial} u_j|^2(p_0) + 1 \leq C + 1.$$

Therefore we have a bound on the full energy density of u_j. This implies that the distance between $u_j(x)$ and $u_j(y)$ in (M, ρ_j) is uniformly bounded for any pair of points $x, y \in M$. In other words, $\operatorname{diam}(M, \rho_j) \leq C'$. We now use the following fact: $K_C = \{(M, \rho) \in \mathcal{R}_g : \operatorname{diam}(M, \rho) \leq C\}$ is compact in $\mathcal{R}_g$, the moduli space, i.e., the diameter is a proper function on the moduli space. Let π be the projection from $\mathfrak{T}_g$ onto $\mathcal{R}_g$. Suppose $P_j \to \infty$ in $\mathfrak{T}_g$. Since $\{\pi(P_j)\}$ lies in a compact subset of $\mathcal{R}_g$ there exists a subsequence $\{j'\}$, and $\{\gamma_{j'}\} \subset \Gamma$ such that $\{\sigma_{j'} := \gamma_{j'}^* \rho_{j'}\}$ converges to a hyperbolic metric σ. The sequence of maps $v_{j'} := \gamma_{j'}^{-1} \circ u_{j'}$ is then a sequence of harmonic maps from (M, ρ_0) to $(M, \sigma_{j'})$ with uniformly bounded first derivatives. It follows that a subsequence converges uniformly to a limiting map. In particular, there are infinitely many members of $\{v_{j'}\}$ which are in the same homotopy class. Since $v_{j'}$ is homotopic to $\gamma_{j'}^{-1}$, this is a contradiction to the assumption that the sequence $\{\gamma_j\}$ goes to infinity in Γ. This completes the proof of the theorem.

§4. Compactification of Teichmüller Space

Thurston gave a compactification $\overline{\Im}_g^{Th}$ of $\Im_g$ by gluing the projective measured foliations onto $\Im_g$. (A good reference for this is Fathi, A., Laudenbach F., and Poenaru, V., Traveaux de Thurston sur les surfaces, Astersque (1979): 66-67.) Thurston's compactification of $\Im_g$ does not depend on the choice of base point within $\Im_g$, and the action of Γ on $\Im_g$ extends to $\overline{\Im}_g^{Th}$. This compactification should be contrasted with Teichmüller's compactification of $\Im_g$: by attaching points at infinity to an embedding of $\Im_g$ into $\mathcal{Q}_0$ (different from that described above), one obtains a compactification $\overline{\Im}_{g,0}^{T}$, which depends on the choice of base point in $\Im_g$. Moreover the action of Γ on $\Im_g$ does not extend continuously to $\partial\overline{\Im}_{g,0}^{T}$. [Kerckhoff, S., The asymptotic geometry of Teichmüller space, Topology 19 (1980), 23-41]. In his thesis [Stanford University, 1986], Michael Wolf shows that the compactification of Teichmüller space obtained by adding points at infinity to the rays of $\mathcal{Q}_0$ using the embedding described in the previous sections agrees with the Thurston compactification of Teichmüller space: Precisely, define

$$B\mathcal{Q}_0 = \{\Phi \in \mathcal{Q}_0 : \|\Phi\|_2 < 1\} ,$$

$$S\mathcal{Q}_0 = \{\Phi \in \mathcal{Q}_0 : \|\Phi\|_2 = 1\} ,$$

$$\overline{B\mathcal{Q}_0} = B\mathcal{Q}_0 \cup S\mathcal{Q}_0 \quad \text{with the } \mathcal{Q}_0 \text{ topology} ,$$

and define a map $\tilde{\mathcal{H}} : \mathfrak{T}_g \to B\mathcal{Q}_0$ given by

$$\tilde{\mathcal{H}}(P) = \frac{4\mathcal{H}(P)}{1 + 4\|\mathcal{H}(P)\|} \quad \left(= \frac{4\Phi_{u_\rho}}{1 + 4\|\Phi_{u_\rho}\|} \right) .$$

Clearly $\tilde{\mathcal{H}}$ is a homemorphism onto $B\mathcal{Q}_0$. Using $\tilde{\mathcal{H}}$ to identify $\mathfrak{T}_g$ with $B\mathcal{Q}_0$, we define a new compactification of $\mathfrak{T}_g$,

$$\overline{\mathfrak{T}}^h_{g,0} = \mathfrak{T}_g \cup S\mathcal{Q}_0 = \overline{B\mathcal{Q}_0} .$$

Then $\overline{\mathfrak{T}}^h_{g,0}$ is homeomorphic to $\overline{\mathfrak{T}}^{Th}_g$. It follows that $\overline{\mathfrak{T}}^h_{g,0} = \overline{\mathfrak{T}}^h_g$ is independent of choice of base point (M, ρ_0), and the action of Γ on $\mathfrak{T}_g$ extends continuously to an action on $\overline{\mathfrak{T}}^h_g$. Of course $\overline{\mathfrak{T}}^h_g \not\approx \overline{\mathfrak{T}}^T_g$. Thus this construction gives an internal ray structure to Thurston's compactification of $\mathfrak{T}_g$ from each point P of $\mathfrak{T}_g$.

§5. Measured Foliations

Let M be a compact C^∞ surface of genus $g > 1$, without boundary. A *measured foliation* F on M is a foliation (with a finite number of singularities) with an invariant transverse measure. This means that if the local charts send leaves of F to horizontal arcs in $\mathbb{R}^2$, the transition functions on $\mathbb{R}^2$ are of the form $\phi_{ij} = \big(f(x,y), c \pm y\big)$ where c

is a constant. The singularities which are allowed are "p-pronged saddles", $p \geq 3$. These are topologically the result of glueing p rectangles together along their horizontal edges. (See Figure below) Note that these singularities are the same topologically as those that occur at $z = 0$ in the line field $z^{p-2}dz^2 > 0$.

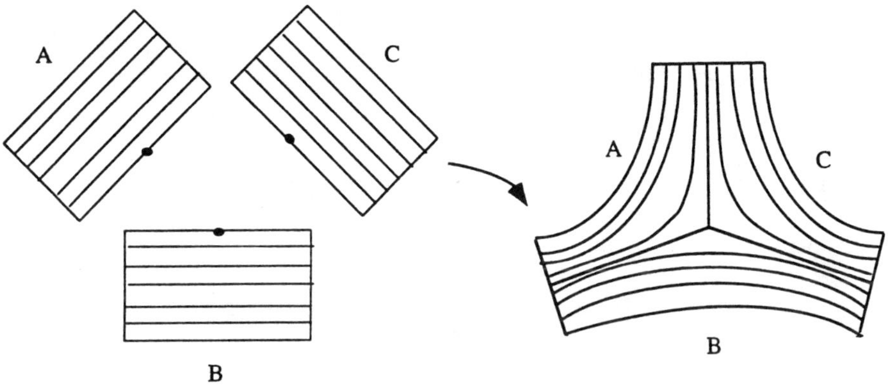

If F is a measured foliation, γ a simple closed curve, then $\int_\gamma F$ is defined to be the total variation of γ in the "y-direction", i.e. the integral of γ with respect to the transverse measure. $i(F,\gamma)$ is the infimum of $\int_{\gamma'} F$ where γ' ranges over all simple closed curves isotopic to γ. $i(F,\gamma)$ is called the *intersection number* of γ with F. Two measured foliation F, F' are *measure equivalent* if, for all simple closed curves γ, $i(F,\gamma) = i(F',\gamma)$. We denote the space of measure classes of foliations by $\mathcal{MF}$. F and F' are *projectively equivalent* if there is a constant b such that $i(F,\gamma) = b\,i(F',\gamma)$ for all γ. The space of projective equivalence classes of foliations is denoted by $\mathcal{PF}$.

Theorem (Thurston). *$\mathcal{MF}$ is homeomorphic to a $6g-6$ dimensional ball; $\mathcal{PF}$ is homeomorphic to a $6g-7$ dimensional sphere.*

Now for any holomorphic quadratic differential $\varphi\, dz^2$ in $\mathcal{Q}_\sigma$ (the space of holomorphic quadratic differentials on (M,σ)), there is a flat metric naturally associated with $\varphi\, dz^2$; locally it is just $|\varphi(z)dz^2|$.

Away from the singularities of $\varphi\, dz^2$ there is a natural conformal coordinate $w = u + iv$ where $dw^2 = \varphi(z)dz^2$. The curves $u =$ constant and $v =$ constant define a pair of foliations with transverse measures $w^*|du|$ and $w^*|dv|$, respectively. These easily extend to where $\varphi(z) = 0$ to give a pair of transverse (singular) measured foliations associated to φdz^2, called the horizontal and vertical foliation of $\varphi\, dz^2$ respectively. In the case where $\varphi\, dz^2$ is determined by a harmonic map, if $z = x + iy$ and $\frac{\partial}{\partial x}, \frac{\partial}{\partial y}$ are tangent to the foliation in a neighborhood $\big($where $\varphi(z) \neq 0\big)$, then $\frac{\partial}{\partial x}$ and $\frac{\partial}{\partial y}$ are the directions of maximal and minimal stretch of the differential dw, and

$$\varphi\, dz^2 = \frac{1}{4}\left(\left\|w_* \frac{\partial}{\partial x}\right\|_\rho^2 - \left\|w_* \frac{\partial}{\partial y}\right\|_\rho^2\right) dz^2$$

with the coefficients of dz^2 determining the transverse measure to the foliations $x =$ constant, $y =$ constant.

Theorem (Hubbard and Masur [HM]). *Given a measured foliation F and a Riemann surface M, there is exactly one quadratic differential on M whose vertical foliation is measure equivalent to F.*

Since $\mathcal{H} : \Im_g \to \mathcal{Q}_\sigma$ is a homeomorphism, the vector space $\mathcal{Q}_\sigma$ provides coordinates for $\Im_g$ centered at σ, as well as some distinguished submanifolds in $\Im_g$. In particular, let Φ_0 be a nonzero holomorphic quadratic differential on (M, σ). Then $t\Phi_0$, $t > 0$ is a ray in $\mathcal{Q}_\sigma$ with vertex at 0. Via the homeomorphism $\mathcal{H}$, $t\Phi_0$ determines a family of metrics ρ_t with $\mathcal{H}(\rho_t) = t\Phi_0$, $\rho_0 = \sigma$ so that ρ_t is divergent in $\Im_g$. On the other hand, $t\Phi_0$ also determines a family of measured foliations on M. Let $F_v(t\Phi_0)$ and $F_h(t\Phi_0)$ denote the vertical and the horizontal measured foliations determined by $t\Phi_0$ respectively.

§6. Asymptotic Relationship between $\{\rho_t\}$ and $\{F_v(t\Phi_0)\}$

In this section we describe an asymptotic relationship between $\{\rho_t\}$ and $\{F_v(t\Phi_0)\}$ as $t \to \infty$ by studying the example of a hyperbolic cylinder M with boundary in which both ρ_t and $F_v(t\Phi_0)$ can be explicity described.

EXAMPLE. Let (M, σ) be the rectangle $[-1,1] \times [0,1]$ in the xy-plane with the metric $ds^2 = dx^2 + dy^2$, and (M, ρ_t) the rectangle $[-\cosh^{-1} t, \cosh^{-1} t] \times [0,1]$ in the uv-plane with the metric $ds^2 = du^2 + \frac{\cosh^2 u}{t^2} dv^2$. Here the top and bottom of each rectangle are identified. Then $K_{\rho_t} = -1$ and $K_\sigma = 0$.

$$w(x,y) = (u,v), \quad u = u(x), \quad v = y$$

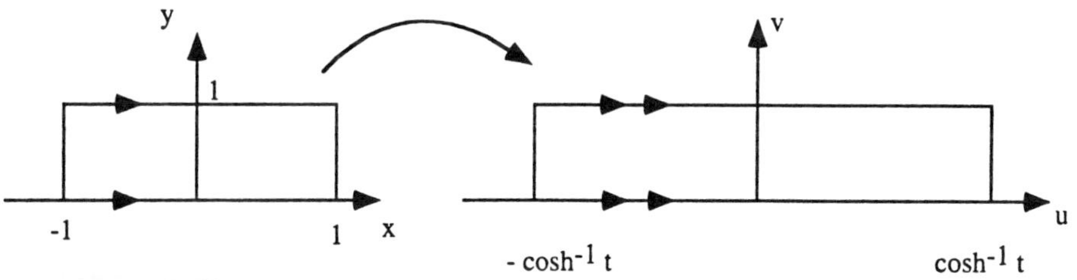

$$ds_\sigma^2 = dx^2 + dy^2 \qquad\qquad ds_{\rho_t}^2 = du^2 + \frac{\cosh^2 u}{t^2} dv^2$$

The metric ρ_t is normalized so that the curves $u = \pm\cosh^{-1} t$ have ρ_t-length 1, while the curve $u = 0$ has ρ_t-length $\varepsilon = \frac{1}{t}$. Let us find a harmonic map $w : (M, \sigma) \to (M, \rho_t)$ with the boundary condition that $w(\pm 1, y) = (\pm\cosh^{-1} t, y)$. By the symmetry of the situation, the desired harmonic map should be of the form $w = u + iv$, $v = y$, $u = u(x)$. In this case the Euler-Lagrange equation becomes

$$\begin{cases} u''(x) = \dfrac{1}{2t^2} \sinh 2u \\ u(0) = 0\,, \quad u(1) = \cosh^{-1} t\,. \end{cases}$$

Let the solution be

$$\begin{cases} u(x) = x\cosh^{-1} t + \beta(x) \\ \beta(0) = 0 = \beta(1) \end{cases}$$

so that $\beta(x)$ represents the deviation of the map from being affine. Then

$$\beta''(x) = \frac{1}{2t^2}\sinh 2u \geq 0 \quad \text{for} \quad u \geq 0 \ .$$

Hence $\beta(x)$ is convex in x, and so $\beta \leq 0$. Now

$$\begin{aligned} 0 \leq \int_0^1 \beta'' &= \beta'(1) - \beta'(0) = u'(1) - u'(0) \\ &= \frac{1}{2t^2}\int_0^1 \sinh 2u\, dx \leq \frac{1}{2t^2}\int_0^1 \sinh(2x\cosh^{-1} t) dx \quad (\because\ \beta < 0) \\ &= \frac{1}{2t^2}\frac{1}{2\cosh^{-1} t}\cosh(2x\cosh^{-1} t)\Big|_0^1 \\ &= \frac{1}{\cosh^{-1} t} - \frac{1}{t^2\cosh^{-1} t} \to 0 \quad \text{as} \quad t \to \infty \ . \end{aligned}$$

Therefore we have $\beta(0) = \beta(1)$, $\beta \leq 0$, and $0 \leq \beta'(1) - \beta'(0) \to 0$ which imply $\beta'(1) \geq 0 \geq \beta'(0)$ and therefore

$$|\beta'(1)| + |\beta'(0)| \to 0 \quad \text{as} \quad t \to \infty \ .$$

This together with convexity of β implies that

$$\beta \to 0 \quad \text{as} \quad t \to \infty \ .$$

So for t very large, $w = u + iv$ deviates very little from an affine stretch. We can rephrase this in terms of vertical measured foliations as follows.

The holomorphic quadratic differential $\mathcal{H}(\rho_t) = \Phi_t$ assoicated to ρ_t is given by

$$\Phi_t = \frac{1}{4}\left(\left\|u_*\frac{\partial}{\partial x}\right\|^2_{\rho_t} - \left\|v_*\frac{\partial}{\partial y}\right\|^2_{\rho_t}\right) dz^2 \ .$$

It has a vertical foliation given by the curves $x =$ constant in the xy-plane, and its vertical transverse measure is

$$\frac{1}{2}\left(\left\|u_*\frac{\partial}{\partial x}\right\|_{\rho_t}^2 - \left\|v_*\frac{\partial}{\partial y}\right\|_{\rho_t}^2\right)^{\frac{1}{2}} |dx| \ ,$$

the coefficient of which, being both real and holomorphic, is a constant. Hence given a horizontal line segment $\gamma : a \leq x \leq b$, the vertical measure of this line is

$$\begin{aligned} i(F_v(\Phi_t), \gamma) &= \int_a^b \frac{1}{2}\left(\left\|u_*\frac{\partial}{\partial x}\right\|_{\rho_t}^2 - \left\|v_*\frac{\partial}{\partial y}\right\|_{\rho_t}^2\right)^{\frac{1}{2}} dx \\ &= \int_a^b \frac{1}{2}\left\|u_*\frac{\partial}{\partial x}\right\|_{\rho_t}\left(1 - \frac{\left\|v_*\frac{\partial}{\partial y}\right\|_{\rho_t}^2}{\left\|u_*\frac{\partial}{\partial x}\right\|_{\rho_t}^2}\right)^{\frac{1}{2}} dx \ , \end{aligned}$$

where

$$\begin{aligned} \frac{\left\|v_*\frac{\partial}{\partial y}\right\|_{\rho_t}^2}{\left\|u_*\frac{\partial}{\partial x}\right\|_{\rho_t}^2} &= \frac{\left\|\frac{\partial}{\partial v}\right\|_{\rho_t}^2}{\frac{\partial u}{\partial x}} = \frac{\frac{\cosh^2 u}{t^2}}{(\cosh^{-1} t + \beta')^2} = \frac{\cosh^2(x\cosh^{-1} t + \beta)}{t^2(\cosh^{-1} t + \beta')^2} \\ &\sim \frac{e^{2x\log t}}{t^2(\cosh^{-1} t + \beta')^2} \sim \frac{t^{2x}}{t^2(\cosh^{-1} t + \beta')^2} \to 0 \quad \text{as} \quad t \to \infty \ . \end{aligned}$$

On the other hand, the ρ_t-length of $w(\gamma)$ is

$$\ell_{\rho_t}(w(\gamma)) = \int_a^b \left\|u_*\frac{\partial}{\partial x}\right\|_{\rho_t} dx \ .$$

Therefore

$$\frac{i(F_v(\Phi_t), \gamma)}{\frac{1}{2}\ell_{\rho_t}(w(\gamma))} \to 1 \ .$$

By constancy of Φ_t with respect to dz^2 (t, fixed), we have

$$i(F_v(\Phi_t), \gamma) = i(c(t)|dx|, \gamma) = c(t)(b - a) \ ,$$

where $c(t)$ is a constant independent of γ. Then the above asymptotic formula implies

$$\frac{1}{b-a}\ell_{\rho_t}(w(\gamma)) \to c \ .$$

Therefore we can conclude that w deviates very little from an affine map: Roughly speaking, vertical lines in xy-plane that are equally spaced with respect to the $w^*(\rho_t)$ metric will converge to vertical lines that are equally spaced with respect to the transverse measure $|dx|$. Moreover we can show that given any arc γ, the ratio of half its ρ_t-length to its measure against the vertical foliation of Φ_t goes to 1 as $t \to \infty$. This means that the asymptotic ratio of ρ_t-lengths of two arcs γ_1 and γ_2 is determined by the ratio of their measures against any Φ_t foliation, which is a topological object on the cylinder. Later we will generalize this situation to compact surfaces of genus $g \geq 2$.

§7. Thurston's and Wolf's Compactifications

First we describe the Thurston compactification $\overline{\Im}_g^{Th}$ of $\Im_g$. ([FLP]). Let $\mathcal{S}$ be the set of free homotopy classes of simple, closed, homotopically nontrival curves on M, a C^∞ compact surface of genus g. Let $\mathcal{MF}^*$ be the space of nontrivial measured foliations on M up to isotopy and Whitehead moves. The following figure illustrates a Whitehead move.

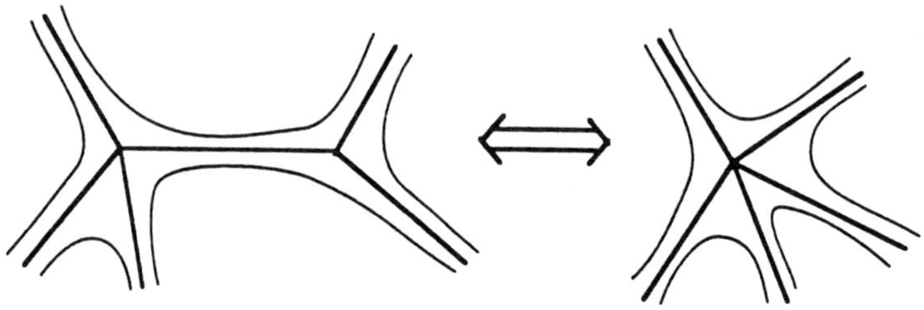

Let $\pi : \mathbb{R}_+^{\mathcal{S}} - \{0\} \to \mathbb{P}(\mathbb{R}_+^{\mathcal{S}})$ be the natural projection onto the projective space of functionals $\mathbb{P}(\mathbb{R}_+^{\mathcal{S}})$. And let ρ be a hyperbolic metric on M. Then for each class $[\gamma] \in \mathcal{S}$, there exists a unique ρ-geodesic

$\gamma \in [\gamma]$ representing $[\gamma]$ on (M,ρ) with ρ-length $\ell_\rho(\gamma)$. This gives rise to a map

$$\ell : \Im_g \to \mathbb{R}_+^{\mathcal{S}} - \{0\} \quad \text{such that} \quad \ell(\rho) = \ell_\rho : \mathcal{S} \to \mathbb{R}_+ \ .$$

Hence $\pi \circ \ell$ is a map from $\Im_g$ into $\mathbb{P}(\mathbb{R}_+^{\mathcal{S}})$. In fact both ℓ and $\pi \circ \ell$ are injections. Using the intersection number of γ with measured foliations, we can also assign a functional $I(F)$ to each nonzero measured foliation $F \in \mathcal{MF}^*$: we define a map

$$I : \mathcal{MF}^* \to \mathbb{R}_+^{\mathcal{S}} - \{0\} \quad \text{such that} \quad I(F)(\gamma) = i(F,\gamma) \ .$$

Then I is also an injection and $\mathcal{PF} = \pi \circ I(\mathcal{MF}^*) \subset \mathbb{P}(\mathbb{R}_+^{\mathcal{S}})$. Finally the Thurston compactification is

$$\overline{\Im}_g^{Th} = \pi \circ \ell(\Im_g) \cup \mathcal{PF}$$

with the topology induced from the topology of $\mathbb{P}(\mathbb{R}_+^{\mathcal{S}})$. Thurston showed that $\mathcal{PF} \approx S^{6g-7}$ and that $\overline{\Im}_g^{Th}$ is a compact manifold with boundary, homeomorphic to a closed ball with boundary sphere $\mathcal{PF}$. By construction, $\overline{\Im}_g^{Th}$ does not depend on the choice of base point $\sigma \in \Im_g$, and since the mapping class group Γ_g acts continuously on $\mathcal{MF}$, its action on $\Im_g$ extends to $\overline{\Im}_g^{Th}$.

Let us now introduce another compactification $\overline{\Im}_g^{h}$ of $\Im_g$. Define

$$B\mathbb{Q}_\sigma = \{\Phi \in \mathbb{Q}_\sigma : \|\Phi\|_2 < 1\} \ , \quad S\mathbb{Q}_\sigma = \{\Phi \in \mathbb{Q}_\sigma : \|\Phi\|_2 = 1\} \ ,$$
$$\overline{B\mathbb{Q}_\sigma} = B\mathbb{Q}_\sigma \cup S\mathbb{Q}_\sigma \quad \text{with the } \mathbb{Q}_\sigma \text{ topology} \ .$$

And we define a map

$$\tilde{\mathcal{H}} : \Im_g \to B\mathbb{Q}_\sigma \quad \text{by} \quad \tilde{\mathcal{H}}(\rho) = \frac{4\mathcal{H}(\rho)}{1 + 4\|\mathcal{H}(\rho)\|_2} \ .$$

Clearly $\tilde{\mathcal{H}}$ is a homeomorphism onto $B\mathcal{Q}_\sigma$. Using $\tilde{\mathcal{H}}$ to identify $\Im_g$ with $B\mathcal{Q}_\sigma$, we define a new compactification of $\Im_g$,

$$\overline{\Im}_g^h(\sigma) = \Im_g \cup S\mathcal{Q}_\sigma = \overline{B\mathcal{Q}_\sigma} .$$

REMARK.

(1) The goal of the remaining sections is to prove $\overline{\Im}_g^h(\sigma) \approx \overline{\Im}_g^{Th}$ so that $\overline{\Im}_g^h(\sigma)$ is independent of choice of base point σ. This requires us to study asymptotic properties of the Beltrami differentials for harmonic maps. It would be interesting to prove directly that $\overline{\Im}_g^h(\sigma)$ is independent of σ without showing $\overline{\Im}_g^h(\sigma) \approx \overline{\Im}^{Th}$.

(2) Suppose $\tilde{\mathcal{H}}_i$ is the map from $\Im_g$ to $B\mathcal{Q}_{\sigma_i}$, $i = 1, 2$. Then $\overline{\Im}_g^h(\sigma)$ being independent of σ implies that $\tilde{\mathcal{H}}_2 \circ (\tilde{\mathcal{H}}_1)^{-1}$ extends as a homeomorphism of $\overline{B\mathcal{Q}_{\sigma_1}}$ onto $\overline{B\mathcal{Q}_{\sigma_2}}$.

§8. Estimates on Stretching

The *Beltrami differential* ν of w is defined by

$$\nu = \frac{\bar{\partial} w}{\partial w} \quad , \text{ hence } \quad |\nu|^2 = \frac{|\bar{\partial} w|^2}{|\partial w|^2} .$$

Classically, much of Teichmüller theory has been based on the study of the Beltrami equation $w_{\bar{z}} = \mu w_z$, where the parameter μ belongs to some infinite dimensional function or tensor space: an appropriate equivalence relation is often required to pass from the parameter solution space to Teichmüller space. This parametrization should be contrasted with that of $\Im_g$ via $\mathcal{Q}_\sigma$, where the parameter space is finite dimensional and no equivalence relation need be imposed.

Proposition. *Let $\Phi_0 \in SQ_\sigma$ and let $\rho_t = \mathcal{H}^{-1}(t\Phi_0)$, $t > 0$. Let $\nu(t)$ be the family of Beltrami differentials determined by the harmonic maps $w_t : (M,\sigma) \to (M,\rho_t)$, $\nu = \dfrac{\bar\partial w_t}{\partial w_t}$.*

(1) *If $\Phi_0(p) \neq 0$, then $|\nu(t)(p)| \uparrow 1$.*

(2) $\dfrac{\|\frac{\partial}{\partial y}\|_{\rho_t}}{\|\frac{\partial}{\partial x}\|_{\rho_t}} \downarrow 0$ *as $t \to \infty$, away from zeroes of Φ_0.*

(3) $\dfrac{\|\frac{\partial}{\partial x}\|_{\mathcal{H}(\rho_t)}}{\frac{1}{2}\|\frac{\partial}{\partial x}\|_{\rho_t}} \uparrow 1$ *as $t \to \infty$, away from zeroes of Φ_0.*

Proof.

(1) Recall that

$$\begin{aligned}
\triangle \log |\bar\partial w| &= |\bar\partial w|^2 - |\partial w|^2 - 1\ , \\
\triangle \log |\partial w| &= |\partial w|^2 - |\bar\partial w|^2 - 1\ , \quad \text{and hence} \\
\triangle \log |\nu(t)| &= 2\big(|\bar\partial w_t|^2 - |\partial w_t|^2\big) \\
&= 2|\partial w| \cdot |\bar\partial w| \cdot \left(\frac{|\bar\partial w|}{|\partial w|} - \frac{|\partial w|}{|\bar\partial w|}\right) \qquad (w_t \text{ is abbreviated as } w.) \\
&= 2|t\Phi_0|\left(|\nu(t)| - \frac{1}{|\nu(t)|}\right) .
\end{aligned}$$

Write $g = \log|\nu(t)|$. Then

$$\begin{aligned}
\Delta g &= 2t|\Phi_0| \quad (e^g - e^{-g}) \\
&= 4t|\Phi_0| \sinh g\ .
\end{aligned}$$

Recall that $J = |\partial w|^2 - |\bar\partial w|^2 > 0$ by Chapter I, §8. Hence

$$|\partial w| > 0\ , \quad \text{and} \quad g \leq 0\ .$$

Since $t|\Phi_0| = |\partial w| \cdot |\bar\partial w|$, we have

$$\nu(t) = \frac{t|\Phi_0|}{|\partial w|^2}\ .$$

Note that $|\Phi_0|$ is independent of t. Therefore it is not difficult to conclude that $\frac{dg}{dt}$ is nonsingular and

$$\left|\frac{dg}{dt}\right| < \infty .$$

Laplacian being independent of time t, we get

$$\triangle\Big(\frac{dg}{dt}\Big) = 4|\Phi_0| \sinh g + 4t(|\Phi_0| \cosh g)\frac{dg}{dt} .$$

Since $g \le 0$, we have

$$\triangle\Big(\frac{dg}{dt}\Big) \le 4t(|\Phi_0| \cosh g)\frac{dg}{dt} .$$

Then by the strong minimum principle on $\frac{dg}{dt}$, we obtain $\min\big(\frac{dg}{dt}\big) > 0$ and so $|\nu(t)|$ is an increasing function with $0 \le |\nu(t)| < 1$. Now let $M_\varepsilon = M \sim \bigcup_{\Phi_0(p)=0} B_\varepsilon(p)$. Then

$$\begin{aligned}
\int_{M_\varepsilon} \frac{(1-|\nu(t)|^2)^{\frac{1}{2}}}{\mathrm{Area}(M_\varepsilon)} d\nu_\sigma &= \int_{M_\varepsilon} \frac{(|\partial w|^2 - |\overline{\partial} w|^2)^{\frac{1}{2}}}{|\partial w| A(M_\varepsilon)} d\nu_\sigma \\
&\le \Big(\int \frac{|\partial w|^2 - |\overline{\partial} w|^2}{A(M_\varepsilon)} d\nu_\sigma\Big)^{\frac{1}{2}} \Big(\int \frac{d\nu_\sigma}{|\partial w|^2 A(M_\varepsilon)}\Big)^{\frac{1}{2}} \\
&\le \Big(\frac{4\pi(g-1)}{A(M_\varepsilon)}\Big)^{\frac{1}{2}} \Big(\int \frac{|\overline{\partial} w|^2 d\nu_\sigma}{|\partial w|^2 A(M_\varepsilon)}\Big)^{\frac{1}{4}} \\
&\quad \Big(\int \frac{d\nu_\sigma}{|\overline{\partial} w|^2 |\partial w|^2 A(M_\varepsilon)}\Big)^{\frac{1}{4}} \\
&\le c \cdot c_\varepsilon \Big(\int \frac{d\nu_\sigma}{t^2 |\Phi_0|^2}\Big)^{\frac{1}{4}} \qquad \Big(\text{because } \frac{|\overline{\partial} w|^2}{|\partial w|^2} < 1\Big) \\
&= c t^{-\frac{1}{2}} .
\end{aligned}$$

Hence $|\nu(t)| \to 1$ a.e. on M_ε as $t \to \infty$.

Next we recall that g is superharmonic on M_ε for all t. So $\lim_{t\to\infty} \log|\nu(t)|$ exists and equals $\log|\nu(\infty)|$. Clearly $\log|\nu(\infty)|$ is still superharmonic. Since $\log |\nu(\infty)| = 0$ a.e. on M_ε, we conclude that

$$\lim_{t\to\infty} |\nu(t)| = |\nu(\infty)| = 1 \quad \text{on} \quad M_\varepsilon \quad \text{for any} \quad \varepsilon > 0 .$$

(2) By construction, the foliations of $\mathcal{H}(\rho_t) = t\Phi_0$ are fixed for all t, but the measures of the measured foliations associated to $t\Phi_0$ change. Let $\frac{\partial}{\partial x}, \frac{\partial}{\partial y}$ be a frame field on (M, σ) (away from the zeroes of Φ_0) tangent to the horizontal and vertical foliations respectively. Then $\frac{\partial}{\partial x}, \frac{\partial}{\partial y}$ are also the directions of maximum and minimum stretch of the differential map dw_t, and so in these coordinates,

$$t\Phi_0 = \frac{1}{4}\left(\left\|\frac{\partial}{\partial x}\right\|_{\rho_t}^2 - \left\|\frac{\partial}{\partial y}\right\|_{\rho_t}^2\right) dz^2 .$$

Moreover, $$|\partial w|^2 - |\bar\partial w|^2 = \text{Jacobian} = \left\|\frac{\partial}{\partial x}\right\|_{\rho_t} \left\|\frac{\partial}{\partial y}\right\|_{\rho_t} ,$$

$$|\partial w|^2 + |\bar\partial w|^2 = \frac{1}{2}\text{energy} = \frac{1}{2}\left(\left\|\frac{\partial}{\partial x}\right\|_{\rho_t}^2 + \left\|\frac{\partial}{\partial y}\right\|_{\rho_t}^2\right) .$$

Therefore $$\nu(t) = \frac{|\bar\partial w|}{|\partial w|} = \frac{\frac{1}{2}\left(\left\|\frac{\partial}{\partial x}\right\|_{\rho_t} - \left\|\frac{\partial}{\partial y}\right\|_{\rho_t}\right)}{\frac{1}{2}\left(\left\|\frac{\partial}{\partial x}\right\|_{\rho_t} + \left\|\frac{\partial}{\partial y}\right\|_{\rho_t}\right)} = \frac{1 - \dfrac{\left\|\frac{\partial}{\partial y}\right\|_{\rho_t}}{\left\|\frac{\partial}{\partial x}\right\|_{\rho_t}}}{1 + \dfrac{\left\|\frac{\partial}{\partial y}\right\|_{\rho_t}}{\left\|\frac{\partial}{\partial x}\right\|_{\rho_t}}}$$

This together with (1) gives (2).

(3) Let $\|\nu\|_{\mathcal{H}(\rho_t)}$ be the norm given by the (singular) flat metric to a vector $\nu \in T_pM$, $\Phi_0(p) \neq 0$. We then notice that

$$\left\|\frac{\partial}{\partial x}\right\|_{\mathcal{H}(\rho_t)}^2 = \frac{1}{4}\left(\left\|\frac{\partial}{\partial x}\right\|_{\rho_t} - \left\|\frac{\partial}{\partial y}\right\|_{\rho_t}\right) .$$

Hence from (2) we obtain the desired result.

Corollary.

(1) $\int_{M_\varepsilon} \|\frac{\partial}{\partial y}\|_{\rho_t} d\nu_\sigma \to 0$ *for any* ε. $\left(M_\varepsilon = M \sim \bigcup_{\Phi_0(p)=0} B_\varepsilon(p)\right)$.

(2) *If* γ *is an arc of the horizontal foliation of* $t\Phi_0$ *not passing through a zero of* Φ_0*, then*

$$0 < c_0 < \ell_{\rho_t}(\gamma) t^{-\frac{1}{2}} < c_1 < \infty \ ,$$

where $\ell_{\rho_t}(\gamma)$ *denotes the* ρ_t*-length of* γ*, and* c_0, c_1 *are constants depending only on* γ.

(3) *If* γ *is an arc of the vertical foliation of* $t\Phi_0$*, then*

$$\ell_{\rho_t}(\gamma) t^{-\frac{1}{2}} \to 0 \quad as \quad t \to \infty \ .$$

Proof.

(1) Since $|t\Phi_0| \le \frac{1}{4}\|\frac{\partial}{\partial x}\|^2_{\rho_t}$ we have

$$\begin{aligned} 4\pi(g-1) &= \int_M J(w_t) d\nu_\sigma = \int_M \left\|\frac{\partial}{\partial x}\right\|_{\rho_t} \left\|\frac{\partial}{\partial y}\right\|_{\rho_t} d\nu_\sigma \\ &\ge \int_M 2t^{\frac{1}{2}} |\Phi_0|^{\frac{1}{2}} \left\|\frac{\partial}{\partial y}\right\|_{\rho_t} d\nu_\sigma \\ &\ge 2t^{\frac{1}{2}} \delta \int_{M_\varepsilon} \left\|\frac{\partial}{\partial y}\right\|_{\rho_t} d\nu_\sigma \ , \end{aligned}$$

which easily gives (1).

(2) $\ell_{\rho_t}(\gamma) = \int_0^1 \|\dot\gamma\|_{\rho_t} ds_\sigma = c\int_0^1 \|\frac{\partial}{\partial x}\|_{\rho_t} ds_\sigma = c\|\frac{\partial}{\partial x}\|_{\rho_t}$. Since $\|\frac{\partial}{\partial x}\|^2_{\rho_t} = 4t|\Phi_0| + \|\frac{\partial}{\partial y}\|^2_{\rho_t}$, we get

$$2t^{\frac{1}{2}}|\Phi_0|^{\frac{1}{2}} \le \left\|\frac{\partial}{\partial x}\right\|_{\rho_t} \le 2t^{\frac{1}{2}}|\Phi_0|^{\frac{1}{2}} + \left\|\frac{\partial}{\partial y}\right\|_{\rho_t} .$$

Consequently,

$$\left\|\frac{\partial}{\partial x}\right\|_{\rho_t} \sim t^{\frac{1}{2}} \quad \text{and} \quad \left\|\frac{\partial}{\partial y}\right\|_{\rho_t} = o\left(\left\|\frac{\partial}{\partial x}\right\|_{\rho_t}\right) = o(t^{\frac{1}{2}}) \ .$$

Thus we obtain (2) and (3) respectively.

§9. Properties of a divergent sequence $\{\rho_n\}$ in $\Im_g$.

Define $\beta : \mathcal{Q}_\sigma \to \mathcal{MF}$ such that $\beta(\Phi)$ is the vertical measured foliation of Φ. Let $\tilde{\beta} = \beta \circ \mathcal{H} \circ \tilde{\mathcal{H}}^{-1} : B\mathcal{Q}_\sigma \to \mathcal{MF}$. Then, for $\tilde{\Phi} \in B\mathcal{Q}_\sigma$,

$$\tilde{\beta}(\tilde{\Phi}) = \beta\Big(\frac{\frac{1}{4}\tilde{\Phi}}{1 - \|\tilde{\Phi}\|_2}\Big) .$$

By a theorem of Hubbard and Masur [HM], β (resp. $\tilde{\beta}$) is a homeomorphism of $\mathcal{Q}_\sigma$ (resp. $B\mathcal{Q}_\sigma$) onto $\mathcal{MF}$. As a consequence, we see that if $\|\tilde{H} \circ \mathcal{H}^{-1}\Phi_n\|_2 \to 1$, then Φ_n converges in the space of projective equivalence classes of $\mathcal{Q}_\sigma$ where two are equivalent if they are postive real numbers of each other

$\Longleftrightarrow$ $\tilde{\mathcal{H}} \circ \mathcal{H}^{-1}\Phi_n$ converges in $B\mathcal{Q}_\sigma$

$\Longleftrightarrow$ $\pi \circ I \circ \beta\Phi_n$ converges in $P(\mathbb{R}^S_+)$.

Lemma. *Let $\rho_n \to \partial\Im_g$. Then for all $[\gamma] \in S$ and for n sufficiently large, there exist k_0, $\eta > 0$ both depending on ρ_n and $[\gamma]$, such that*

$$k_0 i(\beta\mathcal{H}(\rho_n),\ [\gamma]) + \eta \geq \ell_{\rho_n}([\gamma]) \geq i(\beta\mathcal{H}(\rho_n),\ [\gamma])$$

where $k_0 \downarrow 1$ and $\eta\|(\rho_n)\|^{-\frac{1}{2}} \to 0$ as $n \to \infty$.

Proof. First we prove $\ell_{\rho_n}([\gamma]) \geq i(\beta\mathcal{H}(\rho_n),\ [\gamma])$. Let $\gamma : [0,1] \to M$ be a ρ_n-geodesic and let $\dot{\gamma}_h$, $\dot{\gamma}_v$ denote the horizontal and vertical components of $\dot{\gamma}$ relative to the horizontal and vertical foliations. Then

$$\begin{aligned}
\ell_{\rho_n}([\gamma]) = \ell_{\rho_n}(\gamma) &= \int_0^1 \left(\|\dot{\gamma}_h\|^2_{\rho_n} + \|\dot{\gamma}_v\|^2_{\rho_n}\right)^{\frac{1}{2}} ds \\
&\geq \int_0^1 \|\dot{\gamma}_h\|_{\rho_n} ds \geq \int_0^1 \|\dot{\gamma}_h\|_{4\mathcal{H}(\rho_n)} ds \\
&\qquad \Big(\because \Big\|\frac{\partial}{\partial x}\Big\|^2_{\mathcal{H}(\rho_n)} = \frac{1}{4}\Big(\Big\|\frac{\partial}{\partial x}\Big\|^2_{\rho_n} - \Big\|\frac{\partial}{\partial y}\Big\|^2_{\rho_n}\Big)\Big) \\
&= i(\beta\mathcal{H}(\rho_n), \gamma) \geq i(\beta\mathcal{H}(\rho_n),\ [\gamma]) .
\end{aligned}$$

Next we prove $k_0 i(\beta\mathcal{H}(\rho_n),\ [\gamma]) + \eta \geq \ell_{\rho_n}([\gamma])$. Given $[\gamma] \in \mathcal{S}$, let $[\gamma]$ be a piecewise smooth curve such that

(i) all segments of γ are horizontal or vertical,

(ii) γ is vertical near the zeroes of $\mathcal{H}(\rho_n)$, and

(iii) $i(\beta\mathcal{H}(\rho_n), \gamma) = i(\beta\mathcal{H}(\rho_n), [\gamma])$.

By an arguement of [FLP, 5.II.6], conditions (i) and (iii) can always be met. The following figure shows how we

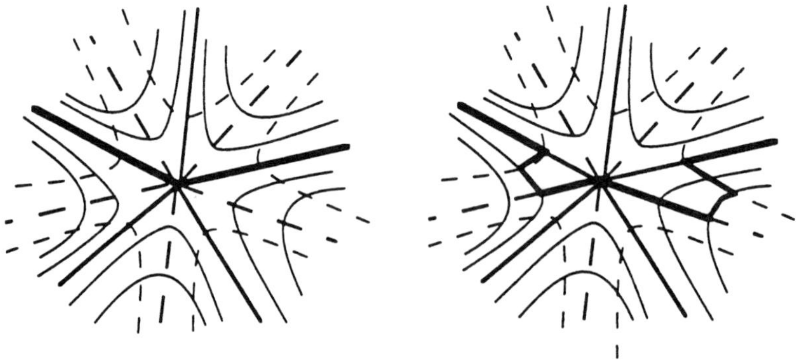

might modify such a curve which does not meet the condition (ii) (i.e., one of its horizontal arcs contains a zero of $\mathcal{H}(\rho_n)$) into a curve which satisfies all three conditions. Thus for each $\Phi_0 \in S\mathcal{Q}_\sigma$, we can find a Φ_0-polygonal representative of $[\gamma]$ whose horizontal arcs lie inside M_ε. Since $S\mathcal{Q}_\sigma$ is compact, we can find $\varepsilon > 0$ depending on $[\gamma]$ such that for any $\Phi \in \mathcal{Q}_\sigma$, there is a Φ-polygonal representative of $[\gamma]$ whose horizontal arcs avoid $\bigcup_{\Phi(p)=0} B_\varepsilon(p)$.

Let $\gamma_h : [a,b] \to M$ be a horizontal arc. Then

$$\begin{aligned}\ell_{\rho_n}(\gamma_h) &= \int_a^b \|\dot{\gamma}_h\|_{\rho_n} ds \\ &= \int_a^b \|\dot{\gamma}_h\|_{4\mathcal{H}(\rho_n)} \cdot \frac{\|\dot{\gamma}_h\|_{\rho_n}}{\|\dot{\gamma}_h\|_{4\mathcal{H}(\rho_n)}} ds\ .\end{aligned}$$

Since by construction γ_h avoids $\bigcup_{\mathcal{H}(\rho_n)(p)=0} B_\varepsilon(p)$, by the proposition of §8,

$$\frac{\|\dot{\gamma}_n\|_{\rho_n}}{\|\dot{\gamma}_h\|_{4\mathcal{H}(\rho_n)}} \downarrow 1$$

uniformly along γ_h. Hence we have

$$\begin{aligned} \ell_{\rho_n}(\gamma_h) &\leq \sup \frac{\|\dot{\gamma}_h\|_{\rho_n}}{\|\dot{\gamma}_h\|_{4\mathcal{H}(\rho_n)}} \int \|\dot{\gamma}_h\|_{4\mathcal{H}(\rho_n)} ds \\ &= k_0 \cdot i(\beta\mathcal{H}(\rho_n), \gamma_h) \end{aligned}$$

with $k_0 \downarrow 1$, depending on ρ_n and $[\gamma]$.

Next let $\gamma_v : [a, b] \to M$ be a vertical arc. Then $i(\beta\mathcal{H}(\rho_n), \gamma_v) = 0$. On the other hand, the corollary of §8 implies that

$$[\ell_{\mathcal{H}^{-1}(t\mathcal{H}(\rho_n)/\|\mathcal{H}(\rho_n)\|)}(\gamma_v)] \cdot t^{-\frac{1}{2}} \to 0 \quad \text{as} \quad t \to \infty .$$

By the compactness of $S\mathbb{Q}_\sigma$, we conclude that there is $\eta > 0$ depending on ρ_n and $[\gamma]$ such that

$$\sum_{\gamma_v} \ell_{\rho_n}(\gamma_v) < \eta \quad \text{and} \quad \eta \cdot \|\mathcal{H}(\rho_n)\|^{-\frac{1}{2}} \to 0 \quad \text{as} \quad \rho_n \to \partial\Im_g .$$

$$\begin{aligned} \ell_{\rho_n}([\gamma]) \leq \ell_{\rho_n}(\gamma) &= \sum_{\gamma_h}(\gamma_h) + \sum_{\gamma_h} \ell_{\rho_n}\ell_{\rho_n}(\gamma_v) \\ &\leq k_0 \sum_{\gamma_h} i(\beta\mathcal{H}(\rho_n),\ \gamma_h) + \eta \\ &= k_0 i(\beta\mathcal{H}(\rho_n),\ \gamma) + \eta = k_0 i(\beta\mathcal{H}(\rho_n),\ [\gamma]) + \eta\ , \end{aligned}$$

with $k_0 \downarrow 1$ and $\eta \cdot \|\mathcal{H}(\rho_n)\|^{-\frac{1}{2}} \to 0$ as $\rho_n \to \partial\Im_g$.

Lemma. *Let $\rho_n \to \partial\Im_g$. Then $\pi \circ \ell(\rho_n)$ converges if and only if $\pi \circ I \circ \beta \circ \mathcal{H}(\rho_n)$ converges, and in the case of convergence, the two sequences have the same limit.*

Proof. The topology of $\Im_g$ is defined by a finite number of curves $\gamma_1, \dots, \gamma_k$ for which, for any $\Phi \subset S\mathbb{Q}_\sigma$, $\sum_j i(\beta\Phi, [\gamma_j]) > \delta > 0$ for some δ. (In fact, the conformal structure of (M, σ) is determined by the lengths of $[\gamma_1], \dots, [\gamma_k]$.)

Suppose $\pi \circ \ell(\rho_n)$ converges. Then there exists a sequence of scalars $\lambda_n > 0$ such that $\lambda_n \rho_n$ converges. This means $\sum_j \ell_{\rho_n}(\gamma_j) \to \infty$, and $\sum_j \lambda_n \ell_{\rho_n}(\gamma_j)$ converges. Then the previous lemma shows that for $[\gamma] \in \mathcal{S}$,

$$\left|\lambda_n \ell_{\rho_n}([\gamma]) - \lambda_n i\big(\beta\mathcal{H}(\rho_n),\, [\gamma]\big)\right| \to 0\ ,$$

since $\lambda_n \eta \to 0$ and $k_0 \to 1$ as $\rho_n \to \partial\mathfrak{T}_g$. Thus $\pi \circ I \circ \beta \circ \mathcal{H}(\rho_n)$ converges to the same limit as $\pi \circ \ell(\rho_n)$. Similarly we can obtain the converse.

§10. Proof of Compactification Theorem

Theorem (Wolf).

$$\overline{\mathfrak{T}}_g^h(\sigma) \approx \overline{\mathfrak{T}}_g^{Th}\ .$$

Proof. Our main tool in this proof will be the theorem of Hubbard and Masur, and the lemma of §9. Recall $\overline{\mathfrak{T}}_g^{Th} = \pi \circ \ell(\mathfrak{T}_g) \cup \mathcal{PF} \subset P(\mathbb{R}_+^{\mathcal{S}})$. Using polar coordinates (θ, r) for $\overline{B\mathcal{Q}}_\sigma$ with $\theta \in S\mathcal{Q}_\sigma$, $r \in [0,1]$, we define a map $\psi : \overline{\mathfrak{T}}_g^{Th} \to \overline{B\mathcal{Q}}_\sigma$ by

$$\psi(y) = \begin{cases} \left(\dfrac{\mathcal{H}(x)}{\|\mathcal{H}(x)\|}\,,\ \dfrac{4\|\mathcal{H}(x)\|}{1+4\|\mathcal{H}(x)\|}\right) & \text{if } \ y = \pi \circ \ell(x),\ x \in \mathfrak{T}_g\ , \\ \left(\lim\limits_n \dfrac{\mathcal{H}(x_n)}{\|\mathcal{H}(x_n)\|}\,,\ 1\right) & \text{if } \ \pi \circ \ell(x_n) \to y \in \partial\mathfrak{T}_g\ . \end{cases}$$

First we show that ψ is well-defined. Suppose x_n, $x'_n \in \mathfrak{T}_g$ such that $\lim \pi \circ \ell(x_n) = \lim \pi \circ \ell(x'_n) = y \in \partial\mathfrak{T}_g \subset P(\mathbb{R}_+^{\mathcal{S}})$. Then the previous lemma implies that $\lim \pi \circ I \circ \beta \circ \mathcal{H}(x_n) = \lim \pi \circ I \circ \beta \circ \mathcal{H}(x'_n)$.

Hence by the Hubbard-Masur theorem (refer to the argument in the first paragraph of §9),

$$\lim \frac{\mathcal{H}(x_n)}{\|\mathcal{H}(x_n)\|} = \lim \frac{\mathcal{H}(x'_n)}{\|\mathcal{H}(x'_n)\|} .$$

Next we claim that ψ is a homeomorphism. First we prove that ψ is continuous. Suppose $\pi \circ \ell(x_n) \to y \in \partial\Im_g$. By definition, ψ is continuous in its first component. Since $\|\mathcal{H}(x_n)\| \to \infty$, ψ is continuous in its second component also. The injectivity of ψ on $\Im_g$ follows from that of $\mathcal{H}$. Suppose y, $y' \in \partial\Im_g$ and $\psi(y) = \psi(y')$. Then there exist x_n, $x'_n \in \Im_g$ such that $\pi \circ \ell(x_n) \to y$ and $\pi \circ \ell(x'_n) = y'$. By hypothesis, $\lim \frac{\mathcal{H}(x_n)}{\|\mathcal{H}(x_n)\|} = \lim \frac{\mathcal{H}(x'_n)}{\|\mathcal{H}(x'_n)\|}$. Then by Hubbard and Masur, $\lim \pi \circ I \circ \beta \circ \mathcal{H}(x_n) = \lim \pi \circ I \circ \beta \circ \mathcal{H}(x'_n)$, and hence by the previous lemma, $y = \lim \pi \circ \ell(x_n) = \lim \pi \circ \ell(x'_n) = y'$.

We now show that $\psi(\overline{\Im_g}^{Th}) = \overline{B\mathcal{Q}_\sigma}$. Obviously, $\psi\big(\pi \circ \ell(\Im_g)\big) = B\mathcal{Q}_\sigma$. So we suppose $\theta \in S\mathcal{Q}_\sigma$ and $\gamma_n\theta \to \theta$. Then $\pi \circ I \circ \tilde{\beta}(\gamma_n\theta) \equiv$ constant, and so the previous lemma says that $\pi \circ I \circ \tilde{\mathcal{H}}^{-1}(\gamma_n\theta)$ converges to a point $y \in \partial\Im_g$. Then by definition of ψ,

$$\psi(y) = \Big(\lim \frac{\mathcal{H}(\tilde{\mathcal{H}}^{-1}(\gamma_n\theta))}{\|\mathcal{H}(\tilde{\mathcal{H}}^{-1}(\gamma_n\theta))\|},\ 1\Big) = (\theta, 1) = \theta .$$

Therefore ψ is surjective.

Finally we claim ψ^{-1} is continuous. Again we have only to show that ψ^{-1} is continuous on $S\mathcal{Q}_\sigma$. Suppose $(\theta_n, \gamma_n) \to (\theta, 1)$. By Hubbard-Masur, $\pi \circ I \circ \tilde{\beta}(\gamma_n\theta_n)$ converges, and so, by the previous lemma, $\pi \circ I \circ \tilde{\mathcal{H}}^{-1}(\gamma_n\theta_n)$ converges to $y \in \partial\Im_g$. By the definition of ψ, however

$$\begin{aligned}\psi\big(\pi \circ \ell \circ \tilde{\mathcal{H}}^{-1}(\gamma_n\theta_n)\big) &= \Big(\frac{\mathcal{H}(\tilde{\mathcal{H}}^{-1}(\gamma_n\theta_n))}{\|\mathcal{H}(\tilde{\mathcal{H}}^{-1}(\gamma_n\theta_n))\|},\ \frac{4\|\mathcal{H}(\tilde{\mathcal{H}}^{-1}(\gamma_n\theta_n))\|}{1 + 4\|\mathcal{H}(\tilde{\mathcal{H}}^{-1}(\gamma_n\theta_n))\|}\Big) \\ &= (\theta_n, \gamma_n)\ , \quad \text{and moreover ,}\end{aligned}$$

$$\begin{aligned}\psi(y) &= \Big(\lim \frac{\mathcal{H}(\tilde{\mathcal{H}}^{-1}(\gamma_n\theta_n))}{\|\mathcal{H}(\tilde{\mathcal{H}}^{-1}(\gamma_n\theta_n))\|}\ ,\ 1\Big) \\ &= (\theta, 1) .\end{aligned}$$

Therefore

$$\psi^{-1}(\theta, 1) = y = \lim \pi \circ \ell \circ \tilde{\mathcal{H}}^{-1}(\gamma_n \theta_n)$$

$$= \lim \psi^{-1}(\theta_n, \gamma_n) .$$

This completes the proof of the theorem.

Chapter III

Harmonic Maps of Kähler Manifolds with Constant Negative Holomorphic Sectional Curvature

Holomorphic maps between hyperbolic complex manifolds have volume-decreasing properties. Toledo, however, obtained examples of harmonic maps between closed surfaces of constant negative curvature, and hence of the Poincaré disk, which are not area-decreasing [To]. He also derived an inequality for the degree of a map from a Riemann surface to a compact Kähler manifold of constant negative holomorphic sectional curvature. This inequality gives a lower bound on the genus of a surface representing a homology class of given degree. Moreover, Toledo proved that a map has maximal degree if and only if it can be deformed to a holomorphic totally geodesic immersion. The proof of this fact is a generalization of the proof that a harmonic degree-one map of a negatively curved surface is a diffeomorphism (theorem in Chapter I §8, or [SY, 3.1]). His results rely on formulas for the Laplacian of the $(1,0)$ and $(0,1)$ energy. Formulas for maps between surfaces were proved in [SY]. In the first three sections we study Toledo's work.

The classical result is in a similar spirit as Kneser's beautiful inequality on the degree of a map between surfaces of genus greater than one [Kn]. Kneser showed that if $f : \sum_g \to \sum_{g'}$, $g, g' > 1$ is a map of degree d then $\chi(\sum_g) \leq d\chi(\sum_{g'})$ and equality holds if and only if f is homotopic to a covering map. This inequality has recently been generalized by M. Gromov to maps between compact manifolds of constant negative curvature. Using the norm defined by Gromov [G], Domic and

Typeset by $\mathcal{A}\mathcal{M}\mathcal{S}$-TEX

Toledo [DT] obtained a more general result than the inequality in the preceeding section. In order to study Domic and Toledo's result, let us first study Gromov norms.

§1. Laplacian of $|\partial f|^2$, $|\overline{\partial} f|^2$

Let S be a Riemann surface with Hermitian metric g and Kähler form ω_S, and let X be an n-dimensional Kähler manifold with Hermitian metric h and Kähler form ω_X. In local coordinates

$$g = g\, dz\, d\overline{z}\ , \qquad \omega_S = \frac{i}{2} g\, dz \wedge d\overline{z}\ ,$$
$$h = h_{\alpha\overline{\beta}}\, dw^\alpha\, d\overline{w}^\beta\ , \qquad \omega_X = \frac{i}{2} h_{\alpha\overline{\beta}}\, dw^\alpha \wedge d\overline{w}^\beta\ .$$

Note that $\omega_S = dA$, the area form of g. Let $f : S \to X$ be a C^∞ map. We write df for the real differential of f and

$$\partial^{1,0} f : T^{1,0}S \to T^{1,0}X\ , \quad \partial^{0,1} f : T^{1,0}S \to T^{0,1}X\ ,$$

for the corresponding components of the complexification of df. In local coordinates $\partial^{1,0} f$, $\partial^{0,1} f$ correspond to w^α_z, $w^\alpha_{\overline{z}}$ respectively. Here the energy density $e(f)$ is defined by

$$e(f) = \frac{1}{2}|df|^2 = \frac{1}{2}\mathrm{tr}\, df\, (df)^t\ .$$

In local coordinates

$$e(f) = \frac{1}{g}\left(h_{\alpha\overline{\beta}}\, w^\alpha_z\, \overline{w}^\beta_{\overline{z}} + h_{\alpha\overline{\beta}}\, w^\alpha_{\overline{z}}\, \overline{w}^\beta_z\right) = e'(f) + e''(f)$$

where $e'(f) = |\partial^{1,0} f|^2$, $e''(f) = |\partial^{0,1} f|^2$. Observe that

$$e'(f) - e''(f) = \frac{1}{g} h_{\alpha\overline{\beta}}\left(w^\alpha_z\, \overline{w}^\beta_{\overline{z}} - w^\alpha_{\overline{z}}\, \overline{w}^\beta_z\right)$$

which can be written as

$$[e'(f) - e''(f)]\omega_S = f^*\omega_X . \tag{1}$$

The harmonic map equation can be written in the Kähler context as

$$w^\beta_{z\overline{z}} + \theta^\beta_{\alpha\gamma}\, w^\alpha_z\, w^\gamma_{\overline{z}} = 0 \tag{2}$$

where

$$\theta = \theta^\beta_{\alpha\gamma}\, dw^\gamma = (\partial N)N^{-1} , \quad N^\beta_\alpha = h_{\alpha\overline{\beta}} , \tag{3}$$

is the connection form of the Hermitian connection on $T^{1,0}X$. If

$$K^{\beta}_{\alpha\gamma\overline{\delta}} = \left(\theta^\beta_{\alpha\gamma}\right)_{\overline{\delta}} , \tag{4}$$

then $\Omega^\beta_\alpha = K^\beta_{\alpha\gamma\overline{\delta}}\, dw^\gamma \wedge d\overline{w}^\delta$ is the curvature form of the Hermitian connection.

In the following theorem we use the notation

$$\beta'(f) = \nabla_1 \partial^{1,0} f , \quad \beta''(f) = \nabla_2 \partial^{0,1} f ,$$

where ∇_1, ∇_2 denote the covariant derivatives on $(T^{1,0}_S)^* \otimes f^*T^{1,0}X$, $(T^{1,0}_S)^* \otimes f^*T^{0,1}X$ respectively. We also write

$$\begin{aligned} q'(f) &= g^{-2} K_{\alpha\overline{\beta}\gamma\overline{\delta}}\left(w^\gamma_z\, \overline{w}^\delta_{\overline{z}} - w^\gamma_{\overline{z}}\, \overline{w}^\delta_z\right) w^\alpha_z\, \overline{w}^\beta_{\overline{z}} , \\ q''(f) &= g^{-2} K_{\alpha\overline{\beta}\gamma\overline{\delta}}\left(w^\gamma_z\, \overline{w}^\delta_{\overline{z}} - w^\gamma_{\overline{z}}\, \overline{w}^\delta_z\right) w^\alpha_{\overline{z}}\, \overline{w}^\beta_z . \end{aligned} \tag{5}$$

In terms of the curvature form Ω we could equivalently write at each $x \in S$,

$$\begin{aligned} q'(f)\omega_S &= \langle f^*\Omega\left(\partial^{1,0} f(v)\right) , \partial^{1,0} f(v)\rangle \\ q''(f)\omega_S &= \langle f^*\Omega\left(\partial^{0,1} f(v)\right) , \partial^{0,1} f(v)\rangle , \end{aligned}$$

where v is a unit vector in $T^{1,0}_x S$ and $\langle \cdot , \cdot \rangle$ is the Hermitian inner product f^*h.

Theorem 1. *If $f : S \to X$ is harmonic, then*

$$\frac{1}{4}\Delta e'(f) = \|\beta'(f)\|^2 + q'(f) + \frac{1}{2}K_S\, e'(f) ,$$
$$\frac{1}{4}\Delta e''(f) = \|\beta''(f)\|^2 - q''(f) + \frac{1}{2}K_S\, e''(f) .$$

Proof. In holomorphic normal coordinates we have

$$g = 1 , \quad g_z = 0 , \quad g_{z\bar{z}} = -\frac{1}{2}K_S \quad \text{at } x , \quad \text{and} \tag{6}$$

$$h_{\alpha\bar{\beta}} = \delta_{\alpha\beta} , \quad h_{\alpha\bar{\beta},\gamma} = \theta^{\beta}_{\alpha\gamma} = 0 ,$$
$$h_{\alpha\bar{\beta},\gamma\bar{\delta}} = \theta^{\beta}_{\alpha\gamma,\bar{\delta}} = K_{\alpha\beta\gamma\bar{\delta}} \quad \text{at } f(x) . \tag{7}$$

We obtain, at x,

$$\begin{aligned} \frac{1}{4}\Delta e' &= g^{-1}e'_{z\bar{z}} = g^{-1}\left(g^{-1}h_{\alpha\bar{\beta}}\, w^{\alpha}_{z}\, \overline{w}^{\beta}_{z}\right)_{z\bar{z}} \\ &= h_{\alpha\bar{\beta}z\bar{z}}\, w^{\alpha}_{z}\, \overline{w}^{\beta}_{\bar{z}} + w^{\alpha}_{zz\bar{z}}\overline{w}^{\alpha}_{\bar{z}} + w^{\alpha}_{zz}\overline{w}^{\alpha}_{\bar{z}\bar{z}} + w^{\alpha}_{z}\, \overline{w}^{\alpha}_{\bar{z}z\bar{z}} \\ &\quad + \frac{1}{2}K_S\, w^{\alpha}_{z}\, \overline{w}^{\alpha}_{\bar{z}} , \end{aligned} \tag{8}$$

where we have used

$$w_{z\bar{z}} = \overline{w}_{\bar{z}z} = 0 \quad \text{at} \quad x \tag{9}$$

which results from (7) and the harmonic map equation (2).

Using the chain rule and simplifying by (7), we get

$$h_{\alpha\bar{\beta},z\bar{z}} = h_{\alpha\bar{\beta},\gamma\bar{\delta}}\, w^{\gamma}_{z}\overline{w}^{\delta}_{\bar{z}} + h_{\alpha\bar{\beta},\bar{\gamma}\delta}\, \overline{w}^{\gamma}_{z}w^{\delta}_{\bar{z}} . \tag{10}$$

Differentiating (2) we obtain in the same way; at x,

$$w^{\beta}_{z\bar{z}z} = -\theta^{\beta}_{\alpha\gamma,\bar{\mu}}\, w^{\alpha}_{z}\, w^{\gamma}_{\bar{z}}\, \overline{w}^{\mu}_{z} , \quad \overline{w}^{\beta}_{\bar{z}z\bar{z}} = -\overline{(\theta^{\beta}_{\alpha\gamma,\bar{\mu}})}\, \overline{w}^{\alpha}_{\bar{z}}\, \overline{w}^{\gamma}_{z} w^{\mu}_{\bar{z}} . \tag{11}$$

By the symmetry relation $\overline{(\theta^{\beta}_{\alpha\gamma,\overline{\mu}})} = \theta^{\alpha}_{\beta\mu,\overline{\gamma}}$ which results from $h_{\alpha\overline{\beta}} = \overline{h_{\beta\overline{\alpha}}}$ and by (7), (10), (11), we get at x

$$\frac{1}{4}\Delta e' = w^{\alpha}_{zz}\overline{w}^{\alpha}_{\overline{z}\,\overline{z}} + K_{\alpha\overline{\beta}\gamma\overline{\delta}}(w^{\gamma}_{z}\overline{w}^{\delta}_{\overline{z}} - w^{\gamma}_{\overline{z}}\overline{w}^{\delta}_{z})w^{\alpha}_{z}\overline{w}^{\beta}_{\overline{z}} + \frac{1}{2}K_S w^{\alpha}_{z}\overline{w}^{\alpha}_{\overline{z}} \,. \tag{12}$$

By (9), the first term is $\left\|\beta'(f)\right\|^2(x)$. The first formula of Theorem 1 follows, because the remaining terms of (12) clearly agree with the corresponding terms of the formula. The second formula is proved in the same way.

Corollary 1. *If $f : S \to X$ is harmonic, then at every point of S where $e' \neq 0$ (respectively $e'' \neq 0$),*

$$\frac{1}{4}\Delta \log e'(f) = \alpha'(f) + \frac{1}{e'(f)}q'(f) + \frac{1}{2}K_S \,,$$
$$\frac{1}{4}\Delta \log e''(f) = \alpha''(f) - \frac{1}{e''(f)}q''(f) + \frac{1}{2}K_S \,,$$

where $\alpha'(f)$, $\alpha''(f) \geq 0$.

Proof. We have

$$\frac{1}{4}\Delta \log e' = \frac{1}{g}(\log e')_{z\overline{z}} = \frac{1}{4}\frac{\Delta e'}{e'} - \frac{1}{g}\frac{e'_z e'_{\overline{z}}}{(e')^2} \,.$$

From this we obtain the first formula with

$$\alpha'(f) = \frac{\|\beta''(f)\|^2}{e'} - \frac{1}{g}\frac{e'_z e'_{\overline{z}}}{(e')^2} \,.$$

Hence

$$\alpha'(f) = \frac{1}{(e')^2}\left[(w^{\alpha}_{z}\,\overline{w}^{\alpha}_{\overline{z}})(w^{\beta}_{zz}\,\overline{w}^{\beta}_{\overline{z}\,\overline{z}}) - (w^{\alpha}_{z}\,\overline{w}^{\alpha}_{\overline{z}\overline{z}})(\overline{w}^{\beta}_{\overline{z}}\,w^{\beta}_{zz})\right] ,$$

which is nonnegative by the Schwarz inequality.

When X has complex dimension one there is only on distinct component of $K_{\alpha\overline{\beta}\gamma\overline{\delta}}$ and this is equal to $-\frac{1}{2}K_X$. Thus we have

Corollary 2. *If X is a Riemann surface and $f : S \to X$ is harmonic, then*

$$\frac{1}{4}\Delta e'(f) = \|\beta'(f)\|^2 - \frac{1}{2}K_X\big(e'(f) - e''(f)\big)e'(f) + \frac{1}{2}K_S\, e'(f)$$
$$\frac{1}{4}\Delta e''(f) = \|\beta''(f)\|^2 + \frac{1}{2}K_X\big(e'(f) - e''(f)\big)e''(f) + \frac{1}{2}K_S\, e''(f)\ .$$

§2. Harmonic Maps which are not Area Decreasing

We first consider the possibility that a harmonic map be area preserving, and obtain the following.

Theorem 2. *Let S and X be compact Riemann surfaces with metrics g and h respectively, and let $f : S \to X$ be a harmonic map that preserves area (i.e., $f^*\omega_X = \pm\omega_X$) and Gaussian curvature (i.e., $f^*K_X = K_S$). Then either*

(1) *f is a holomorphic or anti-holomorphic local isometry, or*

(2) *(S, g) and (X, h) are flat tori.*

Proof. If we assume first $f^*\omega_X = \omega_S$, then $e'(f) - e''(f) \equiv 1$ by (1). Hence Corollary 2 gives

$$\frac{1}{4}\Delta e'(f) = \|\beta'(f)\|^2 \geq 0\ .$$

Therefore e' is constant, and $e'' = e' - 1$ is also constant. Thus

$$0 = e''(f)_{\bar z} = \big(w_{\bar z}\overline{w}_z\big)_{\bar z} = w_{\bar z\bar z}\overline{w}_z\ .$$

We conclude that at any point

$$\text{either}\quad \beta''(f) = 0 \quad\text{or}\quad e''(f) = 0\ .$$

Since $e'(f)$, $e''(f)$ have isolated zeroes or are identically zero (Chapter I, §7), we see that either $e''(f) \equiv 0$ or $\beta''(f) \equiv 0$. In the first case f would be holomorphic and (1) holds. Thus we may assume $\beta''(f) \equiv 0$. Then the second formula of Corollary 2 gives $Ke''(f) \equiv 0$, where $K = f^*K_X = K_S$. Therefore we see that either $K \equiv 0$ or $e''(f) \equiv 0$. Thus both surfaces are flat tori or f is holomorphic, hence also locally isometric since it is both conformal and area-preserving. Finally if f reverses orientation we argue in the same way starting from the second formula of Corollary 2.

Corollary 3. *Let S be a compact Riemann surface with hyperbolic metric, and let $f : S \to S$ be harmonic and area preserving. Then f is a holomorphic or anti-holomorphic isometry.*

Corollary 4. *Let S be a compact Riemann surface with hyperbolic metric, and let $\varphi : S \to S$ be a smooth map of degree one that is not homotopic to a holomorphic map. Then the harmonic diffeomorphism f homotopic to φ is not area decreasing.*

Proof. If f were area decreasing we would have

$$|e'(f) - e''(f)| \leq 1 \quad \text{and} \quad \int (e'(f) - e''(f))\omega_S = \int \omega_S \,,$$

hence $e'(f) - e''(f) \equiv 1$, i.e., f is area-preserving. By the previous corollary f would be holomorphic, contrary to the hypothesis.

REMARK.

(1) There are plenty of maps φ of degree one that are not homotopic to a holomorphic map, for example, any diffeomorphism which is not of finite order in homology, e.g., a "Dehn twist".

(2) A flat torus has plenty of harmonic diffeomorphisms that are area-preserving but neither holomorphic nor anti-holomorphic.

Corollary 5. *There exist harmonic diffeomorphisms of the unit disk with the Poincaré metric that are not area decreasing.*

§3. Maps to Quotients of the Ball

In case the target manifold is a compact Kähler manifold of constant negative holomorphic sectional curvature, we can partially recover the area decreasing property of harmonic maps so that we derive a lower bound on the genus of a surface representing a homology class of given degree.

Let S be a compact surface with hyperbolic metric. Let B^n be the unit ball in $\mathbb{C}^n$ with metric

$$h = 4\frac{(1-\sum w^\alpha \overline{w}^\alpha)(\sum dw^\alpha d\overline{w}^\alpha) + (\sum \overline{w}^\alpha dw^\alpha)(\sum w^\alpha d\overline{w}^\alpha)}{(1-\sum w^\alpha \overline{w}^\alpha)^2} \tag{13}$$

of constant holomorphic sectional curvature -1 and invariant under the group $G^n = SU(1,n)/\text{center}$ of biholomorphic maps of B^n. Finally let Γ be a discrete subgroup of G^n which acts freely on B^n. The quotient manifold $X = B^n/\Gamma$ is then a Kähler manifold of constant holomorphic sectional curvature -1. Our goal is to prove the following theorem:

Theorem 3. *Let S and X be as above, and let $f : S \to X$ be a harmonic map. Then*

$$\left|\int_S f^*\omega_X\right| \leq \int_S \omega_S \,. \tag{14}$$

Moreover equality holds if and only if there is a metric of constant curvature -1 on the surface S with respect to which f is a totally geodesic holomorphic or anti-holomorphic immersion.

To prove this theorem we need the following lemmas.

Lemma 1. *Let D be the Poincaré disk, and let $f : D \to B^n$ be harmonic. Then*

$$q'(f) = \frac{1}{2}e'(f)\big(e'(f) - e''(f)\big) + p(f)\,,$$
$$q''(f) = \frac{1}{2}e''(f)\big(e'(f) - e''(f)\big) - p(f)\,,$$

where $p(f) \geq 0$ and $p(f)(x) = 0$ if and only if $d_x f(T_x D)$ is contained in a complex one-dimensional subspace of $T_{f(x)}B^n$.

Proof of Lemma 1. Let $x \in D$. We may assume $f(x) = 0$ since $q'(f)$ is invariant under G^n. At 0 one finds by direct computation from (13) that

$$K_{\alpha\overline{\beta}\gamma\overline{\delta}} = 4\{\delta_{\alpha\beta}\,\delta_{\gamma\delta} + \delta_{\alpha\delta}\,\delta_{\beta\gamma}\} = \frac{1}{4}\{h_{\alpha\overline{\beta}}\,h_{\gamma\overline{\delta}} + h_{\alpha\overline{\delta}}\,h_{\overline{\beta}\gamma}\}\,.$$

Direct calculation from (5) gives

$$\begin{aligned} q'(f)(x) &= \frac{4}{g^2}\{2(w_z^\alpha\,\overline{w_{\overline{z}}^\alpha})^2 - (w_{\overline{z}}^\gamma\,\overline{w_z^\gamma})(w_z^\alpha\,\overline{w_{\overline{z}}^\alpha}) - (w_z^\alpha\,\overline{w_z^\alpha})(w_{\overline{z}}^\gamma\,\overline{w_{\overline{z}}^\gamma}) \\ &= \frac{1}{4}\{2(e')^2 - e'e'' - \frac{1}{g^2}\langle w_z, w_{\overline{z}}\rangle\langle w_{\overline{z}}, w_z\rangle\} \\ &= \frac{1}{2}\{(e')^2 - e'e''\} + \frac{1}{4g^2}\{|w_z|^2|w_{\overline{z}}|^2 - |\langle w_z, w_{\overline{z}}\rangle|^2\} \end{aligned}$$

from which the first formula follows, letting $p(f)$ be the second term in this expression. From Schwarz's inequality we see that $p(f) \geq 0$.

For the last assertion, let $V = d_x f(T_x D) \subset T_{f(x)}B^n$, and let W be the smallest J-invariant subspace of $T_{f(x)}B^n$ containing V. Then $W \otimes \mathbb{C} = \pi^{1,0}(V \otimes \mathbb{C}) \oplus \pi^{0,1}(V \otimes \mathbb{C}) = \pi^{1,0}(V \otimes \mathbb{C}) \otimes \overline{\pi^{1,0}(V \otimes \mathbb{C})}$, where $\pi^{1,0}$, $\pi^{0,1}$ are the projections of $TB^n \otimes \mathbb{C}$ onto $T^{1,0}B^n$, $T^{0,1}B^n$ respectively. Clearly $\dim_{\mathbb{R}} W \leq 2$ if and only if $\dim_{\mathbb{C}} \pi^{1,0}(V \otimes \mathbb{C}) \leq 1$. Since the latter space is spanned by w_z^α, $w_{\overline{z}}^\alpha$, the equality part of Schwarz's inequality gives that $\dim_{\mathbb{C}} \pi^{1,0}(V \otimes C) \leq 1$ if and only if $p(f)(x) = 0$, and this completes the proof.

Corollary 6. *If $f : D \to B^n$ is harmonic, then*

$$\frac{1}{4}\triangle \log e'(f) = \alpha'(f) + \frac{p(f)}{e'(f)} + \frac{1}{2}(e' - e'' - 1)$$

$$\frac{1}{4}\triangle \log e''(f) = \alpha''(f) + \frac{p(f)}{e''(f)} - \frac{1}{2}(e' - e'' + 1)$$

where α', α'', $p \geq 0$ and p is as in Lemma 1.

Lemma 2. *Let p be an isolated zero of e', and let z be a holomorphic coordinate centered at p. Then there exists a positive integer m_p, a positive constant c, and a function $\rho(z)$ which is nonzero and smooth so that*

$$e'(z) = |z|^{2m_p}(c + \rho(z))$$

in a neighborhood of p. A similar local representation holds near an isolated zero of e''.

Proof. We use equation of harmoinc map (2) to show that the leading term in the Taylor expansion of w_z is holomorphic, i.e.,

$$w_z = Az^m + O(|z|^{m+1})$$

for some positive integer m and some nonzero vector $A = (a^1, \ldots, a^m)$. Thus

$$e'(z) = \frac{h_{\alpha\overline{\beta}}}{g} w_z^\alpha \overline{w}_z^\beta = \frac{h_{\alpha\overline{\beta}}(0)}{g(0)} a^\alpha \overline{a}^\beta |z|^{2m} + O(|z|^{2m+1}) \ .$$

If we let $m_p = m$, $c = \dfrac{1}{g(0)} h_{\alpha\overline{\beta}}(0) a^\alpha \overline{a}^\beta$ and let ρ be the remainder term divided by $|z|^{2m}$ we obtain the desired representation. The bounds on ρ follow from the standard properties of the remainder in Taylor's formula. The representation for e'' is obtained in the same way.

Proof of Theorem 3. To prove the inequality we consider two cases.

(1) f is holomorphic or anti-holomorphic,

(2) f is neither holomorphic nor anti-holomorphic.

Case (1) follows from well-known area decreasing properties of holomorphic maps as follows. Suppose f is holomorphic. Then $e'' \equiv 0$, $f^*\omega_B = e'\omega_S$ and Theorem 1 together with Lemma 1 gives

$$\frac{1}{4}\Delta e' = \|\beta'(f)\|^2 + \frac{1}{2}(e')^2 - \frac{1}{2}e' \tag{16}$$

because $p(f) \equiv 0$. Let x be a point of S where e' attains its maximum. At x we have $\Delta e' \leq 0$, hence $e'(e'-1) \leq 0$ at x. But then we must have $e' \leq 1$ at all points of S, hence $\left|\int f^*\omega_B\right| = \int e'\omega_S \leq \int \omega_S$. If f is anti-holomorphic we get $e' = 0$, $e'' \leq 1$.

Consider now case (2). We can apply Corollary 6 at all points of S except the isolated zeroes of e', respectively e''. Let $S_\varepsilon = S - \bigcup_{e'(x_i)=0} D_\varepsilon(x_i)$ and let $e' = |z|^{2m_i'}(c_i + \rho_i(z))$ given by Lemma 2. Then Green's Theorem easily gives

$$\lim_{\varepsilon \to 0} \int_{S_\varepsilon} (\Delta \log e')\omega_S = -\sum_i 2m_i' \int_{\partial D_\varepsilon} \frac{1}{\varepsilon}\frac{\partial}{\partial r}|z| ds = -4\pi \sum_i m_i' \,.$$

Hence we have from Corollary 6

$$-\pi \sum_i m_i' - \int_S (\alpha' + \frac{p}{e'})\omega_S = \frac{1}{2}\int_S (e' - e'')\omega_S - \frac{1}{2}\int_S \omega_S \,. \tag{17}$$

Since $m_i', \alpha', p \geq 0$ it follows that

$$\int_S f^*\omega_X = \int_S (e' - e'')\omega_S \leq \int_S \omega_S \,, \tag{18}$$

which is one half of the asserted inequality. Arguing in the same way with the zeroes of e'', we get the second half of the inequality, namely

$$\int_S f^*\omega_X \geq -\int_S \omega_S \,. \tag{19}$$

Now we examine when equality can hold in (14). We recall that a map f is totally geodesic if and only if it maps geodesics to geodesics, or equivalently, $\beta(f) \equiv 0$, where β is the second fundamental form of f. In our situation $\beta(f) \equiv 0$ if and only if $\beta'(f) \equiv \beta''(f) \equiv 0$. Note that if f is a holomorphic totally geodesic immersion with respect to some metric of constant negative curvature on S, then $f^*\omega_X$ is the Kähler form of this metric, thus Gauss-Bonnet gives

$$\int_S f^*\omega_X = -2\pi\chi(S) = \int_S \omega_S ,$$

and equality does hold in (14). Similarly, if f is anti-holomorphic and totally geodesic, we get the same equality in (14). Conversely, suppose equality holds in (14). First, if f is holomorphic, we must have $e' \equiv 1$, hence from (16) we see that $\beta'(f) \equiv 0$. Since f is holomorphic $\beta''(f) \equiv 0$, hence the second fundamental form $\beta(f) \equiv 0$ and f is totally geodesic, and an immersion since $e' \equiv 1$, thus $f^*\omega_X$ is never zero. The case that f is anti-holomorphic is handled similary.

Now suppose that equality holds in (14) and f is neither holomorphic nor anti-holomorphic. We must then have equality in (18) or (19). Suppose that equality holds in (18). Then the left hand side of (17) must be identically zero. Since the $m_i' > 0$, this sum must be empty, i.e. $e' > 0$, and we must also have $\alpha' \equiv p \equiv 0$. Hence by Lemma 1 and Corollary 6, we have the following.

Lemma 3. *If $f : S \to X$ is harmonic and $\int_S f^*\omega_X = \int_S \omega_S$, then*

$$e'(x) > 0 \quad \textit{for all} \quad x \in S, \tag{20}$$

$d_x f(T_x S)$ is contained in a complex one-dimensional subspace of $T_{f(x)}X$ for each $x \in S$, and (21)

$$\frac{1}{4}\Delta \log \frac{e'}{e''} = -\alpha'' + e' - e'' . \tag{22}$$

It now follows as in Chapter I, §8 that $e' - e'' \geq 0$. Namely, from (22) we see that $\log \frac{e'}{e''}$ is superharmonic on the set where $e' - e'' < 0$, and from (20) it follows that $\log \frac{e'}{e''}$ is never $-\infty$. If the set where $e' - e'' < 0$ were nonempty, $\log \frac{e'}{e''}$ would take its minimum on the boundary, which is impossible since it is identically zero on the boundary.

We must then have $e' - e'' \geq 0$, hence $\log \frac{e'}{e''} \geq 0$. Arguing as in Chapter I, §8 we see that $e' - e'' > 0$ everywhere on S. Hence $f^*\omega_X$ is never zero and in particular f is an immersion. By (21) we have that for each $x \in S$, $d_x f$ gives an isomorphism between $T_x S$ and a complex line in $T_{f(x)}X$. These isomorphisms give a complex structure on S with respect to which f is holomoprhic. If we give S a hyperbolic metric which is compatible with this complex structure, from the above discussion of equality, it follows that f is totally geodesic. Finally the case of equality in (19) is handled in the same way, thereby completing the proof of Theorem 3.

Theorem 4. *Let S be a compact orientable surface of genus $g > 1$, let X be a compact Kähler manifold of constant holomorphic sectional curvature -1, and let $\varphi : S \to X$ be a smooth map. Then*

$$\left| \int_S \varphi^* \omega_X \right| \leq 4\pi(g-1) \ . \tag{23}$$

Equality holds if and only if a harmonic map homotopic to φ is a totally geodesic holomorphic or anti-holomorphic immersion with respect to some hyperbolic metric on S.

Proof. This follows from Theorem 3, the existence of a harmonic map homotopic to ϕ, and the Gauss-Bonnet Theorem.

Note that $\omega_X = -\frac{4\pi}{n+1} c_1(X)$, where $c_1(X)$ is the first Chern class of X and $n = \dim_{\mathbb{C}} X$. Thus (23) can be re-written as

$$\left| \int_S \varphi^* c_1(X) \right| \leq (n+1)(g-1) \, .$$

§4. Gromov's Pseudonorms

Let X be a topological space and let $C_*(X)$ denote the real singular chain complex of X. (Recall that $C_k(X)$ is the vector space with a basis consisting of all continuous maps of the standard simplex Δ^k into X.) Any k-chain c can be written uniquely as $c = \sum r_i \sigma_i$ where $r_i \in \mathbb{R}$, $\sigma_i : \Delta^k \to X$. Define $\|c\|_1$ by

$$\|c\|_1 = \sum |r_i| \, .$$

If x is a homology class in $H_k(X;\mathbb{R})$ define the *Gromov norm* of x by

$$\|x\|_1 = \inf \left\{ \|c\|_1 \, : \, c \text{ is a singular cycle representing } x \right\} \, .$$

A norm can also be defined on the real cochain complex $C^*(X)$. If c is a k-cochain define

$$\|c\|_\infty = \sup \left\{ |c(\sigma)| \, : \, \sigma \text{ is a singular simplex in } X \right\} \, .$$

If $\alpha \in H^k(X;\, \mathbb{R})$ define $\|\alpha\|_\infty$ by

$$\|\alpha\|_\infty = \inf \left\{ \|c\|_\infty \, : \, c \text{ is a cochain representing } \alpha \right\} \, .$$

It is easily seen that

$$\|x+y\|_1 \leq \|x\|_1 + \|y\|_1 \, , \quad \|\lambda x\|_1 = |\lambda| \, \|x\|_1 \, ,$$

$$|\alpha(x)| \leq \|\alpha\|_\infty \|x\|_1 \, , \tag{1}$$

$$\left\| [\Sigma_g] \right\|_1 = 4(g-1) \quad [G] \, . \tag{2}$$

If $f : X \to Y$ is a continuous map, then

$$\|f_* x\|_1 \leq \|x\|_1 \quad \text{and} \quad \|f^* \alpha\|_\infty \leq \|\alpha\|_\infty . \tag{3}$$

In particular, if M, N are closed oriented manifolds and $[M]$, $[N]$ represent the fundamental classes, then for $f : M \to N$,

$$|\deg f| \, \| [N] \|_1 \leq \| [M] \|_1 . \tag{4}$$

In case $M = N$ and $\| [M] \|_1 \neq 0$, we see that $|\deg f| \leq 1$. Since S^n and T^n admit self maps of all degrees, it follows that $\| [S^n] \|_1 = 0$, $\| [T^n] \|_1 = 0$, $n \geq 1$. More explicity, let $\sigma_j : \Delta^n \to S^n$ be defined by $\sigma_j = f_j \circ g$ where $g : \Delta^n \to S^n$ is the map which identifies $\partial \Delta^n$ to a point, and $f_j : S^n \to S^n$ is a map of degree j. Then $\frac{1}{j}\sigma_j$ represents the fundamental class of S^n. Clearly $\inf_j \|\frac{1}{j}\sigma_j\|_1 = 0$, hence $\| [S^n] \|_1 = 0$. Therefore, Gromov's invariants define a pseudonorm on the real (co) homology but need not be a norm.

Let $\zeta \in H^n(M; \mathbb{R})$ be the fundamental class. Gromov showed that $\|\zeta\|_\infty < \infty$ if and only if $\| [M] \|_1 > 0$ [G]. It is interesting to find manifolds for which $\| [M] \|_1 > 0$. What is known in the literature is the following:

(i) If M is a compact real hyperbolic manifold, $\| [M] \|_1 > 0$ [G], and its value is known [HMu].

(ii) If ω is a nontrivial characteristic class, $\| [\omega] \|_\infty < \infty$ [G].

(iii) If $M = SL(n, \mathbb{R})/SO(n)$, then $\| [M] \|_1 > 0$. [Sa]

Gromov has conjectured that any compact manifold M whose universal cover is a symmetric space of noncompact type has $\| [M] \|_1 > 0$. More generally, Gromov conjectured that for any manifold M with sectional curvature nonpositive and Ricci curvature strictly negative $\| [M] \|_1 > 0$.

Intuitively, Gromov's norm measures the efficiency with which multiples of a homology class can be represented by simplices. A complicated homology class needs many simplices.

§5. Gromov Norms of Hyperbolic Manifolds

For compact hyperbolic manifolds M, $\| [M] \|_1$ was computed by Gromov, and using this, he gave a very quick proof of Mostow's theorem for hyperbolic 3-manifolds.

Let M be a compact hyperbolic manifold and let us consider the quadratic form model for H^n, i.e., the hyperboloid $x_1^2 + \cdots + x_n^2 - x_{n+1}^2 = -1$ in $\mathbb{R}^{n+1}$ with an indefinite metric $dx^2 = dx_1^2 + \cdots + dx_n^2 - dx_{n+1}^2$. Any $k+1$ points $v_0, \dots, v_k$ in $\tilde{M} = H^n$ determine a *straight k-simplex* $\sigma_{v_0,\dots,v_k} : \Delta^k \to H^n$, whose image is the convex hull of $v_0, \dots, v_k$. Out of various ways of defining parametrizations for $\sigma_{v_0,\dots,v_k}$ we have an explicit one for the hyperboloid model. In this model, $v_0, \dots, v_k$ become points in $\mathbb{R}^{n+1}$, so they determine an affine simplex α. The cone from the origin onto α intersects a sheet of the hyperboloid and this cone gives a $1-1$ correspondence between α and $\sigma_{v_0,\dots,v_k}$. Hence we get a *parametrized straight simplex* $\sigma_{v_0,\dots,v_k}$ in H^n which is natural with respect to isometries of H^n. Any singular simplex $\tau : \Delta^k \to M$ can be lifted to a singular simplex $\tilde{\tau}$ in $\tilde{M}$ since Δ^k is simply connected. Let straight $(\tilde{\tau})$ be the straight simplex with the same vertices as $\tilde{\tau}$ and let straight (τ) be the projection of $\tilde{\tau}$ back to M. Straight (τ) does not depend on the lift $\tilde{\tau}$ and straight extends linearly to a chain map $c_*(M) \to c_*(M)$ which is chain homotopic to the identity. Hence τ and straight (τ) represent the same homology class. Clearly $\|\text{straight}(c)\|_1 \leq \|c\|_1$ for all chains c. Therefore, to compute Gromov's invariant of a homology class it suffices to consider only straight simplices.

If M is a complete manifold of nonpositive curvature, we can also consider only the class of "geodesic" simplices. A *geodesic simplex* in M is a singular simplex σ obtained in the following way. Let $v_0, \dots, v_k$ be points of $\tilde{M}$ and let $\tilde{\sigma} : \Delta^k \to \tilde{M}$ be the map

$$(\lambda_0, \dots, \lambda_k) \mapsto \text{center of gravity of the mass distribution}$$
$$\lambda_0 \delta_{v_0} + \cdots + \lambda_k \delta_{v_k} , \text{ where } \delta_v = \text{Dirac measure at } v .$$

Then let $\sigma = \pi \circ \tilde{\sigma}$. Note that the asserted center of gravity exists by a well-known theorem of Cartan. Hence geodesic k-simplices are parametrized by $\tilde{M}/\pi_1(M)$, and their edges are geodesic segments.

A simplex in H^n is called *ideal* if all the vertices are on the sphere at infinity in the Poincaré disk model. A simplex is called *regular* if any permutation of its vertices can be induced by an isometry of H^n.

Theorem ([HMu]). *In H^n, $n \geq 2$, a simplex has finite maximal volume if and only if it is ideal and regular.*

Theorem. *Every closed oriented hyperbolic manifold M^n satisfies the inequality*

$$\| [M] \|_1 > \frac{Vol(M)}{v_n} ,$$

where v_n is the maximal volume of a simplex in H^n, $n \geq 2$.

Proof. Let $\sum r_i \sigma_i$ be any straight cycle representing $[M]$. Then

$$\text{Vol}(M) = \int_{\sum r_i \sigma_i} dV = \sum r_i \int_{\Delta^n} \sigma_i^* \, dV \leq \sum |r_i| v_n .$$

For example, $v_2 = \pi$ by the Gauss-Bonnet formula. From the above theorem it follows that $\| [M] \|_1 > 0$.

By a theorem in [G], if M is a closed oriented manifold, $\| [M] \|_1 = c \operatorname{Vol}(M)$, where c depends only on $\tilde{M}$. Gromov also proves that if $\tilde{M} = H^n$, then $c = \dfrac{1}{v_n}$ so that the inequality of the above theorem is actually equality [T, 6.2].

Corollary. *If $f : M \to N$ is a map between closed oriented hyperbolic manifolds, then*

$$Vol(M) \geq |\deg f| \; Vol(N) \; .$$

Proof. This follows from (4) and the above equality.

Gromov's proof of Mostow's theorem (as presented in [T, 6.3]) for hyperbolic 3-manifolds works also for $n > 3$ since we know that ideal simplices of maximal volume in H^n are automatically regular [HMu] (previous theorem). For the convenience of the reader we state here Mostow's theorem and a very brief outline of Gromov's argument.

Mostow's Theorem. *If M and N are hyperbolic manifolds which are complete with finite volume and homotopy equivalent then they are isometric.*

Gromov's Proof. Let $f : M \to N$ be a homotopy equivalence, and $\tilde{f} : H^n \to H^n$ a lifting of f. Then $\tilde{f}$ induces a continuous map $f^\infty : S^{n-1}_\infty \to S^{n-1}_\infty$ on the sphere at infinity. From the above corollary we get $\operatorname{Vol}(M) = \operatorname{Vol}(N)$. Therefore, whenever $v_0, v_1, \dots, v_n \in S^{n-1}_\infty$ span a regular ideal hyperbolic simplex, then so do $f^\infty(v_0)$, $f^\infty(v_1), \dots, f^\infty(v_n)$. Consider any ideal regular simplex σ together with all images of σ coming from repeated reflections in the faces of σ. The set of vertices of all these images of σ is dense in S^{n-1}_∞. Since f^∞

is determined on this dense set, f^∞ must be the restriction to S^{n-1}_∞ of a unique isometry of H^n which induces the desired isometry $M \to N$.

REMARK. If equality holds in the preceding Corollary then f is homotopic to a covering map.

§6. The Gromov Norm of the Kähler Class of Symmetric Domains

Let D be a bounded symmetric domain of rank p with its Bergman metric. We assume the metric is normalized so that the minimum holomorphic sectional curvature is -1, and thus the maximum holomorphic sectional curvature is $-\frac{1}{p}$. Let ω be the Kähler form of this metric, and let X be a compact manifold with universal cover D. Then ω defines a cohomology class $[\omega] \in H^2(X, \mathbb{R})$. In [DT] Domic and Toledo showed $\| [\omega] \|_\infty = p\pi$ for some classical domains. This result has the following topological corollary. Let f be a continuous map from $\sum_g$, $g > 1$, to X. Then $\left| \int_{\sum_g} f^*\omega \right| \leq 4p(g-1)\pi$. This was proved in [To] for the case $p = 1$, i.e., $D =$ unit ball in $\mathbb{C}^n$, by different methods: harmonic maps and Bochner formula. Extending these methods to higher rank turned out to be impossible, in the sense that the pointwise inequality needed to apply Bochner formula is actually false for $p > 1$. It is thus quite interesting that Gromov's methods can be used to derive the topological corollary. In this section we study [DT].

By the arguments about geodesic simplices in the previous section, to give an upper bound for $\| [\omega] \|_\infty$, it is enough to give an estimate for $\int_\Delta \omega$, Δ a geodesic triangle in D.

Theorem 1.

$$\left|\int_{\Delta}\omega\right| \le p\pi \; .$$

Proof. Observe that if $P \in D$, there exists a (unique) potential P_ρ for the Bergman metric, i.e., a function ρ_P such that $dd^{\mathbb{C}}\rho_P = \omega$, with the following properties:

(a) $\rho_P(P) = 0$,

(b) ρ_P is invariant under $K_P =$ isotropy group of P,

(c) $d^{\mathbb{C}}\rho_P = 0$ on geodesics through P.

Let P, Q, R be the vertices of Δ. Then

$$\begin{aligned}\int_{\Delta}\omega = \int_{\Delta} dd^{\mathbb{C}}\rho_P = \int_{\partial\Delta} d^{\mathbb{C}}\rho_P = \int_{\gamma(Q,R)} d^{\mathbb{C}}\rho_P \\ = \int_{\gamma(Q,R)} d^{\mathbb{C}}(\rho_P - \rho_Q) \; ,\end{aligned}$$

where $\gamma(Q,R)$ is the geodesic segment from Q to R. Now $h_{PQ} = \rho_P - \rho_Q$ is a pluri-harmonic function on D (i.e. $dd^{\mathbb{C}}h_{PQ} = 0$). Since any pluri-harmonic function on a simply connected manifold is the real part of a holomorphic function, $h_{PQ} = \operatorname{Re} H_{PQ}$, H_{PQ} holomorphic on D. Let $k_{PQ} = \operatorname{Im} H_{PQ}$. Then

$$\int_{\Delta}\omega = \int_{\gamma(Q,R)} d^{\mathbb{C}}h_{PQ} = \int_{\gamma(Q,R)} dk_{PQ} = k_{PQ}(R) - k_{PQ}(Q) \; .$$

Thus we have to show that

$$|k_{PQ}(R) - k_{PQ}(Q)| < p\pi \; .$$

We observe that both sides are additive under products, so it is enough to prove the inequality for irreducible domains. Recall that the irreducible bounded symmetric domains are classified as follows [C]:

$$\begin{aligned}
\mathrm{I}_{p,q}:&\quad D_{p,q}=\{Z\in M_{p,q}(\mathbb{C})\ :\ Z^*Z<1\}\ ,\ p\le q\ .\\
\mathrm{II}_p:&\quad \{Z\in D_{p,p}\ :\ Z^t=-Z\}\ .\\
\mathrm{III}_p:&\quad \{Z\in D_{p,p}\ :\ Z^t=Z\}\ .\\
\mathrm{IV}_n:&\quad \{(z_1,\dots,z_n)\in\mathbb{C}^n\ :\ |z_1^2+\cdots+z_n^2|<1 \text{ and}\\
&\qquad 1+|z_1^2+\cdots+z_n^2|^2>2(|z_1|^2+\cdots+|z_n|^2)\}\ .\\
\mathrm{V},\ \mathrm{VI}:&\quad \text{Exceptional domains of dimensions } 1,6,27\ .
\end{aligned}$$

We complete the proof of the inequality for the domain $D_{p,q}$. The same argument goes through to prove Theorem 1 for the domains of type II_p and III_p. $D_{p,q}$ is the noncompact dual of the Grassmannian $G(p,q)$ of complex p-planes in $\mathbb{C}^{p+q}=\{(u,v)\ :\ u\in\mathbb{C}^p,\ v\in\mathbb{C}^q\}$, in fact it is the open submanifold of $G(p,q)$ on which the Hermitian form $|u|^2-|v|^2$ is positive definite, the correspondence being given by $Z\in D_{p,q}\mapsto\{(u,\ Zu) : u\in\mathbb{C}^p\}\in G(p,\ q)$. From this it is easy to see that its group of automorphisms is

$$SU(p,q)=\left\{\begin{pmatrix}A&B\\C&D\end{pmatrix}\in SL(p+q,\mathbb{C}):\right.$$
$$\left.\begin{pmatrix}A^*&C^*\\B^*&D^*\end{pmatrix}\begin{pmatrix}1&0\\0&-1\end{pmatrix}\begin{pmatrix}A&B\\C&D\end{pmatrix}=\begin{pmatrix}1&0\\0&-1\end{pmatrix}\right\}.$$

$SU(p,q)$ acts on $D_{p,q}$ by $Z\longmapsto\dfrac{AZ+B}{CZ+D}$. $D_{p,q}$ has rank 1.

Proposition. *Let $\triangle$ be the geodesic triangle in $D_{p,q}$ with vertices O,Z_0,Z. Then*

$$\int_\triangle\omega=\arg\frac{\overline{\det(1-Z_0^*Z)}}{\det(1-Z_0^*Z)}\ .$$

("arg" is the continuous branch that vanishes when $Z=Z_0$.)

Proof. The potential of the Bergman metric centered at 0 is

$$\rho_0 = -\log\det(1 - Z^*Z) .$$

To obtain ρ_{Z_0} we observe that the following is an element of $SU(p,q)$ that takes Z_0 to 0:

$$\begin{pmatrix} A & B \\ C & D \end{pmatrix} = \begin{pmatrix} (1 - Z_0Z_0^*)^{-\frac{1}{2}} & -(1 - Z_0Z_0^*)^{-\frac{1}{2}}Z_0 \\ -Z_0^*(1 - Z_0Z_0^*)^{-\frac{1}{2}} & (1 - Z_0^*Z_0)^{-\frac{1}{2}} \end{pmatrix} . \tag{5}$$

Then

$$\begin{aligned} \rho_{Z_0} &= -\log\det\left[1 - \{(AZ+B)(CZ+D)^{-1}\}^*(AZ+B)(CZ+D)^{-1}\right] \\ &= -\log\det\left[(CZ+D)^{*-1}(CZ+D)^*(CZ+D)(CZ+D)^{-1} - \right. \\ &\qquad \left. (CZ+D)^{*-1}(AZ+B)^*(AZ+B)(CZ+D)^{-1}\right] \\ &= \log\left|\det(CZ+D)\right|^2 - \log\det\left[(CZ+D)^*(CZ+D) - \right. \\ &\qquad \left. (AZ+B)^*(AZ+B)\right] \\ &= \log\left|\det(CZ+D)\right|^2 + \rho_0 \end{aligned}$$

since $(CZ+D)^*(CZ+D) - (AZ+B)^*(AZ+B) = 1 - Z^*Z$. Thus

$$h_{OZ_0} = -\log\left|\det(CZ+D)\right|^2 = \mathrm{Re}\left[\log\det(CZ+D)^{-2}\right] .$$

Hence

$$\begin{aligned} \int_\Delta \omega &= k_{OZ_0}(Z) - k_{OZ_0}(Z_0) = \arg\det(CZ+D)^{-2}\Big|_{Z_0}^{Z} \\ &= \arg\frac{\det\overline{(CZ+D)}}{\det(CZ+D)}\Bigg|_{Z_0}^{Z} = \arg\frac{\det\overline{(1 + D^{-1}CZ)}}{\det(1 + D^{-1}CZ)}\Bigg|_{Z_0}^{Z} . \end{aligned}$$

Now, using (5),

$$D^{-1}CZ = -(1 - Z_0^*Z_0)^{\frac{1}{2}}\, Z_0^*(1 - Z_0Z_0^*)^{-\frac{1}{2}}Z = -Z_0^*Z ,$$

since $Z_0^* f(Z_0 Z_0^*) = f(Z_0^* Z) Z_0^*$ for any analytic function f. Hence

$$\int_\Delta \omega = \arg \frac{\overline{\det(1 - Z_0^* Z)}}{\det(1 - Z_0^* Z)}\Bigg|_{Z_0}^{Z} = \arg \frac{\det \overline{(1 - Z_0^* Z)}}{\det(1 - Z_0^* Z)} ,$$

as was to be proved.

Theorem 1 for $D_{p,q}$ follows immediately from the following: If $Z_0, Z \in D_{p,q}$ then $Z_0^* Z \in D_{p,p}$. Hence the eigenvalues $d_1, \dots, d_p$ of $Z_0^* Z$ lie in the unit circle. Therefore

$$\Big|\int_\Delta \omega\Big| = \left|\arg \frac{1 - \overline{\lambda}_1}{1 - \lambda_1} + \cdots + \arg \frac{1 - \overline{\lambda}_p}{1 - \lambda_p}\right| < p\pi$$

because

$$\left|\arg \frac{1 - \overline{\lambda}}{1 - \lambda}\right| < \pi \quad \text{for} \quad |\lambda| < 1 .$$

Theorem 2.

$$\big\| [\omega] \big\|_\infty = p\pi .$$

Proof. By Theorem 1 we know that universally $\| [\omega] \|_\infty \le p\pi$. We have only to find, for each D, a discrete torsion-free co-compact Γ in G so that $\| [\omega_X] \|_\infty = p\pi$ for $X = \Gamma \backslash D$: The proportionality principal [G, 2.3] says that $\| [\omega_X] \|_\infty$ is the same for all compact X with universal cover D.

We construct an example for the domain $D_{p,q}$. The other domains of type II_p, III_p are treated in a similar way. We use the following fact from the theory of arithmetic groups [B]: In the Grassmannian model of $D_{p,q}$, change the form to $|u|^2 - \sqrt{2}|v|^2$, i.e., represent $D_{p,q}$ as the open subset of $G(p,q)$ on which $|u|^2 - \sqrt{2}|v|^2$ is positive definite. Let O be the ring of integers in $\mathbb{Q}(i, \sqrt{2})$ and $\Gamma = SL(p+q, O) \cap SU(H)$, where $H = |u|^2 - \sqrt{2}|v|^2$. Then Γ is co-compact in $SU(H) \approx G$.

Now represent the unit disk $D_{1,1} = \left\{ z \in \mathbb{C} : \ |z| < \frac{1}{\sqrt{2}} \right\}$ in the same way: lines in $\mathbb{C}^2$ on which the norm $H_1 = |u|^2 - \sqrt{2}|v|^2$ is positive. The mapping $\psi : D_{1,1} \to D_{p,q}$ defined by

$$\psi(z) = \begin{pmatrix} z & & \\ & \ddots & \\ & & z \\ \hline & 0 & \end{pmatrix}$$

is a totally geodesic mapping whose image has curvature $-\frac{1}{p}$ in the normalized Bergman metric. It is stabilized by a group isomorphic to $SU(H_1)$. Let $\Gamma_1' = SL(2,0) \cap SU(H_1)$. Then Γ_1 is co-compact in $SU(H_1)$. Let Γ_1', Γ' be tortion-free subgroups of finite index in Γ_1, Γ respectively, and let

$$S = D_{1,1}/\Gamma_1' \, , \quad X = D_{p,q}/\Gamma' \, .$$

Then S is a compact Riemann surface of genus $g > 1$, X is a compact manifold covered by $D_{p,q}$, and the above map ψ gives a map $f : S \to X$ with the property that $f^*\omega$ is the Kähler form of a metric of constant curvature $-\frac{1}{p}$ on S. Therefore

$$\int_S f^*\omega = p \int_S \frac{1}{p} f^*\omega = p \int_S -K\, dA = p\, 4\pi (g-1) \, .$$

By the properties of the Gromov's norm $\big((1), (2), (3)\big)$, and Theorem 1,

$$\Big| \int_S f^*\omega \Big| = \big| f^*[\omega_X]([S]) \big| \leq \big\| [\omega_X] \big\|_\infty \, \| [S] \|_1 \leq p\pi 4(g-1) \, .$$

Since equality holds, $\big\| [\omega_X] \big\|_\infty = p\pi$, as was to be proved.

Remark. In the proof of Theorem 2 we saw that totally geodesic complex curves of curvature $-\frac{1}{p}$ give extremals for these inequality, and in §3 ([To]) it was proved that for $p = 1$ these are the only extremals up to homotopy. If $p > 1$, there can be other extremals. For example, if S is a Riemann surface, $X = S \times S$ and f is the graph of any orientation preserving diffeomorphism, then equality also holds. Toledo also pointed out that the use of the Bochner formula, as in [To], easily gives that the only holomorphic extremals are totally geodesic. Thus it would be interesting to find geometric conditions on X that insure that all extremals are homotopic to totally geodesic maps.

Chapter IV
Minimal Surfaces in a Kähler Surface

According to a theorem of Whitney, every smooth n-dimensional manifold M^n can be smoothly embedded in $\mathbb{R}^{2n}$. Viewing $\mathbb{R}^{2n}$ as $\mathbb{C}^n$, one may ask for embeddings which have nice properties relative to the complex structure. The simplest properties relate to the complex tangents. If M has no complex tangents, the embedding is said to be totally real. In general there are global obtructions to finding totally real embeddings. For example, if M is compact, orientable, and totally real, then its Euler number vanishes. In case M has isolated complex tangents Webster obtained an explicit formula for the Euler number. More precisely, in [We1], he considered a compact smooth real two dimensional surface M embedded or immersed in a complex two dimensional manifold N with isolated complex tangents only. He gave a formula relating the Euler number of M, that of its normal bundle, and the sum over all complex tangents of a certain index attached to each complex tangent. If N is Kähler and M is a branched minimal immersion which is not a holomorphic curve, he showed that the complex tangents are isolated and that each has a negative index. Consequently he proved that an S^2 minimally embedded in $\mathbb{C}P^2$ must be a holomorphic curve, hence a standard $\mathbb{C}P^1$ or a conic. We will study [We1] in this chapter.

§1. The Index of an Isolated Complex Tangent

Let M be a real surface, $\tilde{V}$ a complex vector bundle of rank 2 over M, and $V \subset \tilde{V}$ a real subbundle of real rank 2. Let $J, J^2 = -I$, denote the complex multiplication on the underlying real vector bundle of $\tilde{V}$. We have an isolated complex tangent at $p \in M$ if V_q and JV_q coincide for $q = p$ and are transverse for q in a deleted neighborhood of p. Let π_q denote the orthogonal projection of $\tilde{V}_q$ along V_q onto its normal space F_q, relative to some metric on $\tilde{V}$. Let v be a nonzero section of V near p. Then πJv is a local section of F with an isolated zero at p. The index, ind(p) is the degree (i.e., winding number) associated to the fiber coordinate of the map $q \longmapsto \pi Jv(q)$. This is defined up to a sign, which is fixed by choices of local orientation on M and in the fiber of F near p.

Now, in a more special setting, suppose f is a smooth immersion of M into a complex surface N, with tangent bundle TN. Set $\tilde{V} = f^{-1}(TN)$, $V = f_*(TM)$, and F the normal bundle of V in $\tilde{V}$. Here f_* denotes the differential of f thought of as a bundle mapping from TM to $f^{-1}(TN)$. A choice of local orientation of M near p induces naturally a local orientation of TM, and hence of V via f_*. This local orientation of V together with the natural orientation of TN induce a local orientation F by the requirement that $V \oplus F = \tilde{V}$ as oriented bundles. Given an isolated complex tangent at p, we define ind(p) relative to these orientations. It is well defined, since a change in the local orientation of M changes the orientation of both V and F, and is a local bi-holomorphic invariant of M in N near p. If M is compact, we choose on it a smooth tangent vector field v which is nonzero at each complex tangent and has only isolated zeroes. To state our formula we shall make use of the Poincaré-Hopf theorem characterizing the Euler number $\chi(M)$ as the sum of the indices of the zeroes of a vector field tangent to M. Also $\chi(F)$ denotes the sum of the indices of the zeroes of a suitable section of F. Then a comparison of the index sums of v and of $\pi J f_* v$, exactly as in [We2], yields the following, in which M need not be orientable.

Proposition 1. *Let the compact surface M be immersed in the complex surface N with only isolated complex tangents. Then*

$$\chi(M) + \chi(F) = \sum_p ind(p) \ , \tag{1}$$

where the sum extends over all complex tangents to M.

Let us describe ind(p) more explicity. Consider a complex vector bundle $\tilde{V}$ of rank r over a manifold. We suppose that $\tilde{V}$ has a hermitian metric and denote by $g(u,v)$ the real part of the hermitian inner product of two vectors u and v. g is a real inner product on the underlying real vector bundle. We extend g and J complex linearly to $\tilde{V} \otimes \mathbb{C} = V' \oplus V''$, $V'' = \overline{V}'$, $J = iI$ on V'. Thus if u and v are sections of V', then $g(u,v) = 0$ and $g(u,\overline{v})$ is their hermitian inner product. g is compatible with J, $J^* g = g$. We also assume that we have a connection D on $\tilde{V}$ which is compatible with the hermitian metric. It induces a connection, still denoted D, on the real bundle with $D_g = 0$.

Let e_i, $1 \le i \le 2r$, be a local real orthonormal frame field in $\tilde{V}$ and set

$$E_j = \frac{1}{2}(e_{2j-1} - i\, e_{2j}) \ , \quad 1 \le j \le r \ .$$

Then $g(e_i, e_j) = \delta_{ij}$ is equivalent to

$$g(E_i, E_j) = 0 \ , \quad g(E_i, \overline{E}_j) = \frac{1}{2}\delta_{ij} \ . \tag{2}$$

We introduce matrices of connection 1-forms $\xi = (\xi_{ij})$ and $\eta = (\eta_{ij})$

$$DE_i = \xi_{ij} E_j + \eta_{ij} \overline{E}_j, \quad D\overline{E}_i = \overline{\xi}_{ij} \overline{E}_j + \overline{\eta}_{ij} E_j . \tag{3}$$

Covariant differentiation of (2) gives

$$\xi + \overline{\xi}^t = 0 , \quad \eta + \eta^t = 0 . \tag{4}$$

Now let v_i, $1 \le i \le r$, be a local unitary frame field in V', $g(v_i, \overline{v}_j) = \frac{1}{2}\delta_{ij}$. We have

$$Dv_i = \psi_{ij} v_j , \quad \psi + \overline{\psi}^t = 0 . \tag{5}$$

The two frame fields are related by

$$E_i = a_{ij} v_j + b_{ij} \overline{v}_j , \quad \overline{E}_i = \overline{b}_{ij} v_j + \overline{a}_{ij} \overline{v}_j . \tag{6}$$

From (2), we get (in matrix form)

$$0 = ab^t + ba^t , \quad I = a\overline{a}^t + b\overline{b}^t . \tag{7}$$

From (3), (5), (6) we have

$$da + a\psi = \xi a + \eta \overline{b} , \quad db + b\overline{\psi} = \xi b + \eta \overline{a} . \tag{8}$$

Now suppose that $r = 2$ and $V \subset \tilde{V}$ is a real two-plane subbundle with a complex tangent at p. We choose e_i, $i = 1, 2$, to span V and e_α, $\alpha = 3, 4$ to span the normal bundle F. We may further assume that $e_2(p) = Je_1(p)$, $e_4(p) = Je_3(p)$, and that $v_i(p) = E_i(p)$, so that

$$a_{ij}(p) = \delta_{ij} , \quad b_{ij}(p) = 0 . \tag{9}$$

We adapt the unitary frame v_i to the frame E_i as in Chern and Wolfson [CW]. Then we may write

$$E_1 = qv_1 + r\overline{v}_2 , \quad |q|^2 + |r|^2 = |s|^2 + |t|^2 = 1 ,$$

$$E_2 = sv_2 + t\overline{v}_1 , \quad qt + rs = 0 , \quad q(0) = s(0) = 1 , \quad r(0) = t(0) = 0 . \tag{10}$$

With this normalization (8) gives

$$dq + q\psi_{11} = q\xi_{11} + \bar{t}\eta_{12}\ , \quad q\psi_{12} = s\xi_{12}\ ,$$
$$dr - r\psi_{22} = r\xi_{11} + \bar{s}\eta_{12}\ , \quad t\xi_{12} = -r\psi_{12}\ . \tag{11}$$

If we set $Q = \frac{r}{q}$, then (11) gives

$$dQ = Q(\psi_{11} + \psi_{22}) + (q\bar{s} - r\bar{t})q^{-2}\eta_{12}\ . \tag{12}$$

Note that there is a complex tangent precisely where $JE_1 = iE_1$, i.e., where r, or Q, vanishes. If there is an isolated complex tangent at p, we may approximate F near p by the plane spanned by $\operatorname{Re} v_2$, $\operatorname{Im} v_2$, without changing the index. Since

$$e_1 = 2\operatorname{Re}(qv_1 + r\bar{v}_2)\ , \quad Je_1 = 2\operatorname{Re}(iqv_1 - ir\bar{v}_2)\ ,$$
$$\pi Je_1 \approx 2\operatorname{Re}(iqv_1 - ir\bar{v}_2 - iE_1) = 4\operatorname{Re}(-ir\bar{v}_2)\ ,$$

it follows that $\operatorname{ind}(p)$ is the winding number associated to the map

$$p_1 \longmapsto 4i\bar{r}(p_1)\ . \tag{13}$$

§2. Nonholomorphic Minimal Immersions in a Kähler Surface

Now we consider a smooth branched minimal immersion f of M into the Kähler surface (N, g). Even at a branch point we may define the index of a complex tangent as above. We may assume that there is a smooth metric g^0 on M for which f is a weakly conformal branched harmonic immersion. $f^*g = \tilde{c}g^0$, where $\tilde{c} \geq 0$ and has only isolated zeroes corresponding to the branch points.

Proposition 2. *If $f(M)$ is not a holomorphic curve, then it has only isolated complex tangents, each with negative index.*

Proof. We study f near a point p, possibly a branch point, at which the tangent plane V_p is complex. Let

$$E^0 = \frac{1}{2}(e_1^0 - ie_2^0) , \quad \varphi^0 = \theta_1^0 + i\theta_2^0 = \mu \, dz$$

be local frame and coframe fields on M relative to g^0. $z = x + iy$, $z(p) = 0$, is a local isothermal parameter and $\mu \neq 0$. If D^0 denotes the Levi-Civita connection of g^0, then

$$D^0\varphi^0 = -\xi^0 \otimes \varphi^0 , \quad \xi^0 + \overline{\xi}^0 = 0 , \quad d\varphi^0 = \varphi^0 \wedge \xi^0 .$$

We also choose local adapted frames in $\tilde{V} = f^{-1}(TN)$ as in (10), and φ_i the one forms dual to E_i. It follows that

$$f_*E^0 = \lambda E_1 , \quad \text{or} \quad f^*\varphi_1 \equiv \varphi_1 = \lambda\varphi^0 , \quad f^*\varphi_2 \equiv \varphi_2 = 0 \tag{14}$$

for a complex factor λ, $\lambda\overline{\lambda} = \tilde{c}$. N is Kähler and its Levi-Civita connection D coincides with the hermitian connection, so the previous computations apply. We also have

$$D\varphi_i = -\xi_{ji} \otimes \varphi_j - \overline{\eta}_{ji} \otimes \overline{\varphi}_j , \quad d\varphi_i = \varphi_j \wedge \xi_{ji} + \overline{\varphi}_j \wedge \overline{\eta}_{ji} . \tag{15}$$

We write the Jacobian of f as

$$f_* = \varphi^0 f_*E^0 + \overline{\varphi}^0 f_*\overline{E}^0 = \lambda\varphi^0 E_1 + \lambda\overline{\varphi}^0\overline{E}_1 .$$

The second fundamental form of f, a symmetric $\tilde{V}$-valued 2 form, is defined by

$$\mathrm{II}_f = \tilde{D}f_* , \quad \tilde{D} = D^0 \otimes I + I \otimes D .$$

From (14), (15) we get

$$d\lambda + \lambda(\xi_{11} - \xi^0) = \lambda'\varphi^0 , \tag{16}$$

and

$$\lambda\xi_{12} = a\varphi^0 + b\overline{\varphi}^0 , \quad \lambda\eta_{12} = c\varphi^0 + \overline{b}\overline{\varphi}^0 \tag{17}$$

for certain smooth functions a, b, c, λ'. From (16) it follows that λ satisfies an equation of the form $\lambda_{\overline{z}} = K\lambda$. This condition says that λ is almost holomorphic, and by a well-known theorem [Be], since $\lambda \not\equiv 0$, it has the form

$$\lambda = z^\ell \lambda_0 , \quad \ell > 0 , \quad \lambda_0 \neq 0 ,$$

at a branch point p. Also (16) and (17) give

$$\begin{aligned} \mathrm{II}_f &= \lambda'(\varphi^0)^2 E + \overline{\lambda}'(\overline{\varphi}^0)^2 \overline{E}_1 + (\xi_{12} \otimes \varphi_1 + \overline{\eta}_{12} \otimes \overline{\varphi}_1) E_2 \\ &\quad + (\overline{\xi}_{12} \otimes \overline{\varphi}_1 + \eta_{12} \otimes \varphi_1) \overline{E}_2 , \end{aligned} \tag{19}$$

where

$$\xi_{12} \otimes \varphi_1 + \overline{\eta}_{12} \otimes \overline{\varphi}_1 = a(\varphi^0)^2 + 2b\varphi^0\overline{\varphi}^0 + \overline{c}(\overline{\varphi}^0)^2 .$$

The tension field of f is

$$\begin{aligned} \tau_f &= \mathrm{tr}\, \mathrm{II}_f = \mathrm{II}_f(e_1^0, e_1^0) + \mathrm{II}_f(e_2^0, e_2^0) \\ &= 4\mathrm{II}_f(E^0, \overline{E}^0) = 8(bE_2 + \overline{b}\,\overline{E}_2) . \end{aligned} \tag{20}$$

Thus f is harmonic if and only if $b = 0$. In that case (12), (17) imply $Q_{\overline{z}} + KQ = 0$, and hence by [Be] again, we see that either $Q \equiv 0$, or $Q = z^k Q_0$, $k \geq 1$, $Q_0 \neq 0$. If $Q \equiv 0$, then $f(M)$ is a holomorphic curve near $f(p)$. Otherwise p has an isolated complex tangent, and the map (13) has the form $z \longmapsto 4i\overline{r}(z) = 4i\overline{z}^k \overline{Q}_0 \overline{q}$. If follows that $\mathrm{ind}(p) = -k < 0$.

REMARK.

(1) The above proposition was originally motivated by the following observation. If M is immersed in N and has a point of index $+1$ (elliptic point), then there is a smooth one-parameter family of analytic discs with boundary on M and shrinking down to p. This can be used to construct a local variation of M which strictly decreases the area, by Wirtinger's inequality. Hence M cannot be minimal.

(2) At each branch point p' the exponent $\ell = \ell(p')$ in (18) is the branching order. If M does not have a complex tangent at p' we set $\text{ind}(p') = 0$. By Proposition 1 and Proposition 2,

$$\chi(V) + \chi(F) = \sum_p \text{ind}(p) \leq 0 ,$$

and equality holds if and only if V is totally real. Let v be a smooth vector field on the compact surface M, not vanishing at any branch point, and having only isolated zeroes. Then $f_*(v)$ is a section of V. To each zero of v corresponds a zero of $f_*(v)$ of the same index. At a branch point p', $f_*(v)$ picks up an additional zero of index $\ell(p')$. Therefore, summing up over all branch points, we have

$$\chi(M) + \chi(F) + \sum_{p'} \ell(p') = \sum_p \text{ind}(p) \leq 0 .$$

Corollary. *Let M be a compact orientable surface of genus g minimally embedded in $\mathbb{C}P^2$ (with any Kähler metric) with a homological degree $k \in \mathbb{Z}$, $M = k \cdot \mathbb{C}P^1$ in $H_2(\mathbb{C}P^2, \mathbb{Z})$, and $\chi(F) = k^2$. If M is not a complex algebraic curve, then*

$$k^2 \leq 2g - 2 ,$$

with equality holding if and only if M is totally real.

It follows that an embedded minimal S^2 in $\mathbb{C}P^2$ is an algebraic curve of degree k, $(k-1)(k-2) = 2g$, hence is a complex line or conic. If M is an embedded minimal torus, then it is either a nonsingular complex cubic curve or has degree $k = 0$ and is totally real. For the Fubini-Study metric an example of the latter alternative is provided by the Clifford torus.

Chapter V
Stable Minimal Surfaces in Euclidean Space

A minimal surface is called stable if and only if the second variation of the area functional is nonnegative for all compactly supported deformations. The earliest global result for such surfaces is due to Bernstein, who proved that all complete minimal graphs in $\mathbb{R}^3$ (these are automatically stable) are planes. The methods of Schoen, Simon, and Yau [SSY] yield a proof of Bernstein's theorem for stable minimal hypersurfaces M^n in $\mathbb{R}^{n+1}$ provided $\lim_{R\to\infty} R^{-p}\cdot \mathrm{Vol}(B_R)=0$ for some $p\in\left(0, 4+\sqrt{\frac{8}{n}}\right)$. Note that this condition is automatically satisfied by minimal graphs in $\mathbb{R}^{n+1}$, $n\leq 5$. A classification theorem for complete stable minimal surfaces in three dimensional manifolds of nonnegative scalar curvature has been obtained by Fischer-Colbrie and Schoen[FS]. A corollary of their theorem states that all complete oriented stable minimal surfaces in $\mathbb{R}^3$ are planes. This corollary was also proved by do Carmo and Peng [CP].

All the results mentioned so far deal with hypersurfaces. For higher codimensions, Wirtinger [Wi] showed that a holomorphic curve in $\mathbb{C}^n$ is absolutely area minimizing. In view of Wirtinger's result, it is reasonable to ask whether an area minimizing surface in $\mathbb{R}^n$ lies in an even dimensional affine subspace of $\mathbb{R}^n$ and is holomorphic with respect to some orthogonal complex structure on this affine subspace. Morgan [Mo] has shown that the nonzero sum of oriented two-planes in $\mathbb{R}^n$ is area minimizing only if all the planes are complex under some complex structure on the span of the planes. In his paper [Mi] Micallef obtained a partial affirmative answer to the above question for $n\geq 4$. He proved

Typeset by $\mathcal{A}\mathcal{M}\mathcal{S}$-TEX

that any complete oriented parabolic stable minimal surface in $\mathbb{R}^4$ is holomorphic with respect to some orthogonal complex structure on $\mathbb{R}^4$. He also proved that any complete oriented stable minimal surface of finite total curvature and genus zero in $\mathbb{R}^n$, $n \geq 4$, is holomorphic. It is interesting to note that Osserman [O] has constructed families of examples of entire minimal graphs in $\mathbb{R}^4$ which are not holomorphic with respect to any orthogonal complex structure on $\mathbb{R}^4$; these minimal graphs are therefore unstable by Micallef's result.

§1. A Complex Version of the Stability Inequality

The condition that a minimal submanifold M^ℓ of a Riemannian manifold N be stable is expressed by the following inequality (see, for example, [L]) that has to hold for all compactly supported normal vector fields s:

$$\begin{aligned} 0 &\leq I(S,S) \\ &= \int_M \Big(\sum_i \left\| D_{e_i} s \right\|^2 - \sum_{i,j} \langle \nabla'_{e_i} e_j, s\rangle^2 - \sum_i R'(s, e_i, e_i, s) \Big) dV_M \ , \end{aligned}$$

where $e_1, \dots, e_\ell$ are orthonormal vector fields of M, D is the normal connection of M, ∇' is the connection of N, and R' is the curvature tensor of N. If $N = \mathbb{R}^N$, we can simplify the inequality as follows:

$$\int_M \left|(ds)^T\right|^2 \leq \int_M \left|(ds)^N\right|^2 , \tag{1}$$

where ds is the directional derivative of vector valued function s and T, N denote orthogonal projection onto the tangent space and the normal space of M respectively. Inequality (1) is still valid for compactly supported sections of $N_{\mathbb{C}}M = NM \otimes \mathbb{C}$, the complexified normal bundle of M; just add up the stability inequalities for the real and imaginary parts of s. From now on we assume M to be a real two dimensional oriented surface. Let z denote a complex coordinate of M. We can then write $ds = \partial s + \overline{\partial} s$ where $\partial s = (\partial_z s)dz$ and $\overline{\partial} s = (\partial_{\overline{z}} s)d\overline{z}$. Thus (1) may be rewritten in the following way:

$$2\int_M |(ds)^T|^2 \leq \int_M |ds|^2 = \int_M |\partial s|^2 + \int_M |\overline{\partial} s|^2 .$$

Integration by parts yields $\int_M |\partial s|^2 = \int_M |\overline{\partial} s|^2$ because $\mathbb{R}^n$ is flat and s has compact support. Therefore

$$2\int_M |(\partial s)^T|^2 + 2\int_M |(\overline{\partial} s)^T|^2 \leq 2\int_M |\overline{\partial} s|^2 ,$$

that is

$$\int_M |(\partial s)^T|^2 \leq \int_M |(\overline{\partial} s)^N|^2 . \tag{2}$$

Similarly,

$$\int_M |(\overline{\partial} s)^T|^2 \leq \int_M |(\partial s)^N|^2 .$$

§2. A Characterization of Holomorphic Immersions into $\mathbb{R}^{2n}$

Let N^{2n} be a Kähler manifold of real dimension $2n$. By extending the complex structure J to $T_{\mathbb{C}}N = TN \otimes \mathbb{C}$, the complexified tangent bundle of N, we have $T_{\mathbb{C}}N = T_{\mathbb{C}}N^{1,0} \oplus T_{\mathbb{C}}N^{0,1}$, where $T_{\mathbb{C}}N^{1,0}$ and $T_{\mathbb{C}}N^{0,1}$ are the eigenspaces of J corresponding to the eigenvalues i and $-i$ respectively. Since N is Kähler, this splitting of $T_{\mathbb{C}}N$ into $(1,0)$ and $(0,1)$ subbundles is preserved by ∇, the Levi-Civita connection on

TN. The Riemannian metric $\langle\ ,\ \rangle$ on TN can be extended in two ways to $T_{\mathbb{C}}N$:

(i) as a complex bilinear form, to be denoted by $(\ ,\ )$.

(ii) as a Hermitian inner product, to be denoted by $\ll\ ,\ \gg$.

The relationship between these two extensions is given by the formula

$$\ll z, w \gg = (z, \overline{w})\ , \quad z, w \in T_pN \otimes \mathbb{C}\ .$$

Note that $(s,t) \equiv 0$ for all sections s,t of $T_{\mathbb{C}}N^{1,0}$. In other words $T_{\mathbb{C}}N^{1,0}$ is orthogonal to $T_{\mathbb{C}}N^{0,1}$ with respect to $\ll\ ,\ \gg$. When $N = \mathbb{R}^{2n}$, we will use the dot product notation instead of $(\ ,\ )$.

Proposition. *Let $F : M^{2m} \to \mathbb{R}^{2n}$ be an immersion, with $\mathbb{R}^{2n}$ having the usual Euclidean metric. Suppose that there exist vector bundles E and V over M which satisfy conditions (i), (ii) and (iii) below:*

(i) $T_{\mathbb{C}}M \cong E \oplus \overline{E}$, $N_{\mathbb{C}}M \cong V \oplus \overline{V}$.

(ii) *$E \oplus V$ is orthogonal to $\overline{E} \oplus \overline{V}$ with respect to $\ll\ ,\ \gg$.*

(iii) $d : \Gamma(E \oplus V) \to \Gamma\big((E \oplus V) \otimes T^*M\big)$.

Then there exist complex structures $\widetilde{J}$ and J on M^{2m} and $\mathbb{R}^{2n}$ respectively such that $\widetilde{J}$ is orthogonal with respect to the metric induced on M by the immersion, J is orthogonal with respect to the Euclidean inner product on $\mathbb{R}^{2n}$ and F is holomorphic with respect to $\widetilde{J}$ and J.

Proof. Let $\{e_1, \dots, e_m\}$ and $\{e_{m+1}, \dots, e_n\}$ be local frame fields for E and V respectively which are orthonormal with respect to $\ll\ ,\ \gg$. By assumption (iii),

$$de_A = \sum_B \omega_{AB} \otimes e_B$$

for some $n \times n$ matrix of one forms (ω_{AB}), $A, B \in \{1, \dots, n\}$. We define a $2n \times 2n$ complex matrix C by $C^t = (e_1, \dots, e_n, \overline{e}_1, \cdots, \overline{e}_n)$.

Then we have $dC = \omega \otimes C$, where

$$\omega = \begin{pmatrix} (\omega_{AB}) & 0 \\ 0 & (\overline{\omega}_{AB}) \end{pmatrix} .$$

Let

$$J_0 = \begin{pmatrix} -i & & & & & 0 \\ & \ddots & & & & \\ & & -i & & & \\ & & & i & & \\ & & & & \ddots & \\ 0 & & & & & i \end{pmatrix} ,$$

where i and $-i$ each appear n times along the diagonal. Define J by $J = C^{-1}J_0C$. Then $J^2 = -I_{2n}$. Moreover J is a matrix with real entries: by assumption (ii),

$$CC^t = \begin{pmatrix} 0 & I_n \\ I_n & 0 \end{pmatrix} .$$

Therefore (since $C^{-1} = \overline{C}^t$)

$$J - \overline{J} = \overline{C}^t J_0 C + C^t J_0 \overline{C} = \overline{C}^t (J_0 CC^t + CC^t J_0)\overline{C} = 0 .$$

Finally,

$$dJ = -C^{-1}dCC^{-1}J_0C + C^{-1}J_0dC = -C^{-1}(\omega J_0 - J_0\omega)C = 0 .$$

Therefore J is constant along $F(M)$. Thus we can define a complex structure on $\mathbb{R}^{2n}$ by declaring it to be J at any point. J is orthogonal, i.e., $JJ^t = I_{2n}$ because it was defined by means of an orthonormal frame. Note that J preserves E (it multiplies sections of E by i), that is, J preserves $F_*(T_pM)$ at all points $F(p)$ in $F(M)$. Therefore M is a complex submanifold of $\mathbb{R}^{2n}$ with respect to J. The complex structure $\widetilde{J}$ of TM is defined by demanding that F be holomorphic with respect to $\widetilde{J}$ and J, i.e., $\widetilde{J} = F_*^{-1} \circ J \circ F_*$.

We now restrict our attention to immersions $F : M^2 \to \mathbb{R}^4$ of an oriented surface M into Euclidean 4-space. Let $\{e_1, e_2, e_3, e_4\}$ be a local oriented orthonormal frame for $\mathbb{R}^4$ over an open set of $F(M)$ such that $\{e_1, e_2\}$ and $\{e_3, e_4\}$ are local oriented orthonormal frames for the tangent and normal bundles of M respectively. In this case TM and NM can each be given a complex structure, namely, rotation by 90° in the counter-clockwise direction. Let $E = (T_{\mathbb{C}}M)^{1,0}$ and $V = (N_{\mathbb{C}}M)^{1,0}$, where type $(1,0)$ is meant with respect to the complex structure just mentioned. The fibers of E and V are locally spanned by $(e_1 - ie_2)$ and $(e_3 - ie_4)$ respectively. E and V satisfy conditions (i) and (ii) required in the hypotheses of the proposition. Moreover it is not difficult to see that assumption (iii) of the proposition is satisfied if and only if

$$\left[d(e_3 - ie_4)\right] \cdot (e_1 - ie_2) \equiv 0$$

if and only if $\quad \partial(e_3 - ie_4) \cdot F_z \equiv 0 \text{ and } \overline{\partial}(e_3 - ie_4) \cdot F_z \equiv 0$

if and only if $\quad \left[\partial(e_3 - ie_4)\right]^T \equiv 0$

if and only if in a neighborhood of any point in M, there exists a local nonzero section s of V with the property that $(\partial s)^T = 0$. Note that $(\partial s)^T$ is precisely the term on the left in the stability inequality (2).

Corollary 1. *$F : M^2 \to \mathbb{R}^4$ is holomorphic if and only if for any section $V \in \Gamma(N_{\mathbb{C}}M)$ one of $(\partial V^{1,0})^T$ and $(\partial V^{0,1})^T$ vanishes identically.*

§3. Stable Minimal Surfaces of Finite Total Curvature and Genus 0

In this section we prove a result in arbitrary codimension under the condition that our stable surface has genus zero and finite total curvature.

Theorem 1. *Let $F : M^2 \to \mathbb{R}^n$ be an isometric stable minimal immersion of a complete oriented surface M of finite total curvature and genus zero into $\mathbb{R}^n$. Then $F(M)$ lies in an even dimensional affine subspace of $\mathbb{R}^n$ and is holomorphic with respect to some orthogonal complex structure on this affine subspace.*

Before proving Theorem 1 we need the following terminology and lemma. A Riemann surface is said to be *parabolic* if it does not admit positive nonconstant superharmonic function. For example, a minimal surface in $\mathbb{R}^n$ which is an entire graph is parabolic [0], a minimal surface of finite total curvature in $\mathbb{R}^n$ is conformally equivalent to a compact Riemann surface with finitely many punctures and hence is parabolic [CO], and a complete surface of quadratic area growth is parabolic [CY].

Lemma 1. *Let $F : M \to \mathbb{R}^n$ be an isometric stable minimal immersion of a complete oriented parabolic surface M into $\mathbb{R}^n$. If σ is a bounded holomorphic section of $N_{\mathbb{C}}M$ then $(\partial s)^T = 0$.*

Proof. Letting $s = f\sigma$ in the stability inequality (2), with f a C^∞ real valued function of compact support, yields

$$\int_M f^2 \left|(\partial s)^T\right|^2 \leq \frac{1}{2}\int_M |df|^2 \; |\sigma|^2 \leq \frac{C^2}{2}\int_M |df|^2 \, .$$

Hence a theorem of [FS] asserts the existence of $u > 0$ satisfying

$$\Delta u + \frac{2}{C^2}\left|(\partial\sigma)^T\right|^2 u = 0 \, .$$

But then u is a positive superharmonic function and therefore u must be constant. Thus $(\partial\sigma)^T = 0$.

Proof of Theorem 1. As noted above, a complete minimal surface M of finite total curvature in $\mathbb{R}^n$ is conformally equivalent to a compact Riemann surface with finitely many punctures. Moreover the Gauss map extends to the compactified surface $\hat{M}$ as a holomorphic map. Also the normal bundle of M extends to some vector bundle η on $\hat{M}$. Let $\eta_{\mathbb{C}} = \eta \otimes \mathbb{C}$.

By hypothesis, $\hat{M} = S^2$. Moreover, by a theorem of Koszul and Malgrange [KM], $\eta_{\mathbb{C}}$ can be made into a holomorphic vector bundle. Therefore, by a theorem of Grothendieck [Gr], $\eta_{\mathbb{C}}$ splits as the direct sum of holomorphic line bundles $L_1, \ldots, L_{n-2}$

$$\eta_{\mathbb{C}} = L_1 \oplus \cdots \oplus L_p \oplus L_{p+1} \oplus \cdots \oplus L_r \oplus L_{r+1} \oplus \cdots L_{n-2}\ ,$$

where $L_1, \ldots, L_p$ are the positive line bundles, $L_{p+1}, \ldots, L_r$ are the topologically trivial ones, and $L_{r+1}, \ldots, L_{n-2}$ are all negative. By the Riemann Roch theorem we know that each of $L_1, \ldots, L_p, L_{p+1}, \ldots, L_r$ admits a holomorphic section s_j, $j \in \{1, \ldots, r\}$. Simply by restricting the domain of s_j to M, we obtain r linearly independent bounded holomorphic sections of $N_{\mathbb{C}}M$. We know that $c_1(\eta_{\mathbb{C}}) = 0$ because $\eta_{\mathbb{C}}$ is the complexification of a real vector bundle. But $c_1(\eta_{\mathbb{C}}) = c_1(L_1) + \cdots + c_1(L_{n-2})$. Therefore only the following two cases arise.

CASE

(i) $L_1, \ldots, L_{n-2}$ are all trivial. In this case Lemma 1 says that $(\partial s_j)^T = 0$ for all $j \in \{1, \ldots, n-2\}$. But the s_j may be chosen to have no zeroes and then $s_1, \ldots, s_{n-2}$ span $N_{\mathbb{C}}M$. Therefore $(\partial s)^T = 0$ for all sections s of $N_{\mathbb{C}}M$. Thus, for any section s of

$N_{\mathbb{C}}M$, $(ds)^t = (\partial s)^T + \overline{(\partial \overline{s})}^T = 0$. This means that M is totally geodesic and therefore a plane. In particular, Theorem 1 holds.

(ii) $p > 1$ and $r < n-2$. From now on we adopt the following range of indices:

$$1 \le \mu\,,\ \nu \le p\,,\quad p+1 \le A \le n-2\,,$$

$$1 \le j\,,\ k \le r\,,\quad r+1 \le a \le n-2\,.$$

We note that $s_\mu \cdot s_j \equiv 0$. The reason is that

$$\partial_{\overline{z}}(s_\mu \cdot s_j) = (D_{\overline{z}}\, s_\mu) \cdot s_j + s_\mu \cdot (D_{\overline{z}}\, s_j) = 0\,.$$

where D denotes covariant differentiation in $N_{\mathbb{C}}M$. Therefore $s_\mu \cdot s_j$ is constant. But since L_μ is positive, s_μ must vanish somewhere and therefore $s_\mu \cdot s_j \equiv 0$ as claimed. This means that the s_μ lie in $\left[\operatorname{span}\{\overline{L}_1, \dots, \overline{L}_r\}\right]^\perp$, the orthogonal complement of $\operatorname{span}\{\overline{L}_1, \dots, \overline{L}_r\}$. Since $\dim\left[\operatorname{span}\{\overline{L}_1, \dots, \overline{L}_r\}\right]^\perp = n-2-r$, we have $p \le n-2-r$. We also know that each of $L_{r+1}, \dots, L_{n-2}$ admits a meromorphic section s_a, which has only poles. Therefore $s_a \cdot s_A$ is a meromorphic function which either vanishes identically or else has only poles. The latter is impossible and therefore we conclude that $s_a \in \left[\operatorname{span}\{\overline{L}_{p+1}, \dots, \overline{L}_{n-2}\}\right]^\perp$. Since $\dim\big[\operatorname{span}$ $\{\overline{L}_{p+1}, \dots, \overline{L}_{n-2}\}\big]^\perp = p$, $n-2-r \le p$. Combining this with the previous reversed inequality we conclude that $p = n-2-r$ and therefore $\operatorname{span}\{\overline{L}_1, \dots, \overline{L}_p\} = \left[\operatorname{span}\{\overline{L}_1, \dots, \overline{L}_r\}\right]^\perp$. We now claim that $(\partial_z s_j \cdot s_k)dz$ is a holomorphic differential.

$$\begin{aligned}\partial_{\overline{z}}(\partial_z s_j \cdot s_k) &= \partial_{\overline{z}}\, \partial_z s_j \cdot s_k + \partial_z s_j \cdot \partial_{\overline{z}} s_k \\ &= \partial_z(\partial_{\overline{z}}\, s_j \cdot s_k) - \partial_{\overline{z}} s_j \cdot \partial_z s_k + \partial_z s_j \cdot \partial_{\overline{z}} s_k\,.\end{aligned}$$

Since s_j are holomorphic, $\partial_{\overline{z}} s_j$ has only a tangential component and by Lemma 1, $\partial_z s_j$ has only a normal component. Therefore our claim is proved. But since there are no nonzero holomorphic differentials on S^2, $\partial_z s_j \cdot s_k \equiv 0$ for all j, k. From the argument in the previous paragraph we deduce that $\partial_z s_j$ lies in the span of $\{L_1, \dots, L_p\}$.

Let $\xi = L_1 \oplus \cdots \oplus L_p \oplus L_{p+1} \oplus \cdots \oplus L_r \oplus (T_{\mathbb{C}}M)^{1,0}$ (recall that the fibers of $(T_{\mathbb{C}}M)^{1,0}$ are spanned by F_z). We claim that $d : \Gamma(\xi) \to \Gamma(\xi \otimes T^*M)$, that is, d preserves ξ. The above discussion shows that $ds_j \cdot s_k = 0$. Therefore it remains to check that $(dF_z) \cdot F_z = 0$ and $(dF_z) \cdot s_j = 0$. The former follows from $F_z \cdot F_z = 0$ and the latter follows from the minimality of F, that is $F_{z\bar{z}} = 0$, and $(\partial_z s_j)^T = 0$.

A consequence of this claim is that ξ_q, the fiber of ξ over the point q in M, is a fixed subspace of $\mathbb{C}^n$. To see this, let v be a vector in ξ_{q_0} for some fixed $q_0 \in M$, and let $v_\xi(q)$, $v_\xi^\perp(q)$ be the orthogonal projections of v onto ξ_q and $\xi_q^\perp$ respectively. Then $0 = dv = dv_\xi + d_\xi^\perp$. But since d preserves ξ, it also preserve $\xi^\perp$. Therefore $dv_\xi = dv_\xi^\perp = 0$. But $v_\xi^\perp(q_0) = 0$ and therefore $v_\xi^\perp(q) = 0$ for all q. Thus ξ_q is the same for all q and hence $\xi = M \times \Lambda$, where Λ is a subspace of $\mathbb{C}^n$ of dimension $r + 1$.

Now let $T = \Lambda \cap \overline{\Lambda}$. Note that, since $[\operatorname{span}\{\overline{L}_1, \dots, \overline{L}_r\}]^\perp = \operatorname{span}\{L_1, \dots, L_p\}$, we have that $\operatorname{span}\{\Lambda, \overline{\Lambda}\} = \mathbb{C}^n$. But $\dim \operatorname{span}\{\Lambda, \overline{\Lambda}\} = \dim \Lambda + \dim \overline{\Lambda} - \dim T$ and so, $n = 2r + 2 - t$, where $t = \dim T$. Thus $t = r - p$ (since $p + r = n - 2$ from above).

If T is empty, $t = 0$, $r = p$ and Theorem 1 follows by applying the previous proposition with $E = (T_{\mathbb{C}}M)^{1,0}$ and $V = L_1 \oplus \cdots \oplus L_p$. If T is nonempty, then since T is preserved by complex conjugation, $T = W \otimes \mathbb{C}$, where W is a subspace of $\mathbb{R}^n$ of real dimension $t = r - p$. Now note that $(T_{\mathbb{C}}M)^{1,0}$ is not preserved by complex conjugation and so, $M \times W$ is a subbundle of NM which is parallel in $F^*(T\mathbb{R}^n)$. This means that $F(M)$ lies in an affine subspace of $I\!R^n$ orthogonal to W; the dimension of this affine subspace is $n - t = 2p + 2$. It also means that $N_{\mathbb{C}}M$ splits as the Whitney sum $N_{\mathbb{C}}M = M \times T \oplus (M \times T)^{\perp}$. $(M \times T)^{\perp}$ is the complexified normal bundle of M when M is viewed as a surface immersed in the affine subspace orthogonal to W. Now let v be a vector in T. Since $v \in \Lambda$, $v = \sum_{k=1}^{r} a_k s_k + aF_z$. Therefore $v \cdot s_\mu = 0$ for all $\mu \in \{1, \ldots, p\}$. Similarly, since $v \in \overline{\Lambda}$, $v \cdot \overline{s}_\mu = 0$ for all $\mu \in \{1, \ldots, p\}$. Thus T is orthogonal to the fibers of $L_1 \oplus \cdots \oplus L_p \oplus \overline{L}_1 \oplus \cdots \oplus \overline{L}_p$ over any point q in M, that is $L_1 \oplus \cdots \oplus L_p \oplus \overline{L}_1 \oplus \cdots \oplus \overline{L}_p \subset (M \times T)^{\perp}$. But the linear independence of $\{s_1, \ldots, s_p\}$ and the fact that $s_\mu \cdot s_\nu = 0$ for all $\mu, \nu = \{1, \ldots, p\}$ imply that $s_1, \ldots, s_p$, $\overline{s}_1, \ldots, \overline{s}_p$ are linearly independent. Moreover $\dim(M \times T)^{\perp} = 2p$ and therefore

$$L_1 \oplus \cdots \oplus L_p \oplus \overline{L}_1 \oplus \cdots \oplus \overline{L}_p = (M \times T)^{\perp} .$$

Theorem 1 is now proved by applying the proposition with $E = (T_{\mathbb{C}}M)^{1,0}$ and $V = L_1 \oplus \cdots \oplus L_p$.

§4. Stable Minimal Surfaces in $I\!R^4$

We are able to prove a much more general theorem about complete stable surfaces in $I\!R^4$, which we now state.

Theorem 2. *Let $F : M^2 \to I\!R^4$ be an isometric stable minimal immersion of a complete oriented parabolic surface into $I\!R^4$. Then F is holomorphic with respect to some orthogonal complex structure on $I\!R^4$.*

We need two lemmas for the proof of Theorem 2. We recall an observation made in the paragraph following the proof of the proposition in §3: for an immersion $F : M^2 \to I\!R^4$, $N_{\mathbb{C}}M \cong V \oplus \overline{V}$. If v is any vector in $\mathbb{C}^4$, let $v^{1,0}$ and $v^{0,1}$ denote the orthogonal projection of v onto V and $\overline{V}$ respectively.

Lemma 2. *For a minimal immersion $F : M^2 \to \mathbb{R}^4$, $(F_{zz})^{1,0} \otimes (dz)^2$ and $(F_{zz})^{0,1} \otimes (dz)^2$ are holomorphic quadratic differentials with values in V and $\overline{V}$ respectively.*

Proof. Since D preserves V and $\overline{V}$, all we have to check is that $D_{\overline{z}}(F_{zz})^N = 0$. Using $F_z \cdot F_z = 0$ and $F_z \cdot F_{zz} = 0$, we have

$$\begin{aligned}(F_{zz})^N &= F_{zz} - (F_{zz})^T \\ &= F_{zz} - \frac{1}{|F_z|^2}(F_{zz} \cdot F_z)F_{\overline{z}} - \frac{1}{|F_z|^2}(F_{zz} \cdot F_{\overline{z}})F_z \\ &= F_{zz} - \frac{1}{|F_z|^2}(F_{zz} \cdot F_{\overline{z}})F_z \ .\end{aligned}$$

Therefore, by minimality,

$$D_{\overline{z}}(F_{zz})^N = \left\{\partial_{\overline{z}}\left(F_{zz} - \frac{1}{|F_z|^2}(F_{zz} \cdot F_{\overline{z}})F_z\right)\right\}^N = 0 \ .$$

Lemma 3. *If $F : M^2 \to \mathbb{R}^4$ is an isometric minimal immersion and $a \in \mathbb{C}^4$ is a fixed vector then*

$$D_z D_{\overline{z}} a^{1,0} = -\frac{1}{|F_z|^2}(F_{zz} \cdot a^{1,0})(F_{\overline{z}\,\overline{z}})^{1,0} - \frac{1}{|F_z|^2}\big((F_{zz})^{1,0} \cdot a\big)(F_{\overline{z}\,\overline{z}})^{1,0} \ . \tag{4}$$

Proof. This follows from a direct calculation (see [Mi]) which we omit here.

Proof of Theorem 2. Let $s = f\sigma$ in the stability inequality (2) where f is a smooth real valued function with compact support and σ is a complex valued normal section which need not have compact support.

Then

$$\begin{aligned}&\int_M f^2\big|(\partial_z\sigma)^T\big|^2 dx\,dy\\ \le\ &\int_M |f_s|^2|\sigma|^2 dx\,dy + \int_M f f_{\bar z}(D_z\bar\sigma\cdot\sigma)dx\,dy\\ &+\int_M f f_z(\bar\sigma\cdot D_z\sigma)dx\,dy + \int_M f^2(D_z\bar\sigma\cdot D_z\sigma)dx\,dy\ .\end{aligned}$$

Rewriting $ff_{\bar z}$ as $\frac{1}{2}(f^2)_{\bar z}$ and ff_z as $\frac{1}{2}(f^2)_z$ and integrating by parts,

$$\begin{aligned}\int_M f^2\big|(\partial_z\sigma)^T\big|^2 dx\,dy \le \int_M |f_{\bar z}|^2|\sigma|^2 dx\,dy -\\ \int_M f^2\mathrm{Re}(\bar\sigma\cdot D_z D_{\bar z}\sigma)dx\,dy\ .\end{aligned}$$

Since $2|F_z|^2 dx\,dy = dA$, the area element of M, we have

$$|df|^2 = \frac{1}{2|F_z|^2}\{(f_x)^2+(f_y)^2\} = \frac{2}{|F_z|^2}|f_{\bar z}|^2\ .$$

Moreover we can show that

$$\big|(\partial_z\sigma)^T\big|^2 = |\sigma\cdot F_{zz}|^2/|F_z|^2\ .$$

Therefore we get

$$\begin{aligned}&2\int_M f^2\frac{|\sigma\cdot F_{zz}|^2}{|F_z|^4}dA + 2\int_M \frac{f^2}{|F_z|^2}\mathrm{Re}(\bar\sigma\cdot D_z D_{\bar z}\sigma)dA\\ \le\ &\int_M |df|^2|\sigma|^2 dA\ .\end{aligned}$$

Now fix a unit vector $a\in\mathbb{C}^4$ and let $\sigma = a^{1,0}$. Then using (4), we get

$$\int_M f^2 q\,dA \le \int_M |df|^2|a^{1,0}|^2 dA \le \int_M |df|^2 dA\ , \tag{5}$$

where

$$q = -\frac{2}{|F_z|^4}\mathrm{Re}\{(F_{zz})^{1,0}\cdot((F_{zz})^{1,0})a\}\ . \tag{6}$$

A theorem of [FS] says that there exists a smooth function $u > 0$ satisfying the equation

$$\Delta u + qu = 0 \ . \tag{7}$$

Following [FS] we let $w = \log u$. Then w satisfies the equation

$$\Delta w = -q - |dw|^2 \ . \tag{8}$$

Multiplying (8) by f^2, where f is a smooth real valued function with compact support, and integrating over M we obtain

$$2\int_M f(df \cdot dw)dA = \int_M f^2 q \, dA + \int_M f^2 |dw|^2 dA \ . \tag{9}$$

Using the inequality $2|f(df \cdot dw)| \leq \theta f^2 |dw|^2 + \frac{1}{\theta}|df|^2$, $\theta > 0$, we get from (9)

$$\frac{1}{\theta}\int_M |df|^2 dA \geq \int_M f^2 q \, dA + (1-\theta)\int_M f^2 |dw|^2 dA \ . \tag{10}$$

We now let the vector $a \in \mathbb{C}^4$ run over an orthonormal basis $\{a_1, a_2, a_3, a_4\}$ of $\mathbb{C}^4$ and we let q_j stand for the expression in (6) with $a = a_j$. The solution of (7) with $q = q_j$ is denoted by u_j and $w_j = \log u_j$. Summing up over $j \in \{1,2,3,4\}$ the inequality (10) with $q = q_j$ and $w = w_j$ we obtain

$$8\int_m |df|^2 dA \geq \int_M f^2\Big(\sum_j q_j)\Big) dA + \frac{1}{2}\int_M f^2\Big(\sum_j |dw_j|^2\Big) dA \ , \tag{11}$$

where we have set $\theta = \frac{1}{2}$. But from (6) we see that

$$\sum_j q_j = -\frac{2}{|F_z|^4}\mathrm{Re}\{(F_{zz})^{1,0} \cdot (F_{\bar{z}\bar{z}})^{1,0}\} = 0$$

because $b \cdot c = 0$ when b and c are of the same type. Therefore (11) becomes

$$\int_M |df|^2 dA \geq \int_M f^2 r \, dA \ ,$$

where

$$r = \frac{1}{16} \sum_j |dw_j|^2 . \tag{12}$$

Then again by [FS], there exists $v > 0$ with $\Delta v + rv = 0$. But since $r \geq 0$, v is superharmonic, and therefore the parabolicity of M implies that v is constant. It follows that r vanishes identically. From (12) we then deduce that w_j is constant for each j and therefore so is u_j. From (7) with $u = u_j$ and $q = q_j$ we conclude that $q_j \equiv 0$ for each j.

Now at an arbitrary point p of M, either $(F_{zz})^N(p) = 0$ or else we can let $a_1 = \dfrac{(F_{zz})^N(p)}{|(F_{zz})^N(p)|}$. In the latter case, $q_1 = 0$ implies, by (6), that $|(F_{zz})^{1,0}(P)|\,|(F_{zz})^{0,1}(p)| = 0$. This last equality obviously still holds if $(F_{zz})^N(p) = 0$. Therefore, from Lemma 2, we deduce that at least one of $(F_{zz})^{1,0}$ or $(F_{zz})^{0,1}$ vanishes identically. Without loss of generality we may assume that $(F_{zz})^{0,1}$ vanishes identically. But then, if s is a local nonzero section of $V\big(= (N_{\mathbb{C}}M)^{1,0}\big)$, we have, by minimality, that

$$(\partial_z s)^T = -\frac{s \cdot F_{zz}}{|F_z|^2} F_{\bar{z}} = 0 ,$$

where the last equality follows from $0 = (F_{zz})^{0,1} = \dfrac{(F_{zz} \cdot s)\bar{s}}{|s|^2}$. Thus, by Corollary 1 of §2, the proof of Theorem 2 is complete.

Chapter VI

The Existence of Minimal Immersions of 2-Spheres

In [SaU] Sacks and Uhlenbeck developed an existence theory for minimal 2-spheres in compact Riemannian manifolds. Recall that a harmonic map from a sphere is in fact a conformal branched minimal immersion. They proved the existence of minimizing harmonic maps in two cases. If N is a compact Riemannian manifold with $\pi_2(N) = 0$, then every homotopy class of maps from a closed orientable surface M to N contains a minimizing harmonic map. This has also been shown by Lemaire [Le] and Schoen and Yau [SY2]. If $\pi_2(N) \neq 0$, then there exists a generating set for $\pi_2(N)$ consisting of conformal branched minimal immersions of spheres which minimize energy and area in their homotopy classes. Their main result is the proof of the existence of a conformal branched minimal immersion of a sphere when the universal covering space of N is not contractible. When $\pi_2(N) = 0$ this cannot be minimizing for the energy in the single homotopy class of maps. An important tool that is developed is a regularity theorem that a harmonic map with finite energy from the punctured disk into N is C^∞ and harmonic in the entire disk.

One of the difficulties which will arise in their construction of harmonic maps is that the conformal parametrization by the standard sphere is not unique. The group of conformal transformation of S^2 is the group of linear fractional transformations, which is not compact, so that the set of critical maps of the energy integral on $C^1(S^2, N)$ must be noncompact. In some way it is necessary to make a choice of parametrization. This problem is solved by a perturbation technique.

Typeset by $\mathcal{A}\mathcal{M}\mathcal{S}$-TEX

We find the critical maps of a perturbed energy integral for $\alpha > 1$ (E_1 is the usual energy integral plus a constant):

$$E_\alpha(s) = \int_M (1 + |ds|^2)^\alpha d\mu \ ,$$

and check the convergence of these maps as $\alpha \to 1$. E_α satisfies Ljiusternik-Schnirelman theory and a Morse theory if $\alpha > 1$. E_α is not quite invariant under conformal transformations of the sphere, and prefers a parametrization which is carried over when the limit of the critical maps of the perturbed integral are found.

The perturbed integral $\overline{E}_\alpha(s) = \int_M |ds|^{2\alpha} d\mu$ is in many ways easier to deal with than the perturbed integral E_α because of the invariance of $\overline{E}_\alpha$ under expansion from small disks to large disks, which makes explicit calculation easier. However, at some point we need the uniform ellipticity of the Euler-Lagrange equations for the perturbed integral which is satisfied for E_α.

§1. Harmonic Maps from Surfaces

Let M denote a compact orientable surface with a given conformal structure and N^k a C^∞ Riemannian manifold without boundary of dimension $K \geq 2$. Assume that $N \subset \mathbb{R}^k$ is a C^∞ isometric embedding, that M has been given a Riemannian metric compatible with its conformal structure, and that this metric induces the measure $d\mu$ on M. Let $L_1^p(M, N)$ be the Sobolev space of maps $s : M \to N$ whose first derivatives lie in L^p. Recall that a map $s \in L_1^2(M, \mathbb{R}^k) \cap C^0(M, N)$ is harmonic if it is a critical point of the energy functional $E(s) = \int_M |ds|^2 d\mu$. If $N \subset \mathbb{R}^k$ has second fundamental form A, then the Euler-Lagrange equations have the form

$$\Delta s - A(s)(ds, ds) = 0 . \tag{1}$$

Let us now recall the following facts.

Fact 1 (Chapter I, §3). E depends only on the conformal structure of M.

Fact 2 (Chapter I, §5). If s is harmonic then Hopf differential Φ is holomorphic. ($\Phi = \{|s_x|^2 - |s_y|^2 + 2i(s_x, s_y)\}dz^2$ where $z = x + iy$ is a local isothermal parameter on M.)

Fact 3 [ES]. If s is a conformal immersion, then s is harmonic if and only if $s(M)$ is a minimally immersed surface.

Fact 4 [M], [U]. If s is harmonic then $s \in C^\infty(M, N)$.

Fact 5 If s is harmonic and weakly conformal, then s is a branched immersion.

Fact 6 (Chapter I, §5). If $s : S^2 \to N$ is harmonic and $\dim(N) \geq 3$, then s is a C^∞ conformal branched minimal immersion.

The situation for surfaces of genus larger than zero is complicated by the fact that they have many possible conformal structures. The following theorem gives a sufficient condition for a harmonic map from such a surface to be a minimal immersion.

Theorem 1.1. *If s is a critical point of E both with respect to variation of s and the conformal structure on M, then s is a conformal branched minimal immersion.*

Proof. First we show that s is critical with respect to all variations in the metric. Every variation $g(t)$ of the metric $g = g(0)$ on M arises from a composition (See e.g., [EE]) of

(a) the pull-back of g by a C^∞ family $\sigma(t)$ of orientation preserving diffeomorphisms of M,

(b) a smooth curve in the Teichmüller space for M, and

(c) a family of conformal changes in the metric.

The map s is critical for variations of type (b) and (c) by hypothesis and Fact 1 respectively. Obviously the same is true for (a).

Now we show that s is weakly conformal by proving that $\Phi \equiv 0$. Let $d\mu(t)$ be the measure on M induced by $\sigma^*(t)g$, and z a local isothermal parameter on $U \subset M$. Let $g(t) = \big(g_{ij}(t,z)\big)$ and assume that $\big(g_{ij}(t,z)\big) = (\delta_{ij})$ when $t = 0$ or for z near ∂u and that $g_{11}(t,z) = g_{22}(t,z) = 1$ for all (t,z): Then

$$\begin{aligned}
&\frac{d}{dt}\int_M g(t)(ds,\, ds)d\mu(t)\Big|_{t=0} \\
&= \frac{d}{dt}\int_U \big(|s_x(z)|^2 + |s_y(z)|^2 + 2g^{12}(t,z)\big) \\
&\qquad \big(s_x(z),\, s_y(z)\big)\sqrt{1-\big(g^{12}(t,z)\big)^2}\,dx\,dy\Bigg|_{t=0} \\
&= 2\int_U \big(s_x(z),\, s_y(z)\big)\frac{\partial}{\partial t}g^{12}(t,z)\Big|_{t=0} dx\,dy = 0\,.
\end{aligned}$$

Since $\dfrac{\partial}{\partial t}g^{12}(t,z)$ can be chosen arbitrarily, $\big(s_x(z),\, s_y(z)\big) = 0$ for $z \in U$. The same argument applied to $e^{\frac{\pi}{4}i}$ yields

$$\big(s_x(z) + s_y(z)\ ,\ s_x(z) - s_y(z)\big) = \big|s_x(z)\big|^2 - \big|s_y(z)\big|^2 = 0\,.$$

The result follows from Facts 2, 3, 5.

§2. Properties of the Perturbed Problem

For convenience, choose a measure $d\mu$ on M so that $\int_M d\mu = 1$. Let $E_\alpha(s) = \int_M (1 + |ds|^2)^\alpha d\mu$. For $\alpha = 1$, $E_1(s) = 1 + E(s)$ has harmonic maps as critical points. The Sobolev space of maps

$$L_1^{2\alpha}(M, N) = \{s \in L_1^{2\alpha}(M, \mathbb{R}^k) : s(x) \in N\} \subset C^0(M, N)$$

is a C^2 separable Banach manifold for $\alpha > 1$.

Theorem 2.1 [P1]. *E_α is C^2 on the Banach manifold $L_1^{2\alpha}(M, N)$ and satisfies the Palais-Samle condition (C) in a complete Finsler metric on $L_1^{2\alpha}(M, N)$ provided that N is compact.*

REMARK. We say that a function f on a Banach manifold B satisfies the *Palais-Smale condition (C)*, if for any set $S \subset B$ on which $|f(S)|$ is bounded and $|df(S)|$ is not bounded away from zero, the closure $\overline{S}$ contains a critical point of f (a point x where $df_x = 0$).

Theorem 2.2 [P2]. *If f is a C^2 function on a complete, separable C^2 Finsler manifold L, which satisfies the Palais-Smale condition (C) with respect to the Finsler structure, then*

(a) *f takes on its minimum in every component of L,*

(b) if there are no critical values of f in the interval $[a, b]$. then there exists a deformation retraction $\rho : f^{-1}(-\infty, b] \to f^{-1}(-\infty, a]$.

Proposition 2.3. *The critical maps of E_α in $L_1^{2\alpha}(M, N)$ are C^∞ if $\alpha > 1$.*

Proof. By the Sobolev Embedding Theorem, a critical map s is of Hölder class $C^{1-\frac{1}{\alpha}}(M, N)$. By [M,1.11.1], $ds \in L_1^2(M, N)$. The Euler-Lagrange equation is

$$\Delta s + 2(\alpha - 1)\frac{(d^2 s, ds)ds}{1 + |ds|^2} - A(s)(ds, ds) = 0\ . \tag{2}$$

If $\alpha - 1$ is small, we have nice inverses for the linear operator Δ_s : $L^4_s(M,N) \to L^4_0(M,N)$ where

$$\Delta_s u = \Delta u + 2(\alpha - 1)\frac{(d^2u,\ ds)ds}{(1+|ds|^2)} \ . \tag{3}$$

It follows that $s \in L^4_2(M,N) \subset C^1(M,N)$. The smoothness of s follows from Schauder theory.

Proposition 2.4. *Let $\alpha > 1$. In every connected component of $L^{2\alpha}_1(M,N)$ the minimum value of E_α is taken on at some map $s_\alpha \in C^\infty(M,N)$, which also minimizes E_α in its connected component in $C^\infty(M,N)$. There exists a B independent of α such that $\min E_\alpha \le (1+B^2)^\alpha$ in that component.*

Proof. The first statement follows from Proposition 2.3. Let $B = \max_{x\in M}|du(x)|$ for a fixed u in each component. Then $\min E_\alpha \le E_\alpha(u) \le (1+B^2)^\alpha$.

The main difficulty in proving the existence of nontrivial maps of E_α for $\alpha > 1$ is that $N_0 = \{s : s(M) = y \in N\} \cong N$ is the set of trivial critical maps of E_α on which E_α takes its absolute minimum value 1. Recall that at $y \in N_0 \subset L^{2\alpha}_1(M,N)$, we can interpret $T_yL^{2\alpha}_1(M,N) = L^{2\alpha}_1(M,T_yN)$ and $T_yN_0 = \{a : da = 0\}$. Then in a weak L^2 sense we construct a normal bundle to N_0,

$$\mathcal{N} = \bigcup_{y\in N_0} \mathcal{N}_y \subset TL^{2\alpha}_1(M,N)\big|N_0 \ ,$$

$$\mathcal{N}_y = \left\{v \in L^{2\alpha}_1(M,T_yN) : \int_M v\,d\mu = 0\right\} .$$

Define $e : TL^{2\alpha}_1(M,N) \to L^{2\alpha}_1(M,N)$ by the formula $e(s,v)(x) = \exp\big(s(x),\, v(x)\big)$. From the implicit function theorem we get the following.

Lemma 2.5. *$e|\mathcal{N} \to L_1^{2\alpha}(M,N)$ is a diffeomorphism from a neighborhood of the zero section of $\mathcal{N}$ to a neighborhood of $N_0 \subset L_1^{2\alpha}(M,N)$.*

Theorem 2.6. *Given $\alpha > 1$, there exists a $\delta > 0$ depending on α and a deformation retraction*

$$\sigma : E_\alpha^{-1}[1, 1+\delta] \to E_\alpha^{-1}(1) = N_0 \ .$$

Proof. From the above lemma we may conclude that if δ is sufficiently small and $s \in E_\alpha^{-1}[1, 1+\delta]$, then there exist $y \in N$, $v \in \mathcal{N}_y$ such that $s = e(y,v)$ with $\|v\|_\infty$ and $\int_M |dv|^{2\alpha}$ being arbitrarily small. Define the candidate for a retraction σ: $E_\alpha^{-1}[1,1+\delta] \times [0,1] \to L_1^{2\alpha}[M,N]$ by $\sigma(s,t) = e(y,tv)$ for $s = e(y,v)$. Then $\sigma(s,1) = s$, $\sigma(s,t) = y \in N$ and σ is continuous if δ is sufficiently small. Now we can show that $\dfrac{d}{dt}E_\alpha\big(\sigma(s,t)\big) \geq 0$ and thus $\sigma(\cdot, 0)$ is a retraction.

Let $z_0 \in M$ and $q_0 \in N$ be chosen as base points for M and N. $\Omega(M,N)$ will denote the space of base point-preserving maps from M to N. The map $p : C^0(M,N) \to N$ defined by $p(s) = s(z_0)$ is a fibration with fiber $\Omega(M,N)$. This fibration p has a section $N \to N_0 \subset C^0(M,N)$ defined by mapping $q \in N$ to the constant map $s(M) = q$. Therefore the exact homotopy sequence splits and

$$\pi_k\big(C^0(M,N)\big) = \pi_k(N) \oplus \pi_k\big(\Omega(M,N)\big) \ .$$

Theorem 2.7. *If $\Omega(M,N)$ is not contractible, then there exists a $B > 0$ such that for all $\alpha > 1$, E_α has a critical value in the interval $\big(1, (1+B^2)^\alpha\big)$.*

Proof. If $C^0(M,N)$ is not connected, apply Proposition 2.4. Otherwise choose a nonzero homotopy class $\gamma \in \pi_k\big(\Omega(M,N)\big)$. Note that γ is not homotopic to any map $\tilde{\gamma} : S^k \to N_0$. Let $B = \max\limits_{y \in S^k, x \in M} \big|d\gamma(y)(x)\big|$.

Then $E_\alpha\big(\gamma(y)\big) \leq (1+B^2)^\alpha$ for all $y \in S^k$. Suppose that E_α has no critical value in the interval $\big(1,(1+B^2)^\alpha\big)$. Then by Theorem 2.2 there exists a deformation retraction $\rho : E_\alpha^{-1}\big[1,(1+B^2)^\alpha\big] \to E_\alpha^{-1}[1, 1+\delta]$ for all $\delta > 0$. Composing ρ with the deformation retraction σ of Theorem 2.6 produces the deformation retraction

$$\sigma \circ \rho : E_\alpha^{-1}\big[1,(1+B^2)^\alpha\big] \to E_\alpha^{-1}(1) = N_0 \ .$$

But $\sigma \circ \rho \circ \gamma : S^k \to N_0$ is homotopic to γ, a contradiction.

If the covering space of N is not contractible, $\pi_{k+2}(N) = \pi_k\big(\Omega(S^2, N)\big) \neq 0$ for some $k \geq 0$. Hence we have the following.

Proposition 2.8. *IF $M = S^2$ and the universal covering space of N is not contractible, then there exists a B and a critical map of E_α with values in the range $\big(1,(1+B^2)^\alpha\big)$ for $\alpha > 1$.*

§3. Estimates and Extensions

In this section we obtain the main *a prior* estimate needed for convergence and for proving the regularity theorem, Theorem 3.6. As we have seen earlier in the existence theory of harmonic maps, the curvature of N plays an important role. However the curvature and topology of M do not play a role in the *a priori* estimates. To see this, cover M by small disks of radius R on which the metric differs from the ordinary Euclidean metric by terms of order ε. When we expand these disks conformally to be of unit size the integral becomes $E_\alpha(S) = R^{2(1-\alpha)} \int_D (R^2 + |ds|^2)^\alpha d\mu$ where D is the unit disk. Here, the smaller the original disk, the nearer to Euclidean is the metric on the expanded disk. For this reason, *a priori* estimates are uniform in $\alpha \geq 1$, $0 < R \leq 1$.

The Euler-Lagrange equations (2) and (3) appear in the following forms after this conformal dilation:

$$d^*\left(R^2 + |ds|^2\right)^{1-\alpha} ds - \left(R^2 + |ds|^2\right)^{1-\alpha} A(ds, ds) = 0 \tag{4}$$

or

$$\Delta S + 2(\alpha - 1)(d^2 s, ds) ds \left(R^2 + |ds|^2\right)^{-1} - A(ds, ds) = 0 \,. \tag{5}$$

Proposition 3.1. *Let $s : D \to N$ be a critical point of E_α. If $\alpha - 1 > 0$ is sufficiently small depending on $\infty > p > 1$, for all smaller disks $D' \subset D$,*

$$\|ds\|_{D',1,p} \leq k\left(p, D', \|s\|_{D,0,4}\right) \|ds\|_{D,0,4} \,.$$

Proof. Let φ be a smooth function with support in D which is 1 on D', and choose a base point in $\mathbb{R}^k$ so that $\int s = 0$. Multiplying (5) by φ and taking L^p norms for each term, we get an estimate of the form

$$\begin{aligned} \|\Delta(\varphi s)\|_{0,p} \leq 2(\alpha - 1)\|\varphi s\|_{2,p} + \|A\|_{0,\infty} \big\| |d(\varphi s)| \cdot |ds| \big\|_{0,p} \\ + k(\varphi)\|s\|_{1,p} \,. \end{aligned} \tag{6}$$

Let $c(p)$ be the norm of Δ^{-1} as a map from L^p_0 to $L^p_2 \cap L^2_{1,0}$ on the disk. Then from (6) we get

$$\begin{aligned} c(p)^{-1}\|\varphi s\|_{2,p} \leq 2(\alpha - 1)\|\varphi\|_{2,p} + \|A\|_{0,\infty} \| |d(\varphi s)| \cdot |ds| \|_{0,p} \\ + k(\varphi)\|s\|_{1,p} \,. \end{aligned} \tag{7}$$

Letting $p = 2$ and assuming $2(\alpha - 1) < c(2)^{-1}$, we get a bound on $\|s\|_{D'',2,2}$ where $D'' = \{x \in D : \varphi(x) = 1\}$. By the Sobolev inequality, this gives a bound on $\|s\|_{D'',1,p}$ for all p. Repeating (7) for any p with $\operatorname{supp} \varphi \subset D''$, we get a bound on $\|s\|_{2,p}$ in D''.

Main Estimate 3.2. *There exists* $\varepsilon > 0$ *and* $\alpha_0 > 1$ *such that if* $s : D \to N$ *is a smooth critical map of* E_α, $E(s) < \varepsilon$ *and* $1 \le \alpha < \alpha_0$, *then there is an estimate uniform in* $1 \le \alpha < \alpha_0$,

$$\|ds\|_{D',1,p} < C(p, D')\|ds\|_{0,2}$$

for $D' \subset D$ *any smaller disk.*

Proof. By the above Proposition we need only get a bound on $\|ds\|_{D'',0,4}$, $D'' \subset D$. Use (7) with $p = \dfrac{4}{3}$. By the Hölder inequality we estimate the bad quadratic term $\||ds| \cdot |d(\varphi s)|\|_{0,\frac{4}{3}}$ and then by Sobolev we have $\|d(\varphi s)\|_{0,4} \le k'\|\varphi s\|_{2,\frac{4}{3}}$. Now (7) becomes

$$\Big(c(\frac{4}{3})^{-1} - 2(\alpha - 1)\Big)\|\varphi s\|_{2,\frac{4}{3}} \le k'\|A\|_{0,\infty}\|ds\|_{0,2}\|\varphi s\|_{2,\frac{4}{3}} + k(\varphi)\|ds\|_{0,\frac{4}{3}} \,. \qquad (8)$$

Certainly $\|ds\|_{0,\frac{4}{3}} \le \sqrt{E(s)}$ and $(k')^{-1}\|\varphi s\|_{1,4} \le \|\varphi s\|_{2,\frac{4}{3}}$. Therefore if in addition $\big(c(\frac{4}{3})^{-1} - 2(\alpha - 1)\big) - k'\big(\|A\|_{0,\infty}\|ds\|_{0,2}\big) > 0$, we get the desired estimate.

Naturally the above estimate can be made globally on M without the boundary terms containing $k(\varphi)$. Then (8) has the global form

$$\Big(c(\frac{4}{3})^{-1} - 2(\alpha - 1)\Big)\|s\|_{2,\frac{4}{3}} \le k'\|A\|_{0,\infty}\sqrt{E(s)}\|s\|_{2,\frac{4}{3}} \,.$$

Clearly, if $E(s)$ is too small, this has no solutions except $s \equiv \int s$. Hence we get the following.

Theorem 3.3. *There exists* $\varepsilon > 0$ *and* $\alpha_0 > 1$ *such that if* $E(s) < \varepsilon$, $1 \le \alpha < \alpha_0$ *and* s *a critical map of* E_α, *then* $s \in N_0$ *and* $E(s) = 0$.

We will now prove that if $s : D - \{0\} \to N$ is harmonic with finite energy then s is smooth and harmonic in D. In the case that

s is a strict minimum, this was proved by another method by Morrey [M, 4.3]. Notice that due to the conformal equivalence of $D - \{0\}$ with $\mathbb{R}^2 - D$, this theorem can be interpreted as a theorem on the growth at infinity of harmonic maps. We let $D(x_0, R)$ denote the disk of radius R and center x_0; $D(R) = D(0,R)$; $D = D(1)$. Since $\int_D |ds|^2 d\mu < \infty$, $\lim_{R\to 0} \int_{D(R)} |ds|^2 d\mu = 0$. By a conformal expansion to $D(R)$ with $\int_{D(R)} |ds|^2 d\mu < \varepsilon$, we can assume $\int_{D(2)} |ds|^2 d\mu < \varepsilon$. We choose ε later.

Lemma 3.4. *There exists $\varepsilon > 0$, such that if $\int_{D(2)} |ds|^2 d\mu < \varepsilon$, then there exists a constant c such that*

$$\left|ds(x)\right| \cdot |x| < c\Big(\int_{D(2|x|)} |ds|^2 d\mu\Big)^{\frac{1}{2}} \quad \textit{for} \quad x \in D\,.$$

Proof. Define $\tilde{s}(x) = s(x_0 + |x_0|x)$. Then by Sobolev and 3.2 we get

$$\begin{aligned}|ds(x_0)| \cdot |x_0| &= \max_{x \in D(\frac{1}{2})} \left|d\tilde{s}(x)\right| \le \tilde{c}\|ds\|_{D(\frac{1}{2}),1,4} \\ &\le \tilde{c}c\|d\tilde{s}\|_{0,2} \le c\|ds\|_{D(2|x_0|),0,2}\,.\end{aligned}$$

Lemma 3.5. *Let $s : D - \{0\} \to N \subset \mathbb{R}^k$ be a smooth harmonic map such that $E(s) < \infty$. Then*

$$\int_0^{2\pi} \left|s_\theta(z)\right|^2 d\theta = r^2 \int_0^{2x} \left|s_r(z)\right|^2 d\theta\,.$$

Proof. Let $\Phi = w(z)dz^2$ be the Hopf differential. Since $\int_D |w(z)| d\mu \le 2\int_D |ds|^2 d\mu < \infty$, $w(z)$ has a pole of order at most one. A direct computation shows that $\operatorname{Re} w(z)z^2 = \left|s_\theta(z)\right|^2 - |z|^2\left|s_r(z)\right|^2$. Now the lemma follows from Cauchy's theorem.

Theorem 3.6. *If $s : D - \{0\} \to N$ is harmonic with finite energy, then s extends to a smooth harmonic map $s : D \to N$.*

Proof. We approximate s by a radial function q which is piecewise linear in $\log r$. Let $q(2^{-m}) = \dfrac{1}{2\pi}\int_0^{2\pi} s\big(2^{-m},\theta\big) d\theta$. Then $q(r)$ is harmonic for r between 2^{-m+1} and 2^{-m}, $m \geq 1$. Using lemmas 3.4 and 3.5, and the fact that q is the average of s at 2^{-m} and that q does not depend on θ, we can show that $\frac{1}{2}\int_{D(1)} |ds|^2 d\mu \leq \int_{D(1)} |ds - dq|^2 d\mu \leq \frac{1}{2}\int_{C(1)} |s_r|^2 d\theta + \delta \int_{D(1)} |ds|^2 d\mu$, $C(r) = \partial D(r)$. If we translate this inequality by contraction into a disk of any radius, we get for $r \leq 1$,

$$(1 - 2\delta)\int_{D(r)} |ds|^2 d\mu \leq r^2 \int_{C(r)} |ds|^2 d\theta \, .$$

This inequality integrates to yield $\int_{D(r)} |ds|^2 d\mu \leq r^{1-2\delta} \int_D |ds|^2 d\mu$. Apply Lemma 3.4 again to get, for $0 < |x_0| < \frac{1}{2}$,

$$|ds(x_0)| \cdot |x_0| \leq c|2x_0|^{\frac{1}{2}(1-2\delta)} \Big(\int_D |ds|^2 d\mu\Big)^{\frac{1}{2}} \, .$$

This now implies $s \in L_1^{2\alpha}(D, N)$ for $\alpha > 1$, and the proof of Proposition 2.3 applied to equation (1) give regularity.

§4. Convergence Properties of Critical Maps of the Perturbed Problem

In the previous section we obtained existence of critical maps of the perturbed integrals and some uniform estimates. The main result of this section, Theorem 4.7, is that as $\alpha \to 1$, either these converge to a harmonic map, or there is a minimal sphere acting as an obstruction.

Lemma 4.1. *Let s_α, for $\alpha \to 1$, be a sequence of critical maps for E_α, with $E_\alpha(s_\alpha) \leq B$. Then there exists a subsequence $\{\beta\} \subset \{\alpha\}$ such that $s_\beta \to s$ weakly in $L_1^2(M, \mathbb{R}^k)$ and $\lim\limits_{\beta \to 1} E(s_\beta) \geq E(s)$.*

Proof. Every bounded set in the Hilbert space $L_1^2(M, \mathbb{R}^k)$ is weakly compact.

As in the previous section, we assume M has been covered by small disks of radius $R = 2^{-m}$, and we make the additional assumption that each point of M is contained in at most h disks, where h is uniform as $R \to 0$. If we expand these small disks to unit size, the integral appears on the unit disks D in the form $E_\alpha(s) = \int_D \left(R^2 + |ds|^2\right)^\alpha d\mu$. Define $\tilde{E}_\alpha = E_\alpha - R^{2\alpha} \int_D d\mu$.

Lemma 4.2. *Let $s_\alpha : D(R) \to N$ be a sequence of critical maps of E_α for a sequence $\alpha \to 1$, which is weakly convergent in $L^2_1(D(R), \mathbb{R}^k)$. Then there exists $\varepsilon > 0$ such that if $E(s_\alpha) < \varepsilon$ then $s_\alpha \to s$ in $C^1\left(D(\frac{R}{2}), N\right)$ and $s : D(\frac{R}{2}) \to N$ is a smooth harmonic map.*

Proof. We may assume $D(R) = D$. From the Main Estimate 3.2 we have a uniform estimate $\|ds_\alpha\|_{D(\frac{1}{2}),1,4} \leq C(4, D(\frac{1}{2}))\varepsilon$. By Sobolev we have $L^p_2(D(\frac{1}{2}), \mathbb{R}^k) \subset C^1(D(1), \mathbb{R}^k)$, and hence $s_\alpha \to s$ in $C^1(D(\frac{1}{2}), \mathbb{R}^k)$. Equation (4) shows that s is harmonic.

Proposition 4.3. *Let $U \subset M$ be an open set and $s_\alpha : U \to N \subset \mathbb{R}^k$ be a sequence of critical maps of E_α for $\alpha \to 1$, $s_\alpha \to s$ weakly in $L^2_1(U, \mathbb{R}^k)$, $E_\alpha(s_\alpha) < B$. Let $U_m = \{x \in U : D(x, 2^{-m}) \subset U\}$. Then there exists a subsequence $\{\alpha(\ell)\} \subset \{\alpha\}$ and a finite number of points $\{x_{1,m}, \ldots, x_{\ell,m}\}$ where ℓ depends on B and N but not on m, such that*

$$s_{\alpha(\ell)} \to s \text{ in } C^1\left(U_m - \bigcup_{i=1}^{\ell} D(x_{i,m}, 2^{-m-1}), N\right).$$

Proof. Cover U_m by disks $D(x_i, 2^{-m}) \subset U$ such that each point $x \in U$ is covered at most h times and the disks of half the radius cover U_m. Then $\sum_i \int_{D(x_i,2^{-m})} |ds_\alpha|^2 d\mu < Bh$ and for each α there are at most $\dfrac{Bh}{\varepsilon}$ disks on which $\int_{D(x_i,2^{-m})} |ds_\alpha|^2 d\mu > \varepsilon$, where ε is the constant from Lemma 4.2. Except for these $\frac{Bh}{\varepsilon}$ disks, we apply Lemma 4.2 to each disk, and thereby we obtain the desired result.

Theorem 4.4. *Let $U \subset M$ be an open set and $s_\alpha : U \to N$ be a critical map of E_α and $E(s_\alpha) < B$, $\alpha \to 1$ and $s_\alpha \to s$ weakly in $L_1^2(U, \mathbb{R}^k)$. Then there exists a subsequence $\{\beta\} \subset \{\alpha\}$ and a finite number of points $\{x_1, \dots, x_\ell\}$, where ℓ does not depend on U such that $s_\beta \to s$ in $C^1\big(U - \{x_1, \dots, x_\ell\}, N\big)$. Moreover, $s : U \to N$ is a smooth harmonic map.*

Proof. The first part follows from the above proposition by letting $m \to \infty$. Since $s \in C^1\big(U - \{x_1, \dots, x_\ell\}, N\big)$ and $E(s) \leq \lim_{\alpha \to 1} E(s_\alpha) < B$, we can apply Theorem 3.6 to get $s : U \to N$ smooth and harmonic.

We have no assurance that s is not trivial when $U = M$, or that the convergence can be extended over the points $\{x_1, \dots, x_\ell\}$ in the theorem. However in some cases we can directly argue the convergence $s_\alpha \to s$ in the C^1 topology.

Lemma 4.5. *Suppose that the hypotheses of Theorem 4.4 are true and there exists $\delta > 0$ such that $\max_{x \in D(x_i, \delta)} \big|ds_\alpha(x)\big| \leq B < \infty$. Then $s_\alpha \to s$ in $C^1\big(D(x_i, \delta), N\big)$.*

Proof. Choose R such that $\int_{D(x_i, R)} \big|ds_\alpha\big|^2 d\mu \leq \pi R^2 B^2 < \varepsilon$. Then apply Lemma 4.2.

Theorem 4.6. *Let s_α be a sequence of critical maps of E_α for $\alpha \to 1$, $E_\alpha(s_\alpha) < B$ and $s_\alpha \to s$ in $C^1\big(M - \{x_1, \dots, x_\ell\}, N\big)$ but not in $C^1\big(M - \{x_2, \dots, x_\ell\}, N\big)$. Then there exists a nonconstant harmonic map $\tilde{s} : S^2 \to N$ such that*

$$\tilde{s}(S^2) \subset \bigcup_{m \to \infty} \Big(\bigcap_{\alpha \to 1} \bigcup_{\beta \leq \alpha} s_\beta(D(x_1, 2^{-m})) \Big) .$$

Moreover

$$E(s) + E(\tilde{s}) \leq \overline{\lim_{\alpha \to 1}} E(s_\alpha) .$$

Proof. Let $b_\alpha = \max_{x \in D(x_1, 2^{-m})} |ds_\alpha(x)|$ and assume $|ds_\alpha(x_\alpha)| = b_\alpha$, $x_\alpha \in D(x_1, 2^{-m})$. By Lemma 4.5 we may assume $\lim_{\alpha \to 1} b_\alpha = \infty$. Moreover $\lim_{\alpha \to 1} x_\alpha = x_1$. Define $\tilde{s}_\alpha(x) = s(x_\alpha + b_\alpha^{-1} x)$. Then $\tilde{s}_\alpha : D(0, 2^{-m} b_\alpha) \to N$ is a critical map of E_α, $|d\tilde{s}_\alpha(x)| \leq 1$ for $x \in D(0, 2^{-m} b_\alpha)$, and $|d\tilde{s}_\alpha(0)| = 1$. Using Theorem 4.4 and Lemma 4.5, for any $R < \infty$, we can find $\tilde{s}_\alpha \to \tilde{s}$ in $C^1(D(R), N)$, where $\tilde{s} : D(R) \to N$ is smooth and harmonic. Since $|d\tilde{s}(0)| = 1$, $\tilde{s}$ cannot be a map to a point. For a subsequence $\tilde{s}_\beta \to \tilde{s}$ in $C^1(\mathbb{R}^2, N)$,

$$\begin{aligned}
&E(\tilde{s}) + E(s|M - D(x_1, 2^{-m})) \\
&\leq \overline{\lim_{\beta \to 1}} \{ E(\tilde{s}_\beta | D(0, 2^{-m} b_\beta)) + E(s_\beta | M - D(x_1, 2^{-m})) \} \\
&= \overline{\lim_{\beta \to 1}} E(s_\beta) .
\end{aligned}$$

By letting $m \to \infty$, we have $E(\tilde{s}) + E(s) \leq \lim_{\beta \to 1} E(s_\beta)$. But $\mathbb{R}^2 = S^2 - \{p\}$ conformally, and $E(\tilde{s}) < \infty$, so from Theorem 3.6, $\tilde{s}$ extends to a map $\tilde{s} : S^2 \to N$.

Theorem 4.7. *Let s_α be a sequence of critical maps of E_α for $\alpha \to 1$, and $s_\alpha \to s$ weakly in $L_1^2(M, \mathbb{R}^k)$. Then either $s_\alpha \to s$ in $C^1(M, N)$, or there exists a nontrivial harmonic map $\tilde{s} : S^2 \to N$ with $\tilde{s}(S^2) \subset \bigcap_{\alpha \to 1} \overline{\bigcup_{\beta < \alpha} s_\beta(M)}$. Moreover $E(s) + E(\tilde{s}) \leq \overline{\lim_{\alpha \to 1}} E(s_\alpha)$.*

§5. Applications and Results

In this section ε is a uniform constant depending on the second fundamental form of the embedding $N \subset \mathbb{R}^k$ and is assumed to be the minimum of the constants appearing in the Main Estimate 3.2, Theorem 3.3, and Lemma 4.2.

Theorem 5.1. *If N is compact and $\pi_2(N) = 0$, then there exists a minimizing harmonic map in every homotopy class of maps in $C^0(M, N)$.*

Proof. By Proposition 2.4 there exists a minimizing map $s_\alpha : M \to N$ for E_α in a fixed homotopy class. By Theorem 4.4, we can choose a subsequence $\beta \to 1$ such that $s_\beta \to s$ in $C^1\big(M - \{x_1, \dots, x_\ell\}, N\big)$ with $s : M \to N$ harmonic. We claim $s_\beta \to s$ in $C^1(M, N)$.

Let $D(\rho) = D(x_i, \rho)$ and define a modified map $\hat{s}_\beta : D(\rho) \to N$ which agrees with s_β outside $D(\rho)$ and with s at x_i. Let η be a smooth function which is 1 on $r \geq 1$ and 0 on $r \leq \frac{1}{2}$. Let

$$\hat{s}_\beta(x) = \exp_{s(x)}\Big(\eta\big(\frac{|x|}{\rho}\big) \exp^{-1}_{s(x)} \circ\, s_\beta(x)\Big) \,. \tag{9}$$

Then $\hat{s}_\beta \to s$ in $C^1\big(D(\rho), N\big)$, and since $\tilde{E}_\alpha(s) = \int_M \big(1+|ds|^2\big)^\alpha d\mu - 1$,

$$\lim_{\beta \to 1} \tilde{E}_\beta(\hat{s}_\beta) = E\big(s|D(\rho)\big) \,. \tag{10}$$

By assumption s_β and $\hat{s}_\beta$ are homotopic. Since s_β is minimizing, $E_\beta\big(s_\beta|D(\rho)\big) \leq E_\beta\big(\hat{s}_\beta|D(\rho)\big)$. From (10) we have

$$\overline{\lim_{\beta \to 1}}\, E_\beta\big(s_\beta|D(\rho)\big) \leq E\big(s|D(\rho)\big) \leq \rho^2 \pi \|s\|^2_{1,\infty} \,.$$

Hence if ρ is sufficiently small, we can apply Lemma 4.2 to get $s_\beta \to s$ in $C^1\big(D(\rho), N\big)$. We may conclude $s_\beta \to s$ in $C^1(M, N)$. Since s_β minimizes $\tilde{E}_\beta$, s must minimize E in the same homotopy class.

A free homotopy class of (unbased) maps from M to N induces a map between $\pi_1(M)$ and $\pi_1(N)$. The following theorem implies the previous theorem and has exactly the same proof as the previous one.

Theorem 5.2. *Every conjugacy class of homomorphisms from $\pi_1(M)$ into $\pi_1(N)$ is induced by a minimizing harmonic map from M into N.*

Each $\gamma \in \pi_2(N)$ determines a free homotopy class of maps from S^2 into N and two elements γ and γ' in $\pi_2(N)$ determine the same free homotopy class if and only if they belong to the same orbit $\pi_1(N)\gamma = \pi_1(N)\gamma'$ under the usual action of $\pi_1(N)$ on $\pi_2(N)$. We denote by $\Gamma \in \pi_0 C^0(S^2, N)$ the free homotopy class corresponding to $\pi_1(N)\gamma$, and we write $\gamma \in \Gamma$ for any such γ. Given $\Gamma_i = \pi_1(N)\gamma_i$ for $i = 1, 2, 3$, with $\gamma_1 + \gamma_2 = \gamma_3$, then since $\alpha\gamma_1 + \alpha\gamma_2 = \alpha\gamma_3$ we have $\pi_1(N)\gamma_3 \subset \pi_1(N)\gamma_1 + \pi_1(N)\gamma_2$. This relation is the substitution for addition in free homotopy classes. Define

$$\begin{aligned}\#\Gamma &= \min\left\{E(s) : s \in \Gamma \cap L_1^\infty(S^2, N)\right\}\\ &= \lim_{\alpha\to 1}\left\{\min \tilde{E}_\alpha(s) :\ s \in \Gamma \cap L_1^\infty(S^2, N)\right\}.\end{aligned}$$

Note that $\#\Gamma = 0$ if and only if Γ is trivial and that $\#\Gamma > \varepsilon$ otherwise.

Lemma 5.3. *Let $s_\alpha : S^2 \to N$, $\alpha \to 1$, be a sequence of nontrivial critical maps of E_α with $s_\alpha \to s$ in $C^1\big(S^2 - \{p\}, N\big)$. Then $s : S^2 \to N$ is not a map to a point.*

Proof. Computing the variation of $\tilde{E}_\alpha$ at s_α in the radial direction, Sacks and Uhlenbeck obtained

$$\begin{aligned}&\frac{\alpha}{2}\int_{S^+}\left(1 + |ds_\alpha|^2\right)^{\alpha-2}|ds_\alpha|^2\cos\phi\, d\mu\\ \leq\ &\int_{S^-}\left(1 + |ds_\alpha|^2\right)^{\alpha-2}|ds_\alpha|^2(-\cos\phi)d\mu\,,\end{aligned} \tag{11}$$

where S^+, S^- are upper and lower hemispheres respectively. If s is trivial, by Theorem 3.3, s_α cannot approach s in $C^1(S^2, N)$. From Theorem 4.6 we see that there exist expansions $\tilde{s}_\alpha$ of s_α near p which converge to $\tilde{s} : S^2 \to N$. Then

$$\begin{aligned}\frac{1}{2}E(\tilde{s}) &\leq \lim_{\alpha\to 1} E(s_\alpha|D_\alpha) \leq \lim_{\alpha\to 1}\int_{S+} \left|ds_\alpha\right|^2 \cos\phi\, d\mu \\ &\leq \lim_{\alpha\to 1}\Big(-\int_{S-}\left(1+|ds_\alpha|\right)^{\alpha-1}\left|ds_\alpha\right|^2\cos\phi\Big)d\mu \\ &= -\int_{S-}\left|ds\right|^2\cos\phi\, d\mu\ .\end{aligned}$$

Here we use (11) in the last inequality. If $ds = 0$, $\tilde{s}$ is trivial, which is impossible.

Lemma 5.4. *Let $\Gamma \in \pi_0 C^0(S^2, M)$. Then either Γ contains a minimizing harmonic map s or for all $\delta > 0$ there exist nontrivial free homotopy classes $\Gamma_1 = \pi_1(N)\gamma_1$ and $\Gamma_2 = \pi_1(N)\gamma_2$ such that $\Gamma = \pi_1(N)\gamma \subset \gamma_1(N)\gamma_1 + \pi_1(N)\gamma_2$ and $\#\Gamma_1 + \#\Gamma_2 < \#\Gamma + \delta$.*

Proof. By Proposition 2.4 and Theorem 4.4 we can find a sequence $\alpha \to 1$ and maps $s_\alpha \in \Gamma$ which are minimizing for $\tilde{E}_\alpha$ and which converge to s in $C^1\big(S^2 - \{x_1, \dots, x_\ell\}, N\big)$. We can assume that $\lim\limits_{m\to\infty}\overline{\lim\limits_{\alpha\to 1}}\tilde{E}_\alpha\big(s_\alpha|$ $D(x_1, 2^{-m})\big) \geq \varepsilon$, for if not we can apply Lemma 4.2 to remove the singularity x_1 of the convergence of s_α to s. If we can remove all the x_i in this fashion, $s_\alpha \to s$ in $C^1(S^2, N)$ and $s \in \Gamma$ is a harmonic map which minimizes E in Γ. Assuming we cannot do this, pick a small disk $D(\rho)$ around x_1 and use construction (9) to define $\hat{s}_\alpha : D(\rho) \to N$. Let

$$u_\alpha(x) = \begin{cases} s_\alpha(x) & x \in S^2 - D(\rho) \\ \hat{s}_\alpha(x) & x \in D(\rho)\ , \end{cases}$$

$$v_\alpha(x) = \begin{cases} \hat{s}_\alpha \circ f(x) & x \in S^2 - D(\rho) \\ s_\alpha(x) & x \in D(\rho)\ . \end{cases}$$

Here $f : S^2 - D(\rho) \to D(\rho)$ is the conformal reflection leaving the boundary of $D(\rho)$ fixed. Let Γ_1 and Γ_2 be the free homotopy classes of u_α and v_α respectively. The $\pi_1(N)\gamma \subset \pi_1(N)\gamma_1 + \pi_1(N)\gamma_2$. From (10) we can show $\#\Gamma_1 + \#\Gamma_2 < \#\Gamma + \delta$. From the assumption $\lim_{m\to\infty} \overline{\lim_{\alpha\to 1}} \tilde{E}_\alpha (s_\alpha | D(x_1, 2^{-m})) \geq \varepsilon$, we can easily see that $\#\Gamma_2 \neq 0$. If $\#\Gamma_1 = 0$, s would be trivial and then, by Lemma 5.3, there would exists $x_2 \neq x_1$ where $\lim_{m\to\infty} \overline{\lim_{\alpha\to 1}} \tilde{E}_\alpha (s_\alpha | D(x_2, 2^{-m})) \geq \varepsilon$. This must imply that $\#\Gamma_1 \neq 0$ a contradiction. Anyway we can conclude that $\#\Gamma_1 \neq 0$.

Theorem 5.5. *There exists a set of free homotopy classes $\Lambda_i \subset \pi_0 C^0(S^2, N)$ such that elements $\{\lambda \in \Lambda_i\}$ form a generating set for $\pi_2(N)$ as a $Z[\pi_1(N)]$ module (i.e., acted on by $\pi_1(N)$), and each Λ contains a minimizing harmonic map $s_i : S^2 \to N$.*

Proof. Let Λ_i be the homotopy classes containing minimizing harmonic maps. Let $P \subset \pi_2(N)$ be the subgroup generated by $\{\lambda \in \Lambda_i\}$. Suppose $P \neq \pi_2(N)$. Pick a class Γ with $\gamma \in \Gamma$, $\gamma \notin P$ such that if $\#\Gamma' \leq \#\Gamma - \frac{\varepsilon}{2}$, then $\{\gamma' \in \Gamma'\} \subset P$.

By assumption there exist Γ_1 and Γ_2 with $\pi_1(N)\gamma \subset \pi_1(N)\gamma_1 + \pi_1(N)\gamma_2$, $\#\Gamma_1 + \#\Gamma_2 < \#\Gamma + \frac{\varepsilon}{2}$, and $\#\Gamma_j \geq \varepsilon$ for $j = 1, 2$. All this implies $\#\Gamma_j < \#\Gamma - \frac{\varepsilon}{2}$. By assumption the sets $\pi_1(N)\gamma_j$ are both in P, so

$$\pi_1(N)\gamma \subset \pi_1(N)\gamma_1 + \pi_1(N)\gamma_2 \subset P\,.$$

In the next two theorems we look into the situation where the harmonic maps are not necessarily minimizing but may be saddle points. The first theorem treats the case where there are no obstructions to our convergence technique.

Theorem 5.6. *Let $\varepsilon_0 = \min E(s)$ for nontrivial harmonic map $s : S^2 \to N$ and $\varepsilon_0 = \infty$ if this set of harmonic maps is empty. Then $E|E^{-1}[0, \varepsilon_0)$ satisfies a Morse theory for $M \neq S^2$, and for $M = S^2$, $E|E^{-1}[0, 2\varepsilon_0)$ satisfies a Morse theory.*

Proof. The set $\{s_\alpha : s_\alpha$ is critical for E_α, $\tilde{E}_\alpha(s_\alpha) \le \delta < \varepsilon_0\}$ is compact, since by Theorem 4.6, $s_\alpha \to s$ in $C^1(M, N)$ unless there is a minimal harmonic map $\tilde{s} : S^2 \to N$ with image lying in the Hausdorff limit set of $s_\alpha(M)$. In this case $\lim\limits_{\alpha \to 1} \tilde{E}_\alpha(s_\alpha) \ge E(\tilde{s}) + E(s) \ge E(\tilde{s}) \ge \varepsilon_0$. If $M = S^2$, by Lemma 5.3, $E(s) \ge \varepsilon_0$ and hence $\lim\limits_{\alpha \to 1} \tilde{E}_\alpha(s_\alpha) \ge 2\varepsilon_0$ (In case $\tilde{s} \to s$ in $C^1\big(M - \{x_1, x_2\}, N\big)$, we have two nontrivial harmonic maps $\tilde{s}_1$, $\tilde{s}_2$, and then $\lim\limits_{\alpha \to 1} \tilde{E}_\alpha(s_\alpha) \ge \sum\limits_{j=1}^{2} E(\tilde{s}_j) + E(s) \ge 2\varepsilon_0$.) Therefore the results of $[U1]$ are valid.

Theorem 5.7. *If the universal covering space of N is not contractible, then there exists a non-trivial harmonic map* $s : S^2 \to N$.

Proof. We apply Theorem 2.8 and Theorem 3.3 to get a critical map s_α of $\tilde{E}_\alpha$ with $\varepsilon < \tilde{E}_\alpha(s_\alpha) < B$. Then from Theorem 4.4, there is a harmonic map s such that $s_\alpha \to s$ in $C^1\big(S^2 - \{x_1, \cdots, x_\ell\}, N\big)$. If s is not a map to a point, we are finished. If s is constant, then since $\tilde{E}_\alpha(s_\alpha) > \varepsilon$, s_α must fail to converge to s at some points. From Theorem 4.7 there exists a nontrivial harmonic map $\tilde{s}$ with $\tilde{s}(S^2) \subset \bigcap\limits_{\alpha} \bigcup\limits_{\beta < \alpha} s_\beta(S^2)$.

Since the image of a harmonic map from S^2 to N is a conformal branched minimal immersion, the previous theorem yields the main theorem on the existence of minimal spheres. Note that the hypothesis on the universal covering space of N cannot be dropped, for if N has nonpositive curvature then every harmonic map $s : S^2 \to N$ is constant.

Theorem 5.8. *If the universal covering space of N is not contractible, then there exists a nontrivial* C^∞ *conformal branched minimal immersion* $s : S^2 \to N$.

Also, from Theorem 5.5, we get the final result on minimal spheres.

Theorem 5.9. *There exists a finite set of free homotopy classes* $\Lambda_i \in \pi_0 C^0(S^2, N)$ *such that elements* $\{\lambda \in \Lambda_i\}$ *generate* $\pi_2(N)$ *acted on by* $\pi_1(N)$, *and such that each* Λ_i *contains a conformal branched minimal immersion of a sphere having least area among maps of* S^2 *into* N *which lie in* Λ_i.

Chapter VII

Manifolds with Positive Curvature on Totally Isotropic Two-Planes

It was proved in [GM], by means of the theory of harmonic forms, that a compact simply connected n-dimensional Riemannian manifold with positive curvature operator must have the real homology of an n-sphere. Furthermore, Hamilton [Ha1] has used heat equation methods to prove that a compact simply connected four dimensional Riemannian manifold with positive curvature operator must be diffeomorphic to S^4. On the other hand, by the Sphere Theorem [CE], first proved by Rauch and improved by Klingenberg and Berger, we know that if M is simply connected and the sectional curvature of M satisfies $\frac{1}{4} < K_M \leq 1$, then M is homeomorphic to a sphere. Using a new curvature assumption, Micallef and Moore [MM] obtained strengthened versions of these two results. Namely, they proved the Sphere Theorem for manifolds with strictly pointwise $\frac{1}{4}$-pinched sectional curvature, and they showed that every simply connected compact Riemannian manifold with positive curvature operator is homeomorphic to a sphere. More precisely, they proved, via the Sacks-Uhlenbeck theory of minimal S^2's, that a compact simply connected Riemannian manifold of dimension at least four, with positive curvature on totally isotropic two-planes is homeomorphic to a sphere. Very recently, R. Hamilton has extended his heat equation methods to prove that a compact oriented 4-manifold with positive curvature on isotropic 2-planes is diffeomorphic to a connected sum of copies of $S^1 \times M^3$ where M^3 is a 3-manifold of constant

Typeset by $\mathcal{A}_{\mathcal{M}}\mathcal{S}$-TEX

sectional curvature; i.e. a quotient of S^3 by a finite group of isometries acting freely. We describe this important curvature condition in this chapter, and give the proof of the Micallef-Moore theorem. Remarkably, Micallef and Wang(Duke Math J**72**(1993), 649-672) have shown that the connected sum of compact manifolds (of any dimension) with positive curvature on totally isotropic 2-planes also admits such a metric. Thus connected sums of any number of copies of $S^1 \times M^{n-1}$ where M^{n-1} is a constant positive curvature manifold together with constant positive curvature n-manifolds have metrics with positive curvature on totally isotropic 2-planes.

§1. Positive Curvature on Totally Isotropic Two-Planes

Let M be an n-dimensional Riemannian manifold with tangent space T_pM at $p \in M$. The *curvature operator* at p is the self-adjoint linear endomorphism $\mathcal{R} : \wedge^2 T_pM \to \wedge^2 T_pM$ on the second exterior power $\wedge^2 T_pM$ of the tangent space, defined by the formula

$$\langle \mathcal{R}(x \wedge y), u \wedge v \rangle = \langle R(x,y)v, u \rangle \ , \quad \text{for} \quad x, y, u, v \in T_pM \ .$$

As in Chapter V §2, the Riemannian metric $\langle \ , \ \rangle$ is extended to a complex bilinear form $(\ , \ )$ and a Hermitian inner product $\ll \ , \ \gg$ on $T_pM \otimes \mathbb{C}$. Similarly, the Riemannian metric $\langle \ , \ \rangle$ on $\wedge^2 T_pM$ can be extended in two ways to $\wedge^2 T_pM \otimes \mathbb{C}$.

We extend the curvature operator to a complex linear map $\mathcal{R} : \wedge^2 T_pM \otimes \mathbb{C} \to \wedge^2 T_pM \otimes \mathbb{C}$, and to each two plane $\sigma \subset T_pM \otimes \mathbb{C}$ we assign a *complex sectional curvature* $K(\sigma)$, the real number defined by

$$K(\sigma) = \frac{\ll \mathcal{R}(z \wedge w), z \wedge w \gg}{\|z \wedge w\|^2} \ ,$$

where $\{z, w\}$ is a basis for σ.

An element $z \in T_pM \otimes \mathbb{C}$ is said to be *isotropic* if $(z, z) = 0$. A complex linear subspace $V \subset T_pM \otimes \mathbb{C}$ is *totally isotropic* if $(z, z) = 0$ for any $z \in V$.

Definition. *The curvature of a Riemannian manifold M is positive on totally isotropic two-planes if $K(\sigma) > 0$ whenever $\sigma \subset T_pM \otimes \mathbb{C}$ is a totally isotropic two-plane at any $p \in M$.*

REMARK.

(1) The dimension of a totally isotropic subspace of $T_pM \otimes \mathbb{C}$ is no more than $\frac{n}{2}$, and therefore this curvature condition is non-vacuous only for $n \geq 4$.

(2) Just as the condition, "positive sectional curvature" is ideally adapted for studying stability of geodesics, so the condition "positive curvature on totally istropic two-planes" is well adapted for studying stability of minimal surfaces in Riemannian manifolds.

We now investigate some commonly used curvature conditions which imply positive curvature on totally isotropic two-planes. Observe that if σ is a two dimensional totally isotropic subspace of $T_pM \otimes \mathbb{C}$, then there exists a basis $\{z, w\}$ of σ such that

$$z = e_1 + ie_2 \ , \quad w = e_3 + ie_4 \ ,$$

where $e_1, \ldots, e_4$ are real orthonormal tangent vectors. Since

$$z \wedge w = (e_1 \wedge e_3 - e_2 \wedge c_4) + i(e_1 \wedge e_4 + e_2 \wedge e_3) \tag{1}$$

it follows immediately that

$$\begin{aligned} \ll \mathcal{R}(z \wedge w), z \wedge w \gg &= \langle \mathcal{R}(e_1 \wedge e_3 - e_2 \wedge e_4), e_1 \wedge e_3 - e_2 \wedge e_4 \rangle \\ &\quad + \langle \mathcal{R}(e_1 \wedge e_4 + e_2 \wedge e_3), e_1 \wedge e_4 + e_2 \wedge e_3 \rangle \\ &= R_{1331} + R_{2442} + R_{1441} + R_{2332} - 2R_{1234} \ , \quad (2) \end{aligned}$$

where we have used the Bianchi identity.

Suppose now that M is an oriented 4-manifold. Then the real and imaginary parts of (1) are either both self-dual or both anti-self-dual [AHS]. The component of $\mathcal{R}$ which maps the space of self-dual two-forms to itself is $W_+ + \frac{1}{12}s$, where W_+ is the self-dual part of the Weyl tensor W and s is the scalar curvature. Similarly, the component of $\mathcal{R}$ mapping the space of anti-self-dual two-forms to itself is $W_- + \frac{1}{12}s$. Thus for 4-manifolds there is a simple necessary and sufficient condition for positive curvature on totally isotropic two-planes: that the sum of the two lowest eigenvalues of each of the self-adjoint operators $W_+ + \frac{1}{12}s$ and $W_- + \frac{1}{12}s$ be positive. Since W_+ and W_- are trace-free, it is equivalent to the condition that $-W + \frac{1}{6}s$ be positive definite. In particular, the traceless Ricci tensor does not affect curvature on totally isotropic two-planes in dimension 4.

In dimensions ≥ 4, there are a variety of conditions which imply positive curvature on totally isotropic two-planes. The simplest of these is the condition that the curvature operator $\mathcal{R}$ be positive definite, in which case we say that M has *positive curvature operator*. Another condition which implies positive curvature on totally isotropic two-planes is strict pointwise $\frac{1}{4}$-pinching. We say that the sectional curvature of M is *pointwise δ-pinched*, $0 < \delta \leq 1$, if there exsits a positive function $\mathcal{K}$ on M such that

$$\delta \mathcal{K}(p) \leq K(\sigma) \leq \mathcal{K}(p) ,$$

for all real two-plane sections $\sigma \subset T_pM$, and all $p \in M$. The pinching is said to be *global* if $\mathcal{K}$ can be chosen to be constant, and *strict* if one of the inequalities is strict. Suppose that the sectional curvature of M is strictly pointwise $\frac{1}{4}$-pinched. Then each of the first four terms is strictly greater than $\frac{1}{4}\mathcal{K}(p)$, and an inequality of Berger [Bg. p.69] shows that

$$|R_{1234}| \leq \frac{1}{2}\mathcal{K}(p) \ .$$

It follows immediately that

$$\ll \mathcal{R}(z \wedge w), z \wedge w \gg \ > 0 \ ,$$

and hence strict pointwise $\frac{1}{4}$-pinching implies positive curvature on totally isotropic two-planes. Now we state the main theorem.

Main Theorem. *Let M be a compact simply connected n-dimensional Riemannian manifold which has positive curvature on totally isotropic two-planes, where $n \geq 4$. Then M is homeomorphic to a sphere.*

Corollary. *Let M be a compact simply connected n-dimensional Riemannian manifold with $n \geq 4$. If M has positive curvature operator, or M has strictly pointwise $\frac{1}{4}$-pinched sectional curvature, then M is homeomorphic to a sphere.*

REMARK.

(1) The above Corollary holds also for $n = 2, 3$. The case $n = 2$ is well-known, and the case $n = 3$ follows directly from a theorem of Hamilton [Ha2], which in fact shows that M is diffeomorphic to a sphere, under the condition that its Ricci curvature be positive.

(2) Ruh[R] was the first to prove a sphere theorem with only a pointwise pinching assumption. Later, Huisken [Hu] was able to prove a similar result by a heat equation method. Both Ruh and Huisken proved that M is in fact diffeomorphic to a sphere if the curvature is pointwise δ-pinched, where δ is sufficiently close to one ($\delta = \delta(n) \to 1$ as $n = \dim M \to \infty$). It should be mentioned that

for global pinching, one can get a differentiable sphere theorem with δ independent of n [CE, Chapter 7].

The key step in the proof of the Main Theorem consists of showing that a nontrivial branched conformal minimal S^2 in a Riemannian manifold of dimension ≥ 4 with positive curvature on totally isotropic two-planes must have index $\geq \frac{n}{2} - \frac{3}{2}$. This will be proved in the following section.

§2. The Index of Harmonic 2-Spheres in M

Let $\sum$ be a Riemann surface, M a Riemannian manifold, and $f : \sum \to M$ be a nontrivial harmonic map. The tangent bundle TM of M pulls back to a smooth vector bundle f^*TM over $\sum$, and the Riemannian metric and Levi-Civita connection on TM pull back to a metric $\langle\ ,\ \rangle$ and connection ∇ on f^*TM. As we have seen earlier, the fiber metric $\langle\ ,\ \rangle$ extends to the complexified bundle $\mathbb{E} = f^*TM \otimes \mathbb{C}$ in two ways, $(\ ,\ )$ and $\ll\ ,\ ,\gg$. The connection ∇ extends to a complex linear connection on $\mathbb{E}$, which is Hermitian with respect to $\ll\ ,\ \gg$.

Let $\mathcal{A}^{p,q}(\mathbb{E})$ denote the space of (p,q)-forms on $\sum$ with values in $\mathbb{E}$; in terms of a local complex coordinate $z = x + iy$ on $\sum$, $\mathcal{A}^{1,0}(\mathbb{E})$ consists of multiples of dz, whereas $\mathcal{A}^{0,1}(\mathbb{E})$ consists of multiples of $d\overline{z}$. The connection ∇ divides into two components

$$\nabla' : \mathcal{A}^{0,0}(\mathbb{E}) \to \mathcal{A}^{1,0}(\mathbb{E}) , \quad \nabla'' : \mathcal{A}^{0,0}(\mathbb{E}) \to \mathcal{A}^{0,1}(\mathbb{E}) ,$$

and it is well-known [KM], [AHS, Theorem 5.1] that there exists a unique holomorphic structure on $\mathbb{E}$ with respect to which $\nabla'' = \overline{\partial}$, the $\overline{\partial}$-operator on $\mathbb{E}$. With respect to this holomorphic structure, a section W of $\mathbb{E}$ is holomorphic

$$\Longleftrightarrow \quad \nabla'' W \quad \Longleftrightarrow \quad \nabla_{\frac{\partial}{\partial \bar{z}}} W = 0 \, .$$

Obviously, we can regard $\dfrac{\partial f}{\partial z}$ as a local section of $\mathbb{E}$. In this notation, the fact that f is harmonic is expressed by

$$\nabla_{\frac{\partial}{\partial \bar{z}}} \frac{\partial f}{\partial z} = 0 \, ,$$

which asserts that $\dfrac{\partial f}{\partial z}$ is a local holomorphic section of $\mathbb{E}$. On the other hand, f will be conformal if it satisfies

$$\left(\frac{\partial f}{\partial z} \, , \, \frac{\partial f}{\partial z} \right) = 0 \, ,$$

which asserts that $\dfrac{\partial f}{\partial z}$ is isotropic.

Suppose $\sum$ is given a Riemannian metric ds^2 compatible with its conformal structure so that $ds^2 = \lambda^2(dx^2 + dy^2)$. If $V \in \Gamma(f^*TM)$, the second derivative of energy along a deformation of f with deformation field V is given by the *index form* (see Chapter V §1):

$$I(V,V) = 2 \int_{\Sigma} [\| \nabla V \|^2 - \langle \mathcal{K}(V), V \rangle] dA \, , \tag{3}$$

where dA is the area element of $(\sum, ds^2)$,

$$\| \nabla V \|^2 = \frac{1}{\lambda^2} \Big[\Big\langle \nabla_{\frac{\partial}{\partial x}} V \, , \, \nabla_{\frac{\partial}{\partial x}} V \Big\rangle + \Big\langle \nabla_{\frac{\partial}{\partial y}} V \, , \, \nabla_{\frac{\partial}{\partial x}} V \Big\rangle \Big] \, ,$$

and $\mathcal{K}$ is the vector bundle endomorphism of f^*TM defined by

$$\mathcal{K}(V) = \frac{1}{\lambda^2} \Big[R\Big(V, \frac{\partial f}{\partial x}\Big) \frac{\partial f}{\partial x} + R\Big(V, \frac{\partial f}{\partial y}\Big) \frac{\partial f}{\partial y} \Big] \, .$$

The *index* of f is the maximal dimension of a linear subspace of $\Gamma(f^*TM)$ on which the index form (3) is negative definite. The *nullity* of f is the dimension of the subspace of Jacobi fields, an element $U \in \Gamma(f^*TM)$ being called a *Jacobi field* if

$$I(u,v) = 0 \ , \quad \text{for all} \quad V \in \Gamma(f^*TM) \ .$$

Extending the vector bundle endomorphism $\mathcal{K}$ on f^*TM to a complex linear vector bundle endomorphism of $\mathbb{E} = f^*TM \otimes \mathbb{C}$, we can see that the index form (3) extends to a Hermitian symmetric bilinear form on the sections of $\mathbb{E}$. Hence we can define the index and nullity of f for the Hermitian symmetric index form in exactly the same way as above.

Lemma 1. *If* $W \in \Gamma(\mathbb{E})$,

$$I(W,W) = 8 \int_\Sigma \Big[\Big\| \nabla_{\frac{\partial}{\partial \bar z}} W \Big\|^2 - \Big\langle\!\Big\langle \mathcal{R}\Big(W \wedge \frac{\partial f}{\partial z}\Big) \ , \ W \wedge \frac{\partial f}{\partial z} \Big\rangle\!\Big\rangle \Big] dx\, dy \ .$$

Proof. Integrations by parts yield

$$\begin{aligned} \int_\Sigma \Big\| \nabla_{\frac{\partial}{\partial \bar z}} W \Big\|^2 dx\, dy &= \frac{1}{4} \int_\Sigma \Big[\Big\| \nabla_{\frac{\partial}{\partial x}} W \Big\|^2 + \Big\| \nabla_{\frac{\partial}{\partial y}} W \Big\|^2 \Big] dx\, dy \\ &\quad - \frac{i}{4} \int_\Sigma \Big\langle\!\Big\langle R\Big(\frac{\partial f}{\partial x}, \frac{\partial f}{\partial y}\Big) W, W \Big\rangle\!\Big\rangle dx \ , \end{aligned}$$

where the curvature tensor R is extended to be complex linear in all its arguments. Substitution into the Hermitian extension of (3) yields

$$\begin{aligned} I(W,W) = 2 \int_\Sigma \Big[4 \Big\| \nabla_{\frac{\partial}{\partial \bar z}} W \Big\|^2 &+ i \Big\langle\!\Big\langle R\Big(\frac{\partial f}{\partial x}, \frac{\partial f}{\partial y}\Big) W, W, \Big\rangle\!\Big\rangle \\ &- \Big\langle\!\Big\langle R\Big(W, \frac{\partial f}{\partial x}\Big) \frac{\partial f}{\partial x}, W \Big\rangle\!\Big\rangle - \Big\langle\!\Big\langle R\Big(W, \frac{\partial f}{\partial y}\Big) \frac{\partial f}{\partial y}, W \Big\rangle\!\Big\rangle \Big] dx\, dy \ . \end{aligned}$$

Then by the Bianchi identity, we can show that

$$\begin{aligned} &i \Big\langle\!\Big\langle R\Big(\frac{\partial f}{\partial x}, \frac{\partial f}{\partial y}\Big) W, W \Big\rangle\!\Big\rangle - \Big\langle\!\Big\langle R\Big(W, \frac{\partial f}{\partial x}\Big) \frac{\partial f}{\partial x}, W \Big\rangle\!\Big\rangle - \Big\langle\!\Big\langle R\Big(W, \frac{\partial f}{\partial y}\Big) \frac{\partial f}{\partial y}, W \Big\rangle\!\Big\rangle \\ &= -4 \Big\langle\!\Big\langle \mathcal{R}\Big(W \wedge \frac{\partial f}{\partial z}, W \wedge \frac{\partial f}{\partial z}\Big\rangle\!\Big\rangle \ . \end{aligned}$$

This completes the proof.

NOTE. From Lemma 1 it follows that the f_*-image of any holomorphic vector field tangent to $\sum$ is an isotropic complex Jacobi field.

Theorem 1. *Let M be an n-dimensional Riemannian manifold with positive curvature on isotropic two-planes. Then any nontrivial conformal harmonic map $f : S^2 \to M$ has index at least $\frac{n-3}{2}$.*

Proof. We rely very strongly on the fact that our minimal surface has genus zero as we did in §5.3. By a theorem of Grothendieck [Gr], as in §5.3, $\mathbb{E}$ decomposes into a direct sum of homolorphic line bundles,

$$\mathbb{E} = L_1 \oplus \cdots \oplus L_p \oplus L_{p+1} \oplus \cdots \oplus L_r \oplus L_{r+1} \oplus \cdots \oplus L_n \ ,$$

where $c_1(L_i) > 0$ for $1 \le i \le p$, $c_1(L_j) = 0$ for $p+1 \le j \le r$, and $c_1(L_k) < 0$ for $r+1 \le k \le n$. Let $\mathbb{E}_+ = L_1 \oplus \cdots \oplus L_p$, $\mathbb{E}_0 = L_{p+1} \oplus \cdots \oplus L_r$, and $\mathbb{E}_- = L_{r+1} \oplus \cdots \oplus L_n$. Then, by the arguments of §5.3, $\mathbb{E}_+$ is orthogonal to $\mathbb{E}_+ \oplus \mathbb{E}_0$, and $p + r = n$. In particular, $\mathbb{E}_+$ is an isotropic subbundle of $\mathbb{E}$. Moreover, if $z = x + iy$ is the standard coordinate on the Riemann sphere $\mathbb{C} \cup \{\infty\}$, $\frac{\partial f}{\partial z}$ is a holomorphic section of $\mathbb{E}_+$ because it vanishes at ∞ and at the branch points of f.

Let $\mathcal{U}$, $\mathcal{V}$ be complex linear subspaces of $\Gamma(\mathbb{E})$ such that $\mathcal{U}$ is generated by the holomorphic sections of $\mathbb{E}_+$, which are automatically isotropic, and $\mathcal{V}$ is generated by the holomorphic sections with values in a maximal totally isotropic subbundle of $\mathbb{E}_0$. (Existence of holomorphic sections of $\mathbb{E}_+$ and $\mathbb{E}_0$ is guaranteed by the Riemann-Roch Theorem, which says that the complex dimension of holomorphic sections of L_i equals $c_1(L_i) + 1$ if $c_1(L_i) \ge 0$, and 0 if $c_1(L_i) < 0$.) By use of linear algebra, we can show that

$$\begin{aligned}\dim_{\mathbb{C}} \mathcal{V}(q) &= \{w \in T_qM : w \in W(q) \quad \text{for some} \quad W \in \mathcal{V}\} \\ &\geq \frac{r-p-1}{2} .\end{aligned}$$

Define $\mathcal{W} = \mathcal{U} + \mathcal{V}$. It follows that

$$\begin{aligned}\dim_{\mathbb{C}} \mathcal{W}(q) &= \{w \in T_qM : w \in W(q) \quad \text{for some} \quad W \in \mathcal{W}\} \\ &\geq p + \frac{r-p-1}{2} = \frac{n-1}{2} .\end{aligned}$$

A one dimensional subspace of $\mathcal{W}(q)$ is accounted for by $\dfrac{\partial f}{\partial z}(q)$, which is tangent to the sphere. Thus $\mathcal{W}$ contains a linear subspace $\mathcal{W}_0$ of complex dimension at least $\dfrac{n-3}{2}$ consisting of isotropic holomorphic sections of $\mathbb{E}$ such that if $W \in \mathcal{W}_0$, then $W \wedge \dfrac{\partial f}{\partial z}$ is not identically zero. From Lemma 1 it follows that

$$I(W,W) = -8 \int_{S^2} \left\langle\!\left\langle \mathcal{R}\left(W \wedge \frac{\partial f}{\partial z}\right), W \wedge \frac{\partial f}{\partial z} \right\rangle\!\right\rangle dx\, dy \,, \quad W \in \mathcal{W}_0 \,.$$

Note that $W(q)$ and $\dfrac{\partial f}{\partial z}(q)$ span a totally isotropic two-plane in $T_qM \otimes \mathbb{C}$ for almost all $q \in M$. Thus, by assumption, the index form is negative definite on $\mathcal{W}_0$, and hence the index of f is at least $\dfrac{n-3}{2}$, as asserted.

Remark. In case M has positive curvature operator or strictly $\frac{1}{4}$-pinched sectional curvature, a slight modification of the above argument would show that the index of f is at least $n-2$. In the following sections we will see that any compact simply connected Riemannian manifold must contain a nontrivial minimal S^2 of index $\leq n-2$ for purely Morse-theoretic reasons as in [SaU].

§3. Critical Points of Low Index for the α-energy

We now shift our viewpoint to global analysis and the α-energy of Sacks and Uhlenbeck (see the preceding chapter). While their α-energy satisfies condition C of Palais and Smale, its critical points may still be degenerate. Recall, however, the theorem of Morse [Mn] that if f is a C^2 function defined on a compact subset of $\mathbb{R}^n$, then for v in a dense open subset of $\mathbb{R}^n$, the map $x \mapsto f(x) + v \cdot x$ has only nondegenerate critical points. In analogy with this, we will define a perturbed version of the α-energy, to be called the (α, ψ)-energy, which can be chosen to have nondegenerate critical points. Assume that M is isometrically embedded in $\mathbb{R}^N$. Then we can regard $L_1^{2\alpha}(\sum, M)$ as a smooth submanifold of $L_1^{2\alpha}(\sum, R^N)$. Given an L^1 function $\psi : \sum \to \mathbb{R}^N$, the *$(\alpha, \psi)$-energy* is the C^2 function $E_{\alpha,\psi} : L_1^{2\alpha}(\sum, M) \to \mathbb{R}$ defined by

$$E_{\alpha,\psi}(f) = E_\alpha(f) + \int_\Sigma (f \cdot \psi) dA ,$$

where the dot denotes the usual dot product of $\mathbb{R}^N$. $E_{\alpha,\psi}$ still satisfies condition C of Palais and Smale, and is bounded below.

A direct calculation shows that the Euler-Lagrange equation is

$$Q_{\alpha,\psi}(f) = 0 ,$$

where

$$Q_{\alpha,\psi}(f) = 2\alpha(1 + \|df\|^2)^{\alpha-1}\{\Delta f + (\alpha - 1)(1 + \|df\|^2)^{-1} \sum e_i(\|df\|^2) e_i(f) - \sum (A \circ f)\langle e_i(f), e_i(f)\rangle - \frac{1}{2\alpha}(1 + \|df\|^2)^{1-\alpha} P_f(\psi)\} .$$

In this formula, A is the second fundamental form of M in $\mathbb{R}^N$ and $P_f(\psi)(p)$ is the orthogonal projection of $\psi(p)$ into $T_{f(p)}M$.

Lemma 2. *If ψ is C^∞, then any critical point for $E_{\alpha,\psi}$ is C^∞.*

Proof. Straightforward from Proposition 6.2.3.

At a critical point $f \in L_1^{2\alpha}(\sum, M)$ for $E_{\alpha,\psi}$, the second derivative of $E_{\alpha,\psi}$ defines a symmetric bilinear form

$$\delta^2 E_{\alpha,\psi}(f) : T_f L_1^{2\alpha}(\sum, M) \times T_f L^{2\alpha}(\sum, M) \to \mathbb{R} ,$$

which is called the *Hessian* of $E_{\alpha,\psi}$ at f. A relatively straightforward calculation shows that for V, W

$$\begin{aligned}\delta^2 E_{\alpha,\psi}(f)(V,W) &= 4\alpha(\alpha-1)\int_\Sigma (1+\|df\|^2)^{\alpha-2}\langle df, \nabla V\rangle\langle df, \nabla W\rangle dA \\ &+ 2\alpha \int_\Sigma (1+\|df\|^2)^{\alpha-1}\{\langle \nabla V, \nabla W\rangle - \langle \mathcal{K}(V), W\rangle\} dA \\ &+ \int_\Sigma (A \circ f)(V,W) \cdot \psi dA . \end{aligned} \tag{5}$$

The *Jacobi operator* $L_{\alpha,\psi}(f)$ for $E_{\alpha,\psi}$ at a critical point f is the second order differential operator obtained by integrating (5) by parts:

$$\delta^2 E_{\alpha,\psi}(f)(V,W) = \int_\Sigma -\langle L_{\alpha,\psi}(f)(V), W\rangle dA .$$

It is clearly elliptic and self-adjoint when $\alpha - 1$ is small. From the theory of such operators, $L_{\alpha,\psi}(f)$ induces a Fredholm map of index zero $L_{\alpha,\psi}(f) : L_2^{2\alpha}(f^*TM) \to L^{2\alpha}(f^*TM)$. Note that

$$L_{\alpha,\psi}(f)(V) = \delta((Q_{\alpha,\psi})(f))(V) ,$$

the derivative on the right being taken with respect to the variable f.

Lemma 3. *Given any positive constant a, the set $ND^a = \{\psi \in L^{2\alpha}(\sum, \mathbb{R}^N) : Ker(L_{\alpha,\psi}) = 0$ at every critical point f for $E_{\alpha,\psi}$ such that $E_\alpha(f) \leq a\}$ is open and dense.*

Proof. [U2], or [MM, Lemma 2].

Since the C^∞ maps are dense in L^p and ND^a is open, given any constant a, we can find $\psi \in C^\infty(\sum, I\!R^N) \cap ND^a$ of arbitrarily small $L^{2\alpha}$ norm. It follows from Lemma 2 that all the critical maps of $E_{\alpha,\psi}$ are C^∞. Thus the theorem of Uhlenbeck [U3, §5] implies that all the critical points of $E_{\alpha,\psi}$ of α-energy $\leq a$ are weakly nondegenerate and of finite index. It is to functions whose critical points are of this type that the Morse theory on Banach manifolds can be applied. From now on we assume $\sum = S^2$, the sphere with the constant curvature metric of area one.

Lemma 4. *If M is compact and $\pi_k(M) \neq 0$, then there is a critical point for $E_\alpha : L_1^{2\alpha}(S^2, M) \to I\!R$ whose index is $\leq k-2$. Moreover, there is a constant B, independent of α, such that the critical point can be chosen to have α-energy $\leq (1+B^2)^\alpha$.*

Proof. By Lemma 3, there exists $\psi \in C^\infty(\sum, I\!R^N) \cap ND^a$ with $\|\psi\|_{0,2\alpha} < \varepsilon$ for any a and ε such that all the critical points of (α,ψ)-energy $\leq a$ are weakly nondengenerate. Choosing a and ε appropriately, we can apply the proof of Theorem 6.2.7. Hence there exists a critical point $f_{\alpha,\psi}$ for $E_{\alpha,\psi}$ of (α,ψ)-energy $\leq a$. By the Morse theory of [U3] $f_{\alpha,\psi}$ has index $\leq k-2$. Moreover, it follows from Theorem 6.2.7 that there exist positive constants B and C, independent of α, such that $f_{\alpha,\psi}$ can be chosen to satisfy

$$E_{\alpha,\psi}(f_{\alpha,\psi}) \leq (1+B^2)^\alpha + c\|\psi\|_{0,2\alpha} \,. \tag{6}$$

Now choose a sequence $\psi(i) \to 0$ in $L^{2\alpha}$ together with weakly nondengenerate critical points $f_{\alpha,\psi(i)}$ of index $\leq k-2$ satisfying (6). The function E_α is bounded on this sequence while the gradient of E_α is not bounded away from zero. It therefore follows from Palais-Smale condition (C) that there is a critical point f_α in the closure of this sequence. Clearly f_α is nonconstant, has index $\leq k-2$, and satisfies the inequality $E_\alpha(f_\alpha) \leq (1+B^2)^\alpha$.

§4. The Existence of Harmonic Two-Spheres of Low Index

Theorem 2. *If M is a compact Riemannian manifold such that $\pi_k(M) \neq 0$, where $k \geq 2$, then there exists a nonconstant harmonic two-sphere in M of index $\leq k-2$.*

Before proving this theorem, we will state the well-known result that we can remove finitely many points from a nontrivial harmonic two-sphere in M without affecting its index (see e.g.[GL]).

Lemma 5. *Let m be the index of a nontrivial harmonic map $f: S^2 \to M$. Given any finite collection of points $\{p_1, \dots, p_\ell\}$ in M, there is a linear subspace $\mathcal{V}$ of $\Gamma(f^*TM)$ of dimension m, such that: (i) the index form I is negative definite on $\mathcal{V}$, (ii) if $V \in \mathcal{V}$, then V vanishes in an open neighborhood of p_i, for $1 \leq i \leq \ell$.*

Proof of Theorem 2. According to Lemma 4, there is a sequence $\alpha(i) \to 1$, together with critical points $f_{\alpha(i)}$ for $E_{\alpha(i)}$, such that the index $E_{\alpha(i)}$ at $f_{\alpha(i)}$ is $\leq k-2$. Moreover, using Theorem 3.3 of p.140, we can arrange that

$$E_{\alpha(i)}(f_{\alpha(i)}) \leq (1+B^2)^{\alpha(i)}\ , \quad E(f_{\alpha(i)}) \geq \varepsilon\ ,$$

where B and ε are constants which are independent of $\alpha(i)$. Let $\alpha(i) \to 1$. It follows from Theorem 6.4.4 that we can arrange that the $f_{\alpha(i)}$'s converge in C^1 norm on S^2 minus a finite number of points to a constant map or a nontrivial conformal branched minimal sphere $f_1 : S^2 \to M$.

CASE I. The limiting sphere f_1 is nonconstant. It will suffice to show that m = index of f_1 is $\leq k-2$. We consider the diagram of vector bundles

$$\begin{array}{ccccc} f_1^*TM & \longrightarrow & \pi_2^*TM & \longrightarrow & TM \\ \downarrow & & \downarrow & & \downarrow \\ S^2 & \xrightarrow{(id,f_1)} & S^2\times M & \xrightarrow{\pi_2} & M \end{array}$$

where π_2 is the projection on the second factor. Let $\{p_1,\dots,p_\ell\}$ be the point in S^2 at which convergence fails. By the preceding lemma, there exist m linearly independent deformation fields $V_1,\dots,V_m \in \Gamma(f_1^*TM)$, such that

(i) the index form is negative definite on $\operatorname{span}\{V_1,\dots,V_m\}$, and

(ii) $V_1,\dots,V_m$ vanish in neighborhoods of $\{p_1,\dots,p_\ell\}$.

Extend $V_1,\dots,V_m$ to smooth sections $\tilde{V}_1,\dots,\tilde{V}_m$ of π_2^*TM with support in a tubular neighborhood of $(id,f_1)(S^2)$, and let

$$V_r(i) = (id, f_{\alpha(i)})^*(\tilde{V}_r)\ , \quad \text{for} \quad 1\leq r\leq \ell\ .$$

If we regard $V_r(i)$ as a mapping from S^2 to the tangent bundle of M such that $V_r(i)(p)$ lies in the tangent space to M at $f_{\alpha(i)}(p)$, then

$$V_r(i)\to (id,f_1)^*(\tilde{V}_r) = V_r \quad \text{in} \quad C^1\ .$$

Now we need to use the expression for the second variation of the α-energy at a critical point f_α for E_α. Since $Q_\alpha - Q_{\alpha,0}$ vanishes at such a critical point, it follows from (5) that

$$\delta^2E_\alpha(f_\alpha)(U,V) = 4\alpha(\alpha-1)\int_{S^2}(1+\|df_\alpha\|^2)^{\alpha-2}\langle df_\alpha,\nabla u\rangle\langle df_\alpha,\nabla V\rangle dA$$
$$+\,2\alpha\int_{S^2}(1+\|df_\alpha\|^2)^{\alpha-1}\{\langle\nabla U,\nabla V\rangle-\langle\mathcal{K}(U),V\rangle\}dA\ , \tag{7}$$

for all $U, V \in L_1^{2\alpha}((f_\alpha)^*TM)$. As $\alpha \to 1$,

$$(\alpha - 1)\left| \int_{S^2} (1 + \|df_\alpha\|^2)^{\alpha-2} \langle df_\alpha, \nabla V_r(i)\rangle \langle df_\alpha, \nabla V_S(i)\rangle dA \right|$$
$$\leq (\alpha - 1)\left| \int_{S^2} (1 + \|df_\alpha\|^2)^{\alpha-1} \| \nabla V_r(i)\| \, \| \nabla V_S(i)\| dA \right| \to 0 \, ,$$

and hence it follows from (7) that

$$\delta^2 E_{\alpha(i)}(f_{\alpha(i)})(V_r(i), V_s(i))$$
$$\to 2 \int_{S^2} \{\langle \nabla V_r, \nabla V_s\rangle - \langle \mathcal{K}(V_r), V_s\rangle\} dA = I(V_r, V_s) \, ,$$

where I is the index form for the usual energy. Thus when $\alpha(i)$ is sufficiently close to one, the $(m \times m)$-matrix whose (r, s)-entry is $\delta^2 E_{\alpha(i)}(f_{\alpha(i)})(V_r(i), V_s(i))$ must be negative definite. In other words, when $\alpha(i)$ is sufficiently close to one, $f_{\alpha(i)}$ must have index at least m. Hence $m \leq k - 2$, which establishes the theorem in this case.

Case II. The limiting sphere f_1 is constant. Since $E(f_{\alpha(i)}) \geq \varepsilon$, it follows from Theorem 6.4.6 that a non-trivial conformal branched minimal sphere "bubbles off" as $\alpha(i) \to 1$. By the argument in that theorem, we can find a conformal dilation g_i such that the map

$$\tilde{f}_i = f_{\alpha(i)} \circ g_i : \mathbb{C} \to M$$

satisfies $\|d\tilde{f}_i(0)\| = 1$ and $\|d\tilde{f}_i\| \leq 1$ on $\mathbb{C}$. As described in the proof of Theorem 6.4.6, the $\tilde{f}_i$'s converge in C^1 on compact subsets of $\mathbb{C}$ to a nonconstant conformal harmonic map which extends to the entire sphere, $\tilde{f}_1 : S^2 \to M$. To finish the proof, it will suffice to show that $m =$ index of $\tilde{f}_1 \leq k - 2$.

Since the α-energy is not conformally invariant, we need to replace it with a new function $\tilde{E}_i$ defined by

$$\tilde{E}_i(\tilde{f}_i) = E_{\alpha(i)}(f_{\alpha(i)}) = \int_{\mathbb{C}} (1 + \|d\tilde{f}_i\|_i^2)^{\alpha(i)} dA_i \, ,$$

where $dA_i = g_i^*(dA)$, and $\|d\tilde{f}_i\|_i$ is the norm of the differential $d\tilde{f}_i$ measured with respect to $ds_i^2 = g_i^*$ (standard metric on S^2). Similarly, the Hessian of $\tilde{E}_i$ at a critical point can be found by a change of variables in (7). Then, proceeding in exactly the same manner as in Case I, We can show that $m \leq k - 2$.

Proof of the Main Theorem.

Let k be the smallest integer such that $\pi_k(M) \neq 0$. Since M is assumed to be simply connected, $k \geq 2$. By Theorem 2, there is a nontrivial harmonic two-sphere in M whose index is $m \leq k - 2$. On the other hand, it follows from Theorem 1 that $m \geq \frac{n-3}{2}$, and hence $k > \frac{n}{2}$. By the Hurewicz Isomorphism Theorem and Poincaré Duality, it must be that $k = n$, i.e., the lowest nonvanishing homotopy group of M is $\pi_n(M)$. Thus, since M is a CW complex, it follows from Whitehead's theorem that M must be a homotopy sphere. By the resolution of the generalized Poincaré conjecture when $n \geq 4$, we can conclude that M is homeomorphic to a sphere.

REMARK.

(1) The Proof of the Main Theorem shows that if M is a compact oriented (not necessarily simply connected) Riemannian 4-manifold such that $-W + \frac{s}{6}$ (see §1) is positive definite, then $\pi_2(M) = 0$. By comparison, the theory of harmonic forms implies that such a manifold satisfies $H^2(M; \mathbb{R}) = 0$ (see [FU, Appendix C].)

(2) If M satisfies all the hypotheses of the Main Theorem except for being simply connected, it is not necessarily true that the universal cover of M is a sphere, but we can still conclude that

$$\pi_k(M) = 0 \ , \quad \text{for} \quad 2 \leq k \leq \frac{n}{2} \ . \tag{8}$$

This is demonstrated by the standard metric on $S^1 \times S^3$, which is conformally flat and has constant positive scalar curvature, but has universal cover $\mathbb{R} \times S^3$. More generally, if M is an n-dimensional conformally flat manifold, then expanding in terms of a standard basis $z = e_1 + ie_2$, $w = e_3 + ie_4$ for an isotropic two-plane, we find that

$$\ll \mathbb{R}(z \wedge w), z \wedge w \gg = \frac{1}{n-2}[R_{11} + R_{22} + R_{33} + R_{44} - \frac{2}{n-1}s] ,$$

where R_{ij} denotes the Ricci tensor. Hence a compact locally conformally flat Riemannian manifold, for which the sum of the lowest four eigenvalues of the Ricci tensor is $> \frac{2}{n-1}s$ must satisfy (8). See [SY2] for more general results in this direction.

(3) The Main Theorem is a best possible result, in the sense that there are a wide variety of topological types of Riemannian manifolds which have nonnegative curvature on totally isotropic two-planes. Indeed, such manifolds include those which have nonnegative curvature operator (for example, all symmetric spaces of compact type), or those whose curvature is nonstrictly pointwise $\frac{1}{4}$-pinched. For odd-dimensional manifolds there is a stronger curvature condition than nonnegative curvature on totally isotropic two-planes, namely the weakest curvature condition on M which implies that $M \times S^1$ has nonnegative curvature on totally isotropic two-planes: $K(\sigma) \geq 0$ whenever the two-plane σ possesses a basis $\{z, w\}$ such that $(z, z) = (z, w) = 0$. Micallef and Moore obtained a 2-dimensional version of Synge's Theorem: If M is an odd-dimensional compact Riemannian manifold whose sectional curvature is nonstrictly pointwise $\frac{1}{4}$-pinched (but not strictly), then $\pi_2(M) = 0$ [MM, §5].

Chapter VIII
Compact Kähler Manifolds of Positive Bisectional Curvature

Let M be a Kähler manifold of complex dimension n and R its Riemannian curvature tensor. Let σ be a plane (a real two dimensional subspace) in $T_x(M)$, and X, Y an orthonormal basis for σ. The sectional curvature $K(\sigma) = R(X, Y, X, Y)$ is a function defined on the Grassman bundle of planes in the tangent spaces of M. A plane σ is said to be holomorphic if it is invariant by the (almost) complex structure tensor J. The set of J-invariant planes σ is a holomorphic bundle over M with fiber $\mathbb{P}_{n-1}$, the complex projective space of dimension $n-1$. The restriction of the sectional curvature K to this complex projective bundle is called the *holomorphic sectional curvature* and will be denoted by H. In other words, $H(\sigma)$ is defined only when σ is invariant by J, and $H(\sigma) = K(\sigma)$.

Given two J-invariant planes σ and σ' in $T_x(M)$, we define the *holomorphic bisectional curvature* $H(\sigma, \sigma')$ by

$$H(\sigma, \sigma') = R(X, JX, Y, JY) ,$$

where X is a unit vector in σ and Y a unit vector in σ'. It is a simple matter to verify that $R(X, JX, Y, JY)$ depends only on σ and σ'. Since $H(\sigma, \sigma) = H(\sigma)$, the holomorphic bisectional curvature carries more information than the holomorphic sectional curvature. By Bianchi's identity we have

$$R(X, JX, Y, JY) = R(X, Y, X, Y) + R(X, JY, X, JY) .$$

Typeset by $\mathcal{A}_{\mathcal{M}}\mathcal{S}$-TEX

The right hand side is a sum of two sectional curvatures (up to constant factors). Hence the holomorphic bisectional curvature carries less information than the sectional curvature.

The Ricci tensor S may be given by

$$S(X,Y) = \sum_{i=1}^{n} R(X_i, JX_i, X, JY) ,$$

where $(X_1, \dots, X_n, JX_1, \dots, JX_n)$ is an orthonormal basis for $T_x(M)$. Hence it is clear that if holomorphic bisectional curvature is positive (or negative) so is the Ricci tensor.

If M is a submanifold of a Kähler manifold N, then from the Gauss-Codazzi equation we can see that the holomorphic bisectional curvature of M does not exceed that of N.

Frankel conjectured in [F] that

Every compact Kähler manifold M^n of positive sectional (or more generally, positive holomorphic bisectional) curvature is biholomorphic to the complex projective space $\mathbb{P}_n$. The case of dimension two was proved by Androetti-Frankel [F] using a formula for the second variation of arc length and a result for the classification of algebraic surfaces. The case of dimension three was solved by Mabuchi [Ma] using the result of Kobayashi-Ochiai [KO]. Using harmonic maps and the characterization of the complex projective space obtained by Kobayashi-Ochiai [KO1], Siu and Yau [SiY] proved the Frankel conjecture for arbitrary dimension. According to the result of Kobayashi-Ochiai the complex projective space of dimension n is characterized by the fact that its first Chern class equals $\lambda c_1(F)$ for some $\lambda \geq n+1$ and some positive holomorphic line bundle F over it. Since by the result of Bishop-Goldberg [BG] the second Betti number of a compact Kähler manifold M of positive bisectional curvature is 1, to prove the Frankel conjecture it suffices to show that $c_1(M)$ is λ times a generator of $H^2(M, \mathbb{Z})$ for some $\lambda \geq 1 + \dim M$. This can be done by proving that a generator of the free part of $H_2(M, \mathbb{Z})$ can be represented by a rational curve, because the tangent bundle of M restricted to the rational curve splits into a direct sum of holomorphic line bundles over the rational curve according to the result of Grothendieck which we have seen in the preceeding chapters. The existence of a rational curve is obtained in the following way. By the result of Sacks-Uhlenbeck and its improved formulation by Meeks-Yau [MY], the infimum of the energy of maps from S^2 to M representing the generator of $\pi_2(M)$ can be achieved by a sum of stable harmonic maps f_i from S^2 to M $(1 \leq i \leq m)$. The key step in [SiY] is to show that each f_i is either holomorphic or conjugate holomorphic. The known methods of proving the complex-analyticity of a harmonic map use the formula for the Laplacian of the energy function [SY1, To, Wo] or a variation of it [Si]. Siu-Yau use instead the second variation formula of the energy function. In this second variation formula a 2-parameter variation has to be used to imitate the situation of holomorphic deformation. After this key step, holomorphic deformations of rational curves in M are used to show that $m = 1$. This is done by proving that in case $m > 1$ one can holomorphically deform the images of some holomorphic f_i and some conjugate holomorphic f_j so that they are tangential to each other at some point. By removing a disc centered at the point of contact from each and joining the two disc boundaries by a suitable surface, one can obtain a map from S^2 to M with energy smaller than the minimum energy. Thus $m = 1$ and the image of f_1 is a rational curve representing a generator of the free part of $H_2(M, \mathbb{Z})$.

The algebraic counterpart of the Frankel conjecture is Hartshorne's conjecture: Every irreducible n-dimensional non-singualr projective variety with ample tangent bundle defined over an algebraically closed field of characteristic ≥ 0 is isomorphic to the projective space. This conjecture, proved by Mori [Mr], is stronger than the Frankel conjecture.

§1. Energy, $\overline{\partial}$-Energy, and ∂-Energy

Suppose M, N are compact Kähler manifolds whose Kähler metrics are respectively,

$$ds_M^2 = 2 \text{ Re } h_{\alpha\overline{\beta}} dz^\alpha d\overline{z}^\beta \quad \text{and} \quad ds_N = 2 \text{ Re } g_{i\overline{j}} dw^i d\overline{w}^j \ ,$$

where the summation convention is used. The $\overline{\partial}$-energy density of f is defined by

$$|\overline{\partial} f|^2 = g^{i\overline{j}} f^\alpha_{\overline{i}} \overline{f^\beta_{\overline{j}}} h_{\alpha\overline{\beta}} \ ,$$

and the ∂-energy density of f is defined by

$$|\partial f|^2 = g^{\overline{i}j} f^\alpha_j \overline{f^\beta_i} h_{\alpha\overline{\beta}} \ ,$$

where $f^\alpha_{\overline{i}} = \dfrac{\partial f^\alpha}{\partial \overline{w}^i}$ and $f^\alpha_j = \dfrac{\partial f^\alpha}{\partial w^j}$. Therefore energy density $e(f)$ of f, which is defined as the trace of $f^*(ds_M^2)$ with respect to ds_N^2, is equal to $|\overline{\partial} f|^2 + |\partial f|^2$. Now assume $\dim_{\mathbb{C}} N = 1$. The pullback of the Kähler form of M under f is

$$\sqrt{-1} h_{\alpha\overline{\beta}} df^\alpha \wedge d\overline{f}^\beta = \sqrt{-1} h_{\alpha\overline{\beta}} \Big(\frac{\partial f^\alpha}{\partial w} \overline{\frac{\partial f^\beta}{\partial w}} - \frac{\partial f^\alpha}{\partial \overline{w}} \overline{\frac{\partial f^\beta}{\partial \overline{w}}} \Big) dw \wedge d\overline{w} \ .$$

Hence

$$\int_N |\partial f|^2 - \int_N |\overline{\partial} f|^2 = \int_N \sqrt{-1} h_{\alpha\overline{\beta}} df^\alpha \wedge d\overline{f}^\beta$$

which is equal to the Kähler class $\omega(M)$ of M evaluated at the homology class $[f(N)]$ defined by $f : N \to M$ (review §3.1). It follows that

$$\int_N |\bar\partial f|^2 = \frac{1}{2}\int_N e(f) - \frac{1}{2}\omega(M)[f(N)] ,$$
$$\int_N |\partial f|^2 = \frac{1}{2}\int_N e(f) + \frac{1}{2}\omega(M)[f(N)] .$$

As a consequence, the energy-minimizing maps from N to M are precisely the same as the $\bar\partial$-energy-minimizing maps, as was first observed by Lichnerowicz [Li].

§2. The Second Variation Formula

Suppose $\dim_{\mathbb{C}} N \geq 1$. Let $f(t) : N \to M$, $t \in \mathbb{C}$, $|t| < \varepsilon$, be a family of smooth maps parametrized by an open disc in $\mathbb{C}$. In order to compute $\dfrac{\partial^2}{\partial t\,\partial \bar t}|\bar\partial f|^2$ at a point P of N, we choose local holomorphic coordinate systems at P and $Q = f(P)$ such that

$$dg_{i\bar j} = 0 \quad \text{at} \quad P$$

and

$$dh_{\alpha\bar\beta} = 0 \ , \quad \partial_\gamma \partial_\delta h_{\alpha\bar\beta} = 0 \quad \text{at} \quad Q \ .$$

Direct computation yields

$$\begin{aligned}\frac{\partial^2}{\partial t\,\partial \bar t}|\bar\partial f|^2 &= 2\mathrm{Re}\ g^{\bar i j}\Big(\frac{\partial^2}{\partial t\,\partial \bar t} f^\alpha_{\bar i}\Big)\overline{f^\beta_{\bar j}} h_{\alpha\bar\beta} + g^{\bar i j}\Big(\frac{\partial}{\partial t} f^\alpha_{\bar i}\Big(\overline{\frac{\partial}{\partial t} f^\beta_{\bar j}}\Big)\Big) h_{\alpha\bar\beta} \\ &\quad + g^{\bar i j}\Big(\frac{\partial}{\partial \bar t} f^\alpha_{\bar i}\Big(\overline{\frac{\partial}{\partial \bar t} f^\beta_{\bar j}}\Big)\Big) h_{\alpha\bar\beta} + g^{\bar i j} f^\alpha_{\bar i}\overline{f^\beta_{\bar j}}(\partial_\mu \partial_{\bar\nu} h_{\alpha\bar\beta})\frac{\partial f^\mu}{\partial \bar t}\overline{\frac{\partial f^\nu}{\partial \bar t}} \\ &\quad + g^{\bar i j} f^\alpha_{\bar i}\overline{f^\beta_{\bar j}}(\partial_\mu \partial_{\bar\nu} h_{\alpha\bar\beta})\frac{\partial f^\mu}{\partial t}\overline{\frac{\partial f^\nu}{\partial t}} \ .\end{aligned}$$

Consider the vector field ξ on N defined by

$$\begin{aligned}\xi^{\bar{i}} &= g^{\bar{i}j}\Big(\frac{D}{\partial t}\frac{\partial}{\partial \bar{t}} f^\alpha\Big)\overline{f^\beta_{\bar{j}}}h_{\alpha\bar{\beta}}\\ &= g^{\bar{i}j}\Big(\frac{\partial^2}{\partial t\,\partial \bar{t}} f^\alpha + {}^M\Gamma^\alpha_{\beta\gamma}\frac{\partial f^\beta}{\partial t}\frac{\partial f^\gamma}{\partial \bar{t}}\Big)\overline{f^\beta_{\bar{j}}}h_{\alpha\bar{\beta}}\,,\end{aligned}$$

where $\dfrac{D}{\partial t}$ is the covariant differentiation with respect to the connection of the tangent bundle of M and ${}^M\Gamma^\alpha_{\beta\gamma}$ is the Christoffel symbol of M. The divergence of ξ at P is

$$\begin{aligned}\nabla_{\bar{i}}\xi^{\bar{i}} &= g^{\bar{i}j}\Big(\frac{\partial^2}{\partial t\,\partial \bar{t}} f^\alpha_{\bar{i}}\Big)\overline{f^\beta_{\bar{j}}}h_{\alpha\bar{\beta}} + g^{\bar{i}j}\Big(\frac{\partial^2}{\partial t\,\partial \bar{t}}\Big)\overline{\partial_i f^\beta_{\bar{j}}}h_{\alpha\bar{\beta}}\\ &\quad + g^{\bar{i}j}(\partial_\mu\,\partial_{\bar{\nu}} h_{\alpha\bar{\beta}})\overline{f^\nu_i}\,\overline{f^\beta_{\bar{j}}}\frac{\partial f^\mu}{\partial t}\frac{\partial f^\alpha}{\partial \bar{t}}\,.\end{aligned}$$

Now assume that f is harmonic at $t=0$. Then at P and $t=0$

$$g^{\bar{i}j}\overline{\partial_i f^\beta_{\bar{j}}}h_{\alpha\bar{\beta}} = 0$$

and hence at $t=0$

$$\begin{aligned}\frac{\partial^2}{\partial t\,\partial \bar{t}}\int_N |\bar{\partial} f|^2 &= \int_N g^{\bar{i}j}\Big(\frac{D}{\partial t} f^\alpha_{\bar{i}}\Big)\Big(\overline{\frac{D}{\partial t} f^\beta_{\bar{j}}}\Big)h_{\alpha\bar{\beta}} + \int_N g^{\bar{i}j}\Big(\frac{D}{\partial \bar{t}} f^\alpha_{\bar{i}}\Big)\Big(\overline{\frac{D}{\partial \bar{t}} f^\beta_{\bar{j}}}\Big)h_{\alpha\bar{\beta}}\\ &\quad + \int_N g^{\bar{i}j} f^\alpha_{\bar{i}}\overline{f^\beta_{\bar{j}}} R_{\mu\bar{\nu}\alpha\bar{\beta}}\frac{\partial f^\mu}{\partial \bar{t}}\frac{\partial f^\nu}{\partial \bar{t}} + \int_N g^{\bar{i}j} f^\alpha_{\bar{i}}\overline{f^\beta_{\bar{j}}} R_{\mu\bar{\nu}\alpha\bar{\beta}}\frac{\partial f^\mu}{\partial t}\overline{\frac{\partial f^\nu}{\partial t}}\\ &\quad - 2\,\mathrm{Re}\int_N g^{\bar{i}j} R_{\mu\bar{\nu}\alpha\bar{\beta}}\overline{f^\nu_i}\,\overline{f^\beta_{\bar{j}}}\frac{\partial f^\mu}{\partial \bar{t}}\frac{\partial f^\alpha}{\partial t}\,,\end{aligned}$$

where $R_{\mu\bar{\nu}\alpha\bar{\beta}} = \partial_\mu\partial_{\bar{\nu}}h_{\alpha\bar{\beta}} - h^{\bar{\gamma}\delta}\partial_\mu h_{\alpha\bar{\gamma}}\overline{\partial_\nu h_{\beta\bar{\delta}}}$ is the curvature tensor of M.

§3. Complex-Analyticity of Energy-Minimizing Maps

Suppose M is a compact Kähler manifold with positive holomorphic bisectional curvature. Let $f_0 : \mathbb{P}_1 \to M$ be an energy-minimizing map.

Proposition 1. *If $f_0^* c_1(M)$ evaluated at $\mathbb{P}_1$ is nonnegative (respectively nonpositive), then f_0 is holomorphic (respectively conjugate holomorphic).*

Proof. Since the proof of the other case is similar, we prove here only the holomorphic case.

Let T_M be the holomorphic tangent bundle of M, let w be a local coordinate of $\mathbb{P}_1$, and let $\frac{D}{\partial \overline{w}}$ be the covariant differentiation, in the antiholomorphic direction, of local cross sections of $f_0^* T_M$ with respect to the connection of T_M. Let $\mathcal{F}$ be the sheaf of germs of local cross sections s of $f_0^* T_M$ with $\frac{D}{\partial \overline{w}} s = 0$. Clearly $\mathcal{F}$ is an analytic sheaf over $\mathbb{P}_1$. Using the Cauchy kernel and the standard classical iteration process of Korn-Lichtenstein, we can show that for an arbitrary point P of $\mathbb{P}_1$ there exist local cross sections $s_1, \ldots, s_m$ at P such that $\frac{D}{\partial \overline{w}} s_i \equiv 0$, $1 \leq i \leq m$, and $s_1(P), \ldots, s_m(P)$ form a basis of the fiber $f_0^* T_M$ at P, where $m = \dim_{\mathbb{C}} M$. Therefore $\mathcal{F}$ is locally free and the holomorphic vector bundle associated to $\mathcal{F}$ is topologically isomorphic to $f_0^* T_M$.

Now we can regard $f_0^* T_M$ as a holomorphic vector bundle over $\mathbb{P}_1$ (after its identification with the holomorphic vector bundle associated to $\mathcal{F}$). By the theorem of Grothendieck [Gr] again, $f_0^* T_M$ is a direct sum of holomorphic line bundles $L_1, \ldots, L_m$ over $\mathbb{P}_1$. Since the first Chern class of $f_0^* T_M$ evaluated at $\mathbb{P}_1$ is nonnegative, it follows that for some i, $c_1(L_i)$ evaluated at $\mathbb{P}_1$ is nonnegative. By the theorem of Riemann-Roch, we can find a nontrivial global holomorphic section $s = \sum s^\alpha \frac{\partial}{\partial z^\alpha}$ of L_i (and hence of $f_0^* T_M$) over $\mathbb{P}_1$. Construct a smooth family of smooth maps $f(t) : \mathbb{P}_1 \to M$, $t \in \mathbb{C}$, $|t| < \varepsilon$, such that

$f(0) = f_0$ and at $t = 0$,

$$\frac{\partial}{\partial \bar{t}} f^\alpha(t) = 0 \quad \text{and} \quad \frac{\partial}{\partial t} f^\alpha(t) = s^\alpha \ .$$

Since s is a holomorphic cross section of $f_0^* T_M$, it follows that at $t = 0$

$$\frac{D}{\partial t}\frac{\partial}{\partial \overline{w}} f^\alpha = \frac{D}{\partial \overline{w}}\frac{\partial f^\alpha}{\partial t} = \frac{D}{\partial \overline{w}} s^\alpha = 0 \ .$$

Moreover, at $t = 0$,

$$\frac{D}{\partial \bar{t}}\frac{\partial}{\partial \overline{w}} f^\alpha = \frac{D}{\partial \overline{w}}\frac{\partial f^\alpha}{\partial \bar{t}} = 0 \ .$$

From the second variation formula derived in the preceding section, it follows that at $t = 0$

$$\frac{\partial^2}{\partial t\, \partial \bar{t}} \int_{\mathbb{P}_1} |\overline{\partial} f|^2 = \int_{\mathbb{P}_1} \frac{\partial f^\alpha}{\partial \overline{w}} \overline{\frac{\partial f^\beta}{\partial \overline{w}}} R_{\mu\overline{\nu}\alpha\overline{\beta}} \frac{\partial f^\mu}{\partial t} \overline{\frac{\partial f^\nu}{\partial t}} (\sqrt{-1} dw \wedge d\overline{w}) \ . \quad (1)$$

Since f_0 is also $\overline{\partial}$-energy minimizing, we have at $t = 0$

$$\frac{\partial^2}{\partial t\, \partial \bar{t}} \int_{\mathbb{P}_1} |\overline{\partial} f|^2 \geq 0 \ .$$

Since the holomorphic bisectional curvature of M is positive, i.e.,

$$R_{\mu\overline{\nu}\alpha\overline{\beta}} \xi^\mu \overline{\xi}^\nu \eta^\alpha \overline{\eta}^\beta < 0$$

for (ξ^μ), $(\eta^\alpha) \in \mathbb{C}^m - 0$, it follows from (1) that

$$\frac{\partial f_0^\alpha}{\partial \overline{w}} = 0$$

on $\mathbb{P}_1 - Z$ for all α, where Z is the zero-set of s. Therefore the map f_0 is holomorphic, because Z is a finite subset of $\mathbb{P}_1$.

§4. Existence of Energy-Minimizing Maps

Let M be a compact Riemannian manifold. For each C^1 map $f : S^2 \to M$, we let $E([f])$ be the infimum of the sum of the energy of maps whose sum is homotopic to f. The following proposition is proved by using the methods of [SaU] and [MY].

Proposition 2. *For every C^1 map $f : S^2 \to M$ there exist energy-minimizing maps $f_i : S^2 \to M$, $1 \le i \le m$, such that the sum of f_i is homotopic to f and $E([f]) = \sum_{i=1}^{m} E(f_i)$.*

Proof. By Theorem 6.3.3, any nonconstant harmonic map or any homotopically nontrivial C^1 map from S^2 into M has energy $> c$ for some fixed positive constant c. Let k be the smallest nonnegative integer such that

$$E(f) \le \frac{kc}{2} .$$

We are going to prove the proposition by induction on k. The case $k = 0$ is true. We want to prove the case $k = n + 1$ under the assumption that the case $k = n$ is true. By Theorem 6.2.2 and Proposition 6.2.3, for $\alpha > 1$ we can find smooth maps $f_\alpha : S^2 \to M$ which minimize the functional $E_\alpha(g) = \int_{S^2}(1 + |dg|^2)^\alpha$ over the space of all C^1 maps homotopic to f. Let x_α be a point of S^2 so that

$$|df_\alpha|^2(x_\alpha) = \max_{S^2} |df_\alpha|^2 .$$

If $\max_{S^2} |df_\alpha|^2$ is uniformly bounded in α, then by Lemma 6.4.5 the f_α converges to an energy-minimizing map and the assertion is proved. If $\max_{S^2} |df_\alpha|^2$ is unbounded in α, then by Theorem 6.4.6 the map $f_\alpha | D_\alpha$ for some disc D_α with center at x_α converges to $\tilde{g}|D$, where $\tilde{g} : S^2 \to M$ is a harmonic map and D is identified with D_α by a suitable conformal map on S^2. We can assume that $f_\alpha|\partial D_\alpha$ has arbitrarily small length.

As in [MY], we can construct a map $f_\alpha^1 : S^2 \to M$ so that $E(f_\alpha^1)$ is arbitrarily close to $E(f_\alpha) - E(\tilde{g})$ and f is homotopic to the sum of f_α^1 and $\tilde{g}$ as follows. Fix discs D_α and D so that $f_\alpha|D_\alpha$ converges smoothly to $\tilde{g}|D$ (under suitable identification of D_α and D) and the energy of $\tilde{g}$ over D is close to the energy of $\tilde{g}$. We define maps $\tilde{f}_\alpha$ from S^2 into M by extending $f_\alpha|S^2 - D_\alpha$ to S^2 in the following way. Since $f_\alpha|\partial D_\alpha$ is close to $\tilde{g}|\partial D$, we can join $f_\alpha(\partial D_\alpha)$ to $\tilde{g}(\partial D)$ by short geodesics and obtain a map h_α from an annulus into M whose image has small area. Then we obtain a map $\tilde{f}_\alpha$ from S^2 into M by putting together the maps $f_\alpha|S^2 - D_\alpha$, h_α, and $\tilde{g}|S^2 - D$.

By approximation we may assume that the map $\tilde{f}_\alpha$ is a smooth immersion from S^2 into M and the area of the image of $\tilde{f}_\alpha$ is close to the sum of the areas of the images of $f_\alpha|S^2 - D_\alpha$, h_α, and $\tilde{g}|S^2 - D$. Choose a suitable diffeomorphism φ of S^2 such that $f'_\alpha = \overline{f}_\alpha \circ \varphi$ is conformal. Observe that the area of the image of a map from a real surface to a Riemannian manifold is the same as the energy when the map is conformal. Since $\tilde{g}$ is conformal and $E(\tilde{g}|S^2 - D)$ is small, the area of the image of the map $\tilde{g}|S^2 - D$ is small. This together with the smallness of the area of the image of h_α implies that $E(f_\alpha^1)$ is close to $E(f_\alpha) - E(\tilde{g})$.

Since $\tilde{g}$ is harmonic, $E(\tilde{g}) > c$ and $E(f_\alpha^1) \leq \frac{nc}{2}$ for α sufficiently close to 1. By induction hypothesis, for α sufficiently close to 1 we can find energy-minimizing maps $f_i : S^2 \to M$, $1 \leq i \leq m-1$, such that the sum of f_i $(1 \leq i \leq m-1)$ is homotopic to f_α^1 and $E([f_\alpha^1]) = \sum_{i=1}^{m-1} E(f_i)$. Set $f_m = \tilde{g}$, then

$$\begin{aligned} E([f]) &= \lim_{\alpha\to 1} E(f_\alpha) = \lim_{\alpha\to 1} E(f_\alpha^1) + E(\tilde{g}) \\ &\geq E([f_\alpha^1]) + E(\tilde{g}) = \sum_{i=1}^{m} E(f_i) \,. \end{aligned}$$

On the other hand, from the definition of $E([f])$, it is clear that $E([f]) \leq \sum_{i=1}^{m} E(f_i)$. Hence $E([f]) = \sum_{i=1}^{m} E(f_i)$. It follows that f_m is energy-minimizing.

§5. Proof of Frankel Conjecture

First we state the following proposition about holomorphic deformation of rational curves. Its proof can be found in [SiY, §5].

Proposition 3. *Let M be a compact complex manifold whose tangent bundle T_M is positive. Let C_0 be a rational curve in M and $f : \mathbb{P}_1 \to C_0$ be its normalization. Then there exists a proper subvariety Z of $\mathbb{P}(T_M)$ with the following property. If $y \in M$ and $\xi \in (T_M)_y - 0$ define an element of $\mathbb{P}(T_M) - Z$, then there exists a holomorphic map $f' : \mathbb{P} \to M$ homotopic to f (when f is regarded as a map from $\mathbb{P}_1$ to M) such that y is a regular point of $f'(\mathbb{P}_1)$ and the tangent vector of $f'(\mathbb{P}_1)$ at y is a nonzero multiple of ξ.*

Suppose M is an m-dimensional compact Kähler manifold of positive bisectional curvature. Since the Ricci curvature of M is positive, it follows from the theorem of Bonnet-Myers that the universal covering of M is compact. Since $\mathbb{P}_m$ has no fixed-point-free automorphism, by replacing M by its universal covering, we can assume without loss of generality that M is simply connected. It follows that $\pi_2(M)$ is isomorphic to $H_2(M, \mathbb{Z})$.

Since the holomorphic bisectional curvature of M is positive, by the result of Bishop-Goldberg [BG] the second Betti number of M is 1. From the universal coefficient theorem it follows that $H^2(M, \mathbb{Z}) \cong \mathbb{Z}$.

There exists a positive holomorphic line bundle F over M whose first Chern class $c_1(F)$ is a generator of $H^2(M, \mathbb{Z})$. Let g be a generator of the free part of $H_2(M, \mathbb{Z})$ such that the value of $c_1(F)$ at g is 1. Let $f : \mathbb{P}_1 \to M$ be a smooth map so that the element in $\pi_2(M)$ defined by f corresponds to g in the isomorphism between $\pi_2(M)$ and $H_2(M, \mathbb{Z})$.

By Proposition 2, there exist energy-minimizing maps $f_i : \mathbb{P}_1 \to M$, $0 \le i \le k$, such that the sum of f_i $(0 \le i \le k)$ is homotopic to f and $E([f]) = \sum_{i=0}^{k} E(f_i)$. By Proposition 1, each f_i is either holomorphic or conjugate holomorphic. So each $f_i(\mathbb{P}_1)$ is a rational curve. Since $c_1(T_M)$ is a positive integral multiple of $c_1(F)$ and the value of $c_1(F)$ at g is 1, it follows that at least one f_i is holomorphic. If $k > 0$, then at least one f_j is conjugate holomorphic. We distinguish now between two cases.

Case 1. $k = 0$.

Let E be the divisor of the differential df_0 of f_0 and let $[E]$ be the line bundle over $\mathbb{P}_1$ associated to E. Then the line bundle $T_M \otimes [E]$ is a subbundle of $f_0^* T_M$ and the quotient bundle $(f_0^* T_M)/(T_M \otimes [E])$ splits into a direct sum of line bundles $Q_2, \dots, Q_m$. Each Q_i $(2 \le i \le m)$ is a positive line bundle. It follows that

$$c_1(f_0^* T_M) = c_1(T_M) + c_1([E]) + \sum_{i=2}^{m} c_1(Q_i) .$$

Hence $c_1(T_M)$ evaluated at g is $\ge n + 1$. That is $c_1(T_M) = \lambda c_1(F)$ for some integer $\lambda \ge n + 1$. By the result of Kobayashi-Ochiai [KO], M is biholomorphic to $\mathbb{P}_m$.

Case 2. $k > 0$.

Without loss of generality we can assume that f_0 is holomorphic and f_1 is conjugate holomorphic. From Proposition 3 we obtain a proper subvariety Z of $\mathbb{P}(T_M)$ for the rational curve $f_1(\mathbb{P}_1)$ of M. Take a point $y \in M$ and a non-zero tangent vector ξ of M at y so that the point of $\mathbb{P}(T_M)$ defined by y and ξ does not belong to $Z \cup Z'$. The map f_0 (respectively f_1) is homotopic to a holomorphic map f_0' (respectively conjugate holomorphic map f_1') from $\mathbb{P}_1$ to M such that y is a regular point of $f_0'(\mathbb{P}_1)$ (respectively $f_1'(\mathbb{P}_1)$) whose tangent vector at y is a multiple of ξ. Since for $\nu = 0, 1$ both $E(f_\nu)$ and $E(f_\nu')$ are equal to the absolute value of the Kähler class of M at $[f_\nu(\mathbb{P}_1)]$, by replacing f_ν by f_ν' we can assume without loss of generality $f_\nu = f_\nu'$.

Choose a local coordinate system $z_1, \ldots, z_n$ at y with y as the origin such that $\xi = \frac{\partial}{\partial z_1}$. Choose a local coordinate system ζ of $\mathbb{P}_1$ so that both $f_0^{-1}(y)$ and $f_1^{-1}(y)$ correspond to $\zeta = 0$ and f_ν $(\nu = 0, 1)$ is of the form

$$z_1 = \zeta\,, \quad z_\mu = f_\mu(\zeta) \qquad (2 \leq \mu \leq m)$$

near $\zeta = 0$. For $\delta > 0$ let Δ_δ be the closed disc in $\mathbb{C}$ of radius δ centered at 0. For δ sufficiently small, we remove $f_\nu(\Delta_\delta)$ from $f_\nu(\mathbb{P}_1)$ $(\nu = 1, 2)$ and replace these two discs by the surface S_δ defined by

$$z_1 = \delta e^{i\theta}\,, \; z_\mu = t f_0(\delta e^{i\theta}) + (1 - t) f_1(\delta e^{i\theta})$$

for $0 \leq \theta \leq 2\pi$ and $0 \leq t \leq 1$. The surface

$$S_\delta \cup \Big(\bigcup_{\nu=1,2} f_\nu(\mathbb{P}_1 - \Delta_\delta) \Big)$$

oriented in the obvious way is the image of a map $\tilde{f}$ from S^2 to M which is homotopic to the sum of f_0 and f_1. Moreover, for δ sufficiently small the area of this surface is strictly less than the combined area of $f_0(\mathbb{P}_1)$ and $f_1(\mathbb{P}_1)$. By smoothing out $\tilde{f}$, and composing $\tilde{f}$ with a suitable diffeomorphism of S^2, we can get a conformal map $\hat{f}$ with

$E(\hat{f}) < E(f_0) + E(f_1)$. The sum of $\hat{f}$ and $f_2, \ldots, f_k$ is homotopic to f and yet

$$E(\hat{f}) + \sum_{i=2}^{k} E(f_i) < \sum_{i=0}^{k} E(f_i) = E([f]) ,$$

which is a contradiction. Hence $k = 0$ and M is biholomorphic to $\mathbb{P}_m$.

References

[AHS] M.F. Atiyah, N.J. Hitchin and I.M. Singer, *Self-duality in four dimensional Riemannian geometry*, Proc. Royal Soc. London, **362(A)** (1978), 425-461.

[Bg] M. Berger, *Sur quelques variétés riemanniennes suffisamment pincées*, Bull. Soc. Math. France, **88** (1960), 57-71.

[Be] L. Bers, *An outline of the theory of pseudo-analytic functions*, Bull. Amer. Math. Soc., **62** (1956), 291-331.

[BG] R.L. Bishop and S.I. Goldberg, *On the other second cohomology group of a Kähler manifold of positive curvature*, Proc. Amer. Math. Soc., **16** (1965), 119-122.

[B] A. Borel, *Compact Clifford-Klein forms of symmetric spaces*, Topology, **2** (1963), 111-122.

[CP] M. do Carmo and C.K. Peng, *Stable minimal surfaces in* $\mathbb{R}^3$ *are planes*, Bull. Amer. Math. Soc., **1** (1979), 903-906.

[C] E. Cartan, *Sur les domaines bornes homogenes de l'espace de n variables complexes*, Abh. Math. Sem. Hamburg, **ii** (1935), 116-162, Oevres Completes, Parie I, 1259-1305.

[CE] J. Cheeger and D. Ebin, *Comparison theorems in Riemannian geometry*, North-Holland, Amsterdam, 1975.

[CY] S.Y. Cheng and S.-T. Yau, *Differential equations on Riemannian manifolds and their geometric applications*, Comm. Pure Appl. Math., **28** (1975), 333-354.

[CG] S.S. Chern and S.I. Goldberg, *On the volume decreasing property of a class of real harmonic mappings*, Amer. J. Math., **97** (1975), 133-147.

[CO] S.S. Chern and R. Osserman, *Complete minimal surfaces in Euclidean n-space*, J. Analyse Math., **19** (1967), 15-34.

[DT] A. Domic and D. Toledo, *The Gromov norm of the Kähler class of symmetric domains*, Math. Ann., **276** (1987), 425-432.

[EE] C.J. Earle and J. Eells, *A fiber bundle description of Teichmüller theory*, J. Diff. Geom., **3** (1969), 19-43.

[EL] J. Eells and L. Lemaire, *A report on harmonic maps*, Bull. London Math. Soc., **10** (1978), 1-68.

[ES] J. Eells and J. Sampson, *Harmonic mappings of Riemannian manifolds*, Amer. J. Math., **86** (1964), 109-160.

[FLP] A. Fathi, F. Laudenbach and V. Poenaru, *Traveaux de Thurston sur les surfaces*, Asterisque, (1979), 66-67.

[FS] D. Fisher-Colbrie and R. Schoen, *The structure of complete stable minimal surfaces in 3-manifolds of nonnegative scalar curvature*, Comm. Pure Appl. Math., **33** (1980), 199-211.

[F] T. Frankel, *Manifolds with positive curvature*, Pacific J. Math., **11** (1961), 165-174.

[FU] D.S. Freed and K. Uhlenbeck, *Instantons and four manifolds*, Math. Sci. Inst. Publ., Springer-Verlag, New York, **1** (1984).

[GM] S. Gallot and D. Meyer, *Opérateur de courbure et Laplacien des formes différentielles d'une variété Riemannienne*, J. Math. PureAppl., **54** (1975), 259-284.

[GT] D. Gilbarg and N.S. Trudinger, *Elliptic partial differential equations of second order*, Springer-Verlag, New York, 1983.

[G] M. Gromov, *Volume and bounded cohomology*, Inst. Hautes Etudes Sci. Publ. Math.

[Gr] A. Grothendieck, *Sur la classification des fibrées holomorphes sur la sphère Riemann*, Amer. J. Math., **79** (1957), 121-138.

[GL] R. Gulliver and H.B. Lawson, *The structure of stable minimal hypersurfaces near a singularity*, Proc. Symposia in Pure Math., **44** (1986), 213-237.

[HMu] V. Haagerup and H.J. Munkholm, *Simplices of maximal volume in hyperbolic n-space*, Acta Math., **147** (1981), 1-11.

[Ha1] R. Hamilton, *Four manifolds with positive curvature operator*, preprint.

[Ha2] R. Hamilton, *Three-manifolds with positive Ricci curvature*, J. Diff. Geom., **17** (1982), 255-306.

[H] E. Heinz, *On certain elliptic differential equations and univalent mappings*, J. Analyse Math., **5** (1956-57), 197-272.

[HM] J. Hubbard and H. Masur, *Quadratic differentials and foliations*, Acta Math., **142** (1979), 221-274.

[Hu] G. Huisken, *Ricci deformation of the metric on a Riemannian manifold*, J. Differential Geom., **21** (1985), 47-62.

[K] S. Kerckhoff, *The asymptotic geometry of Teichmüller space*, Topology, **19** (1980), 23-41.

[Kn] H. Kneser, *Die Kleinste Bedeckungzahl innerhalb einer Klasse von Flächenabbildungen*, Math. Ann., **103** (1930), 347-358.

[KO] S. Kobayashi and T. Ochiai, *Characterizations of complex projective spaces and hyperquadratics*, J. Math. Kyoto Univ., **13** (1973), 31-47.

[KM] J.L. Koszul and B. Malgrange, *Sur certaines fibrées complexes*, Arch. Math., **9** (1958), 102-109.

[L] H.B. Lawson, Jr., *Lectures on minimal submanifolds*, **I**, Publish or Perish, Berkeley, 1980.

[Le] L. Lemaire, *Applications harmoniques des surfaces Riemanniennes*, J. Differential Geom., **13** (1978), 51-87.

[Li] Lichnerowicz, *Applications harmoniques et variétes kahlériennes*, Symp. Math., **3** (1970), 341-402.

[Ma] T. Mabuchi, $\mathbb{C}^3$*-actions and algebraic threefolds with ample tangent bundle*, Nagoya Math. J., **69** (1978), 33-64.

[MY] W. Meeks and S.-T. Yau, *Topology of three dimensional manifolds and the embedding problems in minimal surface theory*, Ann. of Math., **122** (1980), 441-484.

[Mi] M. Micallef, *Stable minimal surfaces in Euclidean space*, J. Differential Geom., **19** (1984), 57-84.

[MM] M. Micallef and J.D. Moore, *Minimal two-spheres and the topology of manifolds with positive curvature on totally isotropic two-planes*, Ann. of Math.

[Mn] J. Milnor, *Lectures on the h-cobordism theorem*, Mathematical Notes, **1**, Princeton Univ. Press, Princeton, New Jersey, 1965.

[MG] F. Morgan, *On the singular structure of two-dimensional area minimizing surfaces in* $\mathbb{R}^n$,

[Mr] S. Mori, *Projective manifolds with ample tangent bundles*, Ann. of Math., **110** (1979), 593-606.

[M] C.B. Morrey, *Multiple integrals in the calculus of variations*, Springer-Verlag, New York, 1966.

[O] R. Osserman, *A survey of minimal surfaces*, Van Nostrand Reinhold, New York, 1969.

[P1] R.S. Palais, *Foundations of global non-linear analysis*, Benjamin, New York, 1968.

[P2] R.S. Palais, *Ljusternik-Schnirelman theory on Banach manifolds*, Topology, **5** (1966), 115-132.

[R] E.A. Ruh, *Riemannian manifolds with bounded curvature ratios*, J. Differential Geom., **17** (1982), 643-653.

[SaU] J. Sacks and K. Uhlenbeck, *The existence of minimal immersions of 2-spheres*, Ann. of Math., **113** (1981), 1-24.

[S] R. Schoen, *Analytic aspects of the harmonic map problem*, Math. Sci. Res. Inst. Publ., Springer-Verlag, **?** (198?).

[Sa] R. Savage, *The space of positive definite matrices and Gromov's invariant*, Trans. AMS, **274** (1982), 239-263.

[SU] R. Schoen and K. Uhlenbeck, *A regularity theory for harmonic maps*, J. Differential Geom., **17** (1987), 307-335.

[SSY] R. Schoen, L. Simon and S.-T. Yau, *Curvature estimates for minimal hypersurfaces*, Acta Math., **134** (1975), 275-288.

[SY3] R. Schoen and S.-T. Yau, *The geometry and topology of manifolds of positive scalar curvature.*

[SY2] R. Schoen and S.-T. Yau, *Existence of incompressible minimal surfaces and the topology of three dimensional manifolds with nonnegative scalar curvature*, Ann. of Math., **110** (1979), 127-142.

[SY1] R. Schoen and S.-T. Yau, *Univalent harmonic maps of surfaces*, Invent. Math., **44** (1978), 265-278.

[Si] Y.-T. Siu, *The complex-analycity of harmonic maps and the strong rigidity of compact Kähler manifolds*, Ann. of Math., **112** (1980), 73-111.

[SiY] Y.-T. Siu and S.-T. Yau, *Compact Kähler manifolds of positive bisectional curvature*, Invent. Math., **59** (1980), 189-204.

[T] W. Thurston, *The geometry and topology of three-manifolds*, Princeton University Lecture Note.

[To] D. Toledo, *Harmonic maps from surfaces to certain Kähler manifolds*, Math. Scand., **45** (1979), 13-26.

[U1] K. Uhlenbeck, *Morse theory by perturbation methods with applications to harmonic maps*, Trans. AMS, **267** (1981), 569-583.

[U2] K. Uhlenbeck, *Integrals with nondegenerate critical points*, Bull. Amer. Math. Soc., **76** (1970), 125-128.

[U3] K. Uhlenbeck, *Morse theory on Banach manifolds*, J. Functional Analysis, **10** (1972), 430-445.

[We1] S.M. Webster, *Minimal surfaces in a Kähler surface*, J. Differential Geom., **20** (1984), 463-470.

[We2] S.M. Webster, *The euler and Pontrjagin numbers of an n-manifold in* $\mathbb{C}^n$, Math. Helv., **60** (1985), 193-216.

[Wi] W. Wirtinger, *Eine Determinantenidentität and ihre Anwendung auf analytische Gebilde and Hermitesche Maßbestimmung*, Monatsh. Math. Physik, **44** (1936), 343-365.

[W] M. Wolf, *The Teichmüler theory of harmonic maps*, J. Differential Geom., **29** (1989).

[Wo] J.C. Wood, *Holomorphicity of certain harmonic maps from a surface to complex projective n-space*, J. London Math. Soc., **20** (1979), 137-142.

PART II

Chapter IX

Analytic Aspects of The Harmonic Map Problem

In this chapter we consider some of the basic existence, uniqueness, and regularity theorems for harmonic mappings. We give a proof of the Eells-Sampson theorem concerning the homotopy problem for maps to manifolds of nonpositive curvature, as well as the theorem of Hamilton which deals with the homotopy/Dirichlet problem. This chapter is an expanded and updated version of the article with the same title by R. Schoen [Sc]

1. Formulation of the Basic Problems

Given Riemannian (M^n, g), (N^k, h) and a C^1 map $u : M \to N$, the energy density of u is given by

$$e(u) = \mathrm{Tr}_g(u^*h) = \sum_{\alpha,\beta,i,j} g^{\alpha\beta}(x)h_{ij}(u(x))\frac{\partial u^i}{\partial x^\alpha}\frac{\partial u^j}{\partial x^\beta}$$

where x^α, u^i, $1 \leq \alpha \leq n$, $1 \leq i \leq k$ are local coordinates. The energy functional $E(u)$ is then

$$E(u) = \int_M e(u)\mathrm{d}v$$

where $\mathrm{d}v = (\det g)^{1/2}\mathrm{d}x$ is the volume element of M. From the definition of E it is clear that the natural class of maps u for which E is finite are those which are bounded and have first derivatives in L^2. Call this class of maps $\mathcal{D}$, that is

$$\mathcal{D} = L^\infty(M, N) \bigcap L_1^2(M, N).$$

The first observation one makes is that there is no obvious definition of these spaces because the expression for $e(u)$ depends on the fact that the image of an open set in M is contained in a coordinate chart of N. Since we don't expect maps in $\mathcal{D}$ to be a priori continuous we cannot guarantee this. The simplest way to get around this difficulty is to embed N as a submanifold of a Euclidean space $\mathbb{R}^K$. For convenience we assume that N is isometrically embedding, although one can get by with a smooth embedding. Now we can make sense of $L^\infty(M,N)$ as those members of $L^\infty(M,\mathbb{R}^K)$ which have image lying almost everywhere in N and similarly $L_1^2(M,N)$ is the subset of $L_1^2(M,\mathbb{R}^K)$ consisting of maps having image almost everywhere in N. Notice that N is defined by a system of equations of the form

$$\Phi(u) = (\Phi_1(u), \dots, \Phi_{K-k}(u)) = 0$$

where the Jacobian matrix of Φ has rank $K-k$ at each point u with $\Phi(u)=0$. Thus the space $\mathcal{D}$ is given by

$$\mathcal{D} = L^\infty(M,\mathbb{R}^K)\bigcap L_1^2(M,\mathbb{R}^K)\bigcap\{u : \Phi(u(x)) = 0\, a.e.\, x\}.$$

In this formulation the energy functional becomes the ordinary Dirichlet integral

$$E(u) = \sum_{i=1}^K \int_M |\nabla u^i|^2 dv$$

and the harmonic map problem is to find critical points of E which respect the constraint $\Phi(u) = 0$. The Euler-Lagrange equation in coordinate form is

$$\Delta_M u^i + g^{\alpha\beta}\Gamma^i_{jl}(u(x))\frac{\partial u^j}{\partial x^\alpha}\frac{\partial u^l}{\partial x^\beta} = 0, \quad i = 1, \dots, k \tag{1.1}$$

where Γ^i_{jl} are the Christoffel symbols for the Levi-Civita connection on N; i.e. they are computed as nonlinear combinations of h_{ij} and its first derivatives on N. The Euler-Lagrange equation for the constrained problem is clearly $(\Delta u)^T = 0$ where $(\ )^T$ means projection of a vector into the tangent space of N at $u(x)$. Since the normal component of Δu can be represented in terms of the second fundamental form of N, the equation becomes

$$\Delta_M u^i = g^{\alpha\beta} A^i_{u(x)} \left(\frac{\partial u}{\partial x^\alpha}, \frac{\partial u}{\partial x^\beta} \right), \quad i = 1, \dots, K \tag{1.2}$$

where $A_u(X,Y) \in (T_u N)^\perp$ is the second fundamental form of n given by $A(X,Y) = (D_X Y)^\perp$ where D denotes the directional derivative in $\mathbb{R}^K$ and X, Y are extended arbitrarily as tangent vector fields to N in a neighborhood of $u \in N$. The advantage of (1.2) over (1.1) is that (1.2) makes sense in the weak form for a map in $\mathcal{D}$.

Definition. *A map $u \in \mathcal{D}$ is a (weakly) harmonic map if u satisfies the weak form of (1.2); that is for any $\eta \in C_0^\infty(M, \mathbb{R}^K)$ we have*

$$\int_M \sum_i \nabla \eta^i \cdot \nabla u^i + g^{\alpha\beta} \eta^i(x) A^i_{u(x)} \left(\frac{\partial u}{\partial x^\alpha}, \frac{\partial u}{\partial x^\beta} \right) dv = 0. \tag{1.3}$$

More generally one can ask that u be a critical point of E in $\mathcal{D}$. Observe, however, that since $\mathcal{D}$ is only defined as a subset of the Hilbert space $L_1^2(M, \mathbb{R}^K)$, there is no clear meaning of critical point. We formulate a definition based on utility. There are two basic classes of variations which one would like to perform on a map u. The first is defined by a function $\eta \in C_0^\infty(M, \mathbb{R}^K)$. Given such an η we can consider the map $u + t\eta$. Generally the map $u + t\eta$ will not be in $\mathcal{D}$ because it will not satisfy the constraint $\Phi(u) = 0$. If N is at least a C^1 submanifold we can correct for this deficiency by projecting $u + t\eta$ onto N. Precisely let $\Pi : O \to N$ be the nearest point projection defined on a small neighborhood O of N and observe that since u has bounded image we will have $u(x) + t\eta(x) \in O$ for t sufficiently small, a.e. $x \in M$. Therefore we can get $u_t = \Pi \cdot (u + t\eta)$ and it is not difficult to see that the curve $t \to u_t$ is a C^1 curve in $L_1^2(M, \mathbb{R}^K)$ and hence we can require

$$\frac{\mathrm{d}}{\mathrm{d}t} E(u_t) \bigg|_{t=0} = 0 \quad \text{for every } \eta \in C_0^\infty(M, \mathbb{R}^K).$$

It can be seen without much difficulty (see [SU1]) that u is harmonic if and only if u is critical for these variations. There is a second type of variation which one can consider. This is a variation of the parameter domain M; that is, given a family $F_t : M \to M$ of diffeomorphisms which are equal to the identity outside a compact set of M and with $F_0 = \mathrm{id}$, we can set $u_t = u \circ F_t$ and require that

$$\frac{\mathrm{d}}{\mathrm{d}t} E(u_t)\Big|_{t=0} = 0 \quad \text{for every family } F_t.$$

We will see below that the function $t \to E(u_t)$ is a smooth function of t so that the derivative exists. It is not generally true that the curve $t \to u_t$ is a differentiable curve in $L_1^2(M, \mathbb{R}^K)$.

Definition. *A map $u \in \mathcal{D}$ is a stationary point of E if u is critical with respect to both of the types of variation described above.*

As we have noted, a stationary point is harmonic but the converse is not true (see [BCL] for an example). It is clear that a C^2 harmonic map is a stationary point; in fact we'll see below that a C^0 harmonic map is a stationary point. It is also clear that an energy minimizing map is a stationary point.

We now derive a basic set of equations which holds for stationary maps. First, we consider the metric g on M to be an independent variable, and we denote by $E(u, g)$ the energy of the map u computed with respect to the metric g. Now let X be a smooth vector field with compact support on M, and let F_t denote the one parameter group of diffeomorphisms determined by X. We then have $E(u \circ F_t, g) = E(u, F_{-t} * g)$ since F_t is an isometry from (M, g) to $(M, F_{-t} * g)$. This shows that $E(u_t)$ is a smooth function of t. We denote by $\tau = \tau_{\alpha\beta} dx^\alpha dx^\beta$ the pullback of the metric from N; that is, $\tau_{\alpha\beta} = \frac{\partial u}{\partial x^\alpha} \cdot \frac{\partial u}{\partial x^\beta}$. We then have

$$E(u, g) = \int_M g^{\alpha\beta} \tau_{\alpha\beta} \sqrt{det(g)} dx,$$

and we may calculate $\frac{dE(u_t)}{dt}$ at $t = 0$ and use the stationarity of u to obtain

$$\int_M \langle \tau, \mathcal{L}_X g - \frac{1}{2} \mathrm{Tr}_g(\mathcal{L}_X g) \rangle dv = 0$$

for every smooth compactly supported vector field X, where $\mathcal{L}_X g$ denotes the Lie derivative of g with respect to X. Note that this may be rewritten

$$\int_M \langle \bar{\tau}, \mathcal{L}_X g \rangle dv = 0$$

where $\bar{\tau} = \tau - \frac{1}{2}\mathrm{Tr}_g(\tau)$, and is the weak form of the system of PDE's $g^{\alpha\beta}\bar{\tau}_{\gamma\alpha;\beta} = 0$. The following lemma is essentially a classical result which follows in case the dimension of M is 2.

Lemma 1.1. *Suppose $n = \dim M = 2$, and $u \in \mathcal{D}$ is a stationary point for E. The Hopf differential*

$$\Phi(z) = \left[\left(\| \left(\frac{\partial u}{\partial x} \right) \|^2 - \| \left(\frac{\partial u}{\partial y} \right) \|^2 \right) - 2i \left(\frac{\partial u}{\partial x} \right) \cdot \left(\frac{\partial u}{\partial y} \right) \right] dz^2$$

where $z = x + iy$ is a complex coordinate on M, is holomorphic and hence smooth.

Proof : If the map u is C^2, this result was derived in [CG]. Since the result is local, we can assume that the domain of u is the unit disk D in $\mathbb{C}$, and since the energy is conformally invariant we can assume the metric is euclidean. For any C^∞ function $\eta(x, y)$ with compact support in D, let $X = \eta \frac{\partial}{\partial x}$, and observe that $\mathcal{L}_X g = 2\eta_x dx^2 + \eta_y dx dy$. The equation above then becomes $\int_D \{(\tau_{11} - \tau_{22})\eta_x + 2\tau_{12}\eta_y\} dx dy = 0$. Similarly, we may choose $X = \zeta(x, y)\frac{\partial}{\partial y}$ and derive the equation $\int_D \{(\tau_{11} - \tau_{22})\zeta_y - 2\tau_{12}\zeta_x\} dx dy = 0$. These are the Cauchy-Riemann equations for the L^1 function $\phi(z) = \tau_{11} - \tau_{22} - 2i\tau_{12}$, and hence by Weyl's lemma, $\phi(z)$ is a smooth holomorphic function of z. This completes the proof of Lemma 1.1.

In dimension greater than two, stationary maps enjoy the following monotonicity inequality. For general stationary maps this was derived in [P]. For maps of least energy this was used in the regularity theory in [SU1].

Lemma 1.2. *Suppose u is a stationary map from a domain Ω in $\mathbb{R}^n$ into a manifold N. Then we have for $0 < \sigma < \rho < dist(x_0, \partial\Omega)$*

$$(1.4)\quad \rho^{2-n}\int_{B_\rho(x_0)} e(u)dx - \sigma^{2-n}\int_{B_\sigma(x_0)} e(u)dx = 2\int_{B_\rho(x_0)\setminus B_\sigma(x_0)} |x-x_0|^{2-n}\left|\frac{\partial u}{\partial r}\right|^2 dx$$

where $r = |x - x_0|$.

Proof To derive monotonicity from our calculations, we assume without loss of generality that $x_0 = 0$, and choose the vector field $X = \zeta(x)x^\alpha\frac{\partial}{\partial x^\alpha}$ where $\zeta(x)$ is a smooth function with compact support in $B_\sigma(0)$. We then have

$$\left(\mathcal{L}_X g - \frac{1}{2}\mathrm{Tr}_g(\mathcal{L}_X g)\right)_{\alpha\beta} = (2-n)\zeta\delta_{\alpha\beta} + \zeta_\beta x^\alpha + \zeta_\alpha x^\beta - \zeta_\gamma x^\gamma\delta_{\alpha\beta}.$$

If ζ is chosen to be a function of $r = |x|$, the stationarity then gives

$$(n-2)\int \zeta|\nabla u|^2 dx = \int r\zeta_r(2|\frac{\partial u}{\partial r}|^2 - |\nabla u|^2)dx.$$

Letting ζ approach the characteristic function of $B_\sigma(0)$ this implies

$$(n-2)\int_{B_\sigma}|\nabla u|^2 dx - \int_{\partial B_\sigma}|\nabla u|^2 dv = -2\int_{\partial B_\sigma}|\frac{\partial u}{\partial r}|^2 dv.$$

This may be written

$$\frac{d}{d\sigma}(\sigma^{2-n}E(B_\sigma(0))) = 2\int_{\partial B_\sigma} r^{2-n}|\frac{\partial u}{\partial r}|^2 dv$$

which implies the conclusion of Lemma 1.2 by integration.

We close this introductory section by stating the basic existence questions which we will consider in this chapter.

(1) **Homotopy problem** Assume M is closed and compact. Given $\varphi : M \to N$, find a harmonic map u with u homotopic to φ.

(2) **Dirichlet problem** Assume M is compact with smooth boundary ∂M. Given $\varphi : M \to N$, find a harmonic map with $u \equiv \varphi$ on ∂M.

(3) **Homotopy/Dirichlet problem** Assume M is compact with smooth boundary ∂M. Given $\varphi : M \to N$, find a harmonic map with $u \equiv \varphi$ on ∂M and which is homotopic to φ (rel. ∂M). (i.e. there exsits u_t, $0 \le t \le 1$, with $u_0 = \varphi$, $u_1 = u$, and $u_t \equiv \varphi$ on ∂M.)

2. Solvability of the Dirichlet problem

In the previous section we introduced the class of maps u for which the energy is finite; that is, those maps which have first derivatives in L^2, and are bounded. We called this class of maps $\mathcal{D}$, so that we defined $\mathcal{D} = L^\infty(M,N) \cap L_1^2(M,N)$. In case both M and N are compact, we have $\mathcal{D} = L_1^2(M,N)$. Let us assume that both M and N are compact. Note that the inner product and norm on the Hilbert space $L_1^2(M,\mathbb{R}^K)$ are

$$(f,g)_1 = \int_M [\langle f(x), g(x)\rangle + \langle df(x), dg(x)\rangle] dv_M$$

and

$$|f|_1 = (f,f)_1^{\frac{1}{2}} ,$$

where $\langle df(x), dg(x)\rangle$ denotes the natural inner product in $\mathrm{Hom}(M_x, \mathbb{R}^K)$.

A sequence $\{f_i\}$ in $L_1^2(M,\mathbb{R}^K)$ is said to *converge weakly to* $f \in L_1^2(M,\mathbb{R}^K)$ if for every $g \in L_1^2(M,\mathbb{R}^K)$ we have

$$\lim_{i\to\infty} (f_i, g)_1 = (f,g)_1 \ .$$

Note that $C^\infty(M,N)$ is not dense in $L_1^2(M,N)$, that is, $\overline{C^\infty(M,N)} \neq L_1^2(M,N) \subset L_1^2(M,\mathbb{R}^K)$(see [SU2] for an example). Furthermore $\overline{C^\infty(M,N)}$ need not be weakly closed, so it is not sufficient for solving the harmonic map problem to restrict oneself to maps which are L_1^2 limits of smooth (or continuous) maps. See [B1] for more detailed results on the approximation of maps in $L_1^2(M,N)$ by smooth maps.

We now state well known compactness and lower semicontinuity results whose proof may be found in [M3, Theorems 1.8.1, 3.4.4].

Lemma 2.1. *Let* $\{f_i\}$ *be a sequence in* $L_1^2(M, \mathbb{R}^K)$ *such that* $|f_i|_1 \leq K$ *for all* i *and some positive number* K. *Then* f_i *has a subsequence* $\{f_j\}$ *which converges weakly in* $L_1^2(M, \mathbb{R}^K)$, *pointwise almost everywhere, and strongly in* $L^2(M, \mathbb{R}^K)$ *to some* $f \in L_1^2(M, \mathbb{R}^K)$. *Moreover, we have*

$$E(f) \leq \lim_{j \to \infty} E(f_j) \ .$$

Remark 2.2. *By this lemma we see that bounded subsets of* $L_1^2(M, N)$ *are weakly compact. Indeed, if* $u \in L_1^2(M, \mathbb{R}^K)$ *is the weak limit of* $\{u_i\}$ *with* $|u_i|_1 < K$, *then a.e. pointwise convergence implies* $u \in L_1^2(M, N)$.

Remark 2.3. *In case* M *is 1-dimensional,* $L_1^2(M, N)$ *is an infinite dimensional Hilbert manifold. But if* $\dim M \geq 2$, *this is not true since in this case we cannot locally parametrize* $L_1^2(M, N)$ *on a Hilbert space. Recall that if* $pk > n$, $L_k^p(M^n, \mathbb{R}^K)$ *is embedded in* $C^0(M, \mathbb{R}^K)$, *and this makes it possible to show that* $L_k^p(M^n, \mathbb{R}^K)$ *is a Banach manifold. Note that the 2-dimensional problem is a border line case in this discussion.*

Suppose M is compact with smooth boundary, and N is compact and closed. Given $\varphi \in L_1^2(M, N)$, the Dirichlet problem is to find a harmonic map $u \in L_1^2(M, N)$ with $u - \varphi \in L_{1,0}^2(M, \mathbb{R}^K)$, where $L_{1,0}^2(M, \mathbb{R}^K) = \overline{C_c^\infty(M, \mathbb{R}^K)} \subseteq L_1^2(M, \mathbb{R}^K)$. We now show that the Dirichlet problem is always solvable in that one find a minimizing map with given boundary data. This result does not require any smoothness on N, but only that N be a compact subset of R^K.

Proposition 2.4. *Let* $N \subseteq R^K$ *be a compact set, and let* $\varphi \in L_1^2(M, N)$ *be given. Define* $\mathcal{D}_\varphi = \{v \in L_1^2(M, N) : v - \varphi \in L_{1,0}^2(M, \mathbb{R}^K)\}$ *and let* $E_0 = \inf\{E(v) : v \in \mathcal{D}_\varphi\}$. *There exists a map* $u \in \mathcal{D}_\varphi$ *with* $E(u) = E_0$.

Proof Take a minimizing sequence $\{u_i\}$ in $\mathcal{D}_\varphi$, that is $u_i \in \mathcal{D}_\varphi$ and $E(u_i) \to E_0$. By the weak compactness of bounded subsets of $L_1^2(M, N)$, there exist a subsequence $\{u_i'\}$ and $u \in L_1^2(M, N)$ such that $u_{i'}$ converges weakly to u. We now note that $L_{1,0}^2(M, \mathbb{R}^K)$ is weakly closed since $L^2_{1,0}(M, \mathbb{R}^K)$ is a closed linear subspace of $L^2_1(M, \mathbb{R}^K)$. Hence $u - \varphi \in L^2_{1,0}(M, \mathbb{R}^K)$ since $u_{i'} - \varphi \in L^2_{1,0}(M, \mathbb{R}^K)$. Therefore $u \in \mathcal{D}_\varphi$. By the lower semicontinuity of energy

$$E(u) \leq \underline{\lim}\, E(u_i) = E_0 \ .$$

Thus u minimizes the energy and this completes the proof.

Example 1. The classical Dirichlet problem arises in case $\varphi \in C^0(\partial M)$ is a real valued function. Then there exists a unique harmonic function $u : M \to \mathbb{R}$ such that $u = \varphi$ on ∂M.

Example 2. Let $\mathbb{R}^n \to \mathbb{R}^{n+1}$ be the injection $i(x^1, \ldots, x^n) = (x^1, \ldots, x^n, 0)$. Let us consider the following Dirichlet problem: Find a harmonic map $u : B^n \to S^n \subset \mathbb{R}^{n+1}$ such that $u\big|_{\partial B^n} = i\big|_{\partial B^n}$, that is u maps the boundary of B^n to the equator of S^n. We will seek rotationally symmetric u only.

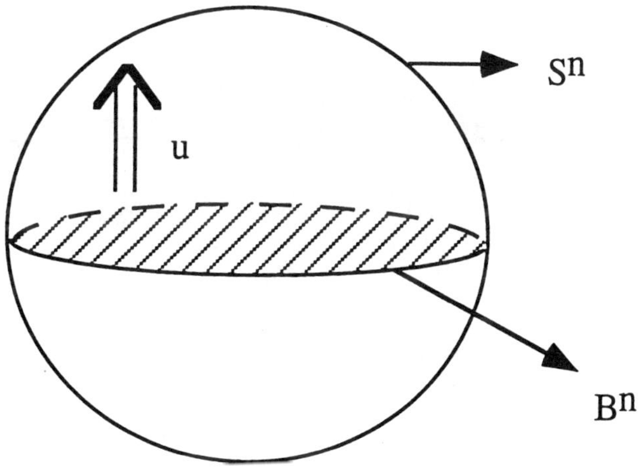

Let (r, θ) be polar coordinates on B^n and (ρ, φ) geodesic coordinates on S^n such that ρ is the distance from the north pole of S^n and $\varphi \in S^{n-1}$. With respect to these coordinates, the metric on B^n is $dr^2 + r^2 d\theta^2$ and the metric on S^n is $d\rho^2 + (\sin^2 \rho) d\varphi^2$. Clearly we may identify the θ and φ coordinates.

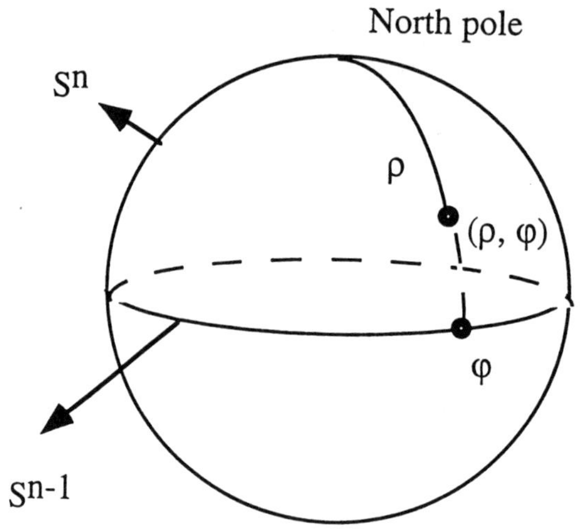

The solution we are looking for is of the form

$$u(r,\theta) = (\rho(r),\theta) \ , \ u \text{ harmonic} \ , \ \rho(1) = \frac{\pi}{2} \ .$$

Note that

$$u^*(d\rho^2 + \sin^2\rho\, d\varphi^2) = (\rho'(r))^2 dr^2 + \sin^2\rho(r)\, d\theta^2 \ .$$

Choose an orthonormal frame $\theta_1, \dots, \theta_{n-1}$ on S^{n-1} such that

$$d\theta^2 = \sum_{i=1}^{n-1} \theta_i^2 \ , \quad \theta_n = dr \ .$$

With respect to this frame, the metric on B^n is

$$r^2\theta_1^2 + \cdots + r^2\theta_{n-1}^2 + \theta_n^2 \ ,$$

and the pull-back of the metric of S^n is

$$\sin^2\rho\,\theta_1^2 + \cdots + \sin^2\rho\,\theta_{n-1}^2 + (\rho'(r))^2\theta_n^2 \ .$$

Hence $|du|^2 = (\rho')^2 + (n-1)\dfrac{\sin^2\rho}{r^2}$ and

$$\begin{aligned} E(u) &= \int_{S^{n-1}} \int_0^1 \Big[(\rho')^2 + (n-1)\frac{\sin^2\rho}{r^2}\Big] r^{n-1} dr\, d\theta \\ &= \mathrm{Vol}(S^{n-1}) \int_0^1 \Big[(\rho')^2 + (n-1)\frac{\sin^2\rho}{r^2}\Big] r^{n-1} dr \ . \end{aligned}$$

For any $\eta(r) \in C_c^\infty((0,1))$,

$$\begin{aligned} 0 &= \frac{d}{dt} E(\rho + t\eta)\Big|_{t=0} \\ &= \mathrm{Vol}(S^{n-1}) \int_0^1 (2\rho'\eta' + 2(n-1)\sin\rho\cos\rho \cdot r^{-2}\eta) r^{n-1} dr \\ &= \mathrm{Vol}(S^{n-1}) \int_0^1 \eta(-2(r^{n-1}\rho')' + (n-1)\sin 2\rho \cdot r^{-2} r^{n-1}) dr \ , \end{aligned}$$

where the first term is obtained by integration by parts. Hence the Euler-Lagrange equation becomes

$$-2(r^{n-1}\rho')' + (n-1)\sin 2\rho \cdot r^{n-3} = 0 \ ,$$

or

$$\frac{1}{r^{n-1}} \frac{d}{dr}\left(r^{n-1}\frac{d\rho}{dr}\right) - \frac{n-1}{2}\frac{\sin 2\rho}{r^2} = 0 \ .$$

If we write $t = \log r$ and $\alpha = 2\rho$, we get

$$\frac{d^2\alpha}{dt^2} + (n-2)\frac{d\alpha}{dt} - (n-1)\sin\alpha = 0 \ ,$$

where $t \in (-\infty, 0]$ and $\alpha(0) = \pi$ since $r \in [0,1]$ and $\rho(1) = \frac{\pi}{2}$. Without loss of generality we may assume $u(0)$ to be the north pole, or $\rho(0) = 0$. Therefore $\lim_{t \to -\infty} \alpha(t) = 0$. Thus the Dirichlet Problem for rotationally symmetric maps reduces to the following ODE problem with boundary conditions:

$$\begin{cases} \dfrac{d^2\alpha}{dt^2} + (n-2)\dfrac{d\alpha}{dt} - (n-1)\sin\alpha = 0 \ , \\ \alpha(0) = \pi \ , \quad \lim_{t \to -\infty} \alpha(t) = 0 \ . \end{cases}$$

We consider three separate cases:

Case (i) $n = 2$: The desired solution is $\alpha(t) = 4\tan^{-1}(e^t)$. Hence $\rho(r) = 2\tan^{-1} r$. It can be seen that the map u is the inverse of stereographic projection.

Case (ii) $3 \leq n \leq 6$: The graph of the solution is

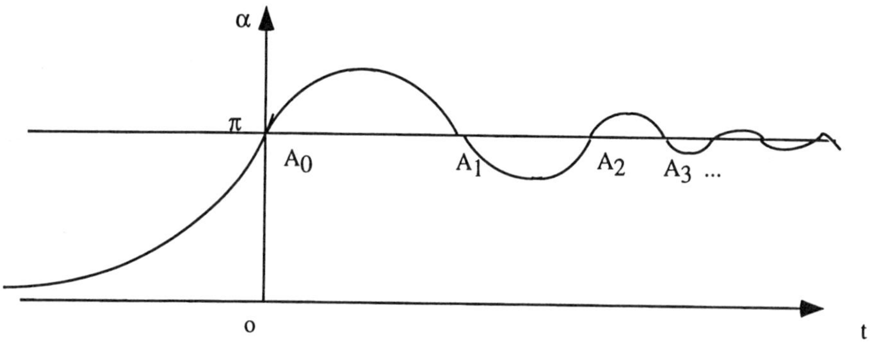

Any translation of the graph moving A_i $(i = 1, 2, \dots)$ to A_0 gives another solution. Therefore we get infinitely many solutions in this case. The graph for the corresponding $\rho(r)$ is

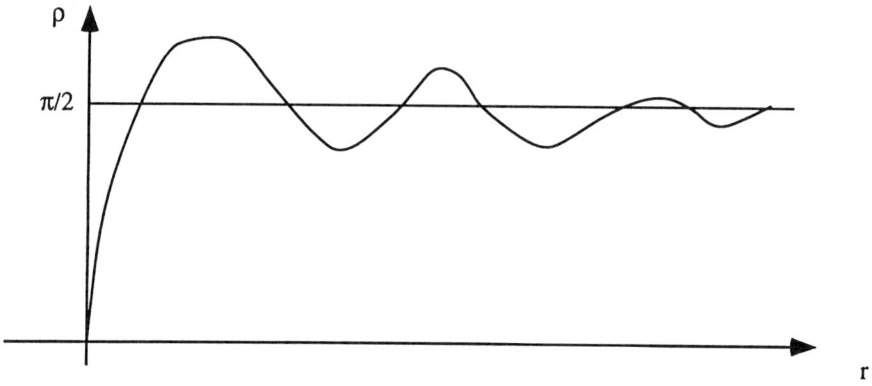

The infinitely many solutions $u_0, u_1, \dots$ converge to the map u_∞ given by $u_\infty(r, \theta) = \left(\frac{\pi}{2}, \theta\right)$. Since the corresponding $\rho_0, \rho_1, \dots$ converge to the constant function $\rho = \frac{\pi}{2}$. Note that

$$E(u_1) < E(u_2) < \cdots < E(u_\infty) .$$

Case (iii) $n \geq 7$: The graph of a solution of the ODE must be

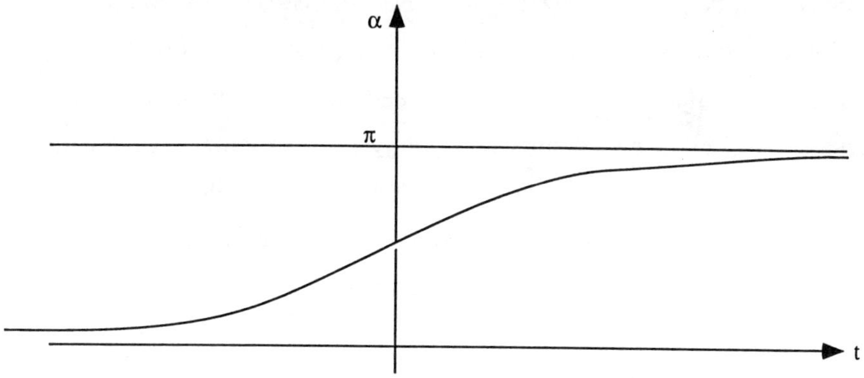

Since this does not satisfy the first boundary condition we do not have a solution in this case.

3. Convexity and Uniqueness Theorems

Suppose M is a compact manifold with boundary, and $N = R^k$ is a Euclidean space. If φ is a map from M to N, then the set defined by $C^1_\varphi(M,N) = \{u \in C^1 : u = \varphi \text{ on } \partial M\}$ is a convex set, in other words, $u_t = tu_1 + (1-t)u_0 \in C^1_\varphi$, $0 \leq t \leq 1$, whenever u_0, $u_1 \in C^1_\varphi(\Omega)$. Furthermore, one can easily see that the energy is strictly convex along these paths. Motivated by this simple case, we derive similar result in $C^1(M,N)$ when N satisfies a certain curvature assumption. In this case the space $C^1(M,N)$ is obviously not a linear space.

Definition. *Let u_0, $u_1 \in C^1(M,N)$. A homotopy u_t, $0 \leq t \leq 1$, is a geodesic homotopy if for any $x \in M$, the curve $t \mapsto u_t(x)$ is a constant speed (with speed depending on x) geodesic in N.*

Proposition 3.1. *Assume N is a complete manifold of nonpositive sectional curvature, and that M is compact with possibly empty boundary. Given two homotopic (rel. ∂M) maps u_0, $u_1 \in C^1(M,N)$ which agree on ∂M, there is a unique geodesic homotopy u_t from u_0 to u_1 through maps u_t which agree with u_0 on ∂M.*

Proof : Let v_t for $0 \leq t \leq 1$ be a chosen homotopy with $v_0 = u_0$, $v_1 = u_1$, and $v_t = u_0$ on ∂M for all t. For any chosen $x \in M$, let $\gamma_x(t)$ be the unique constant speed geodesic in N parametrized on $0 \leq t \leq 1$ with $\gamma_x(0) = u_0(x)$, $\gamma_x(1) = u_1(x)$, and such that γ_x is homotopic to the curve $t \to v_t(x)$. Then define $u_t(x) = \gamma_x(t)$. Elementary Jacobi field arguments imply that the map $u_t \in C^1(M, N)$ for each t. Uniqueness of the geodesic homotopy follows from the uniqueness for geodesics in N with fixed endpoints and fixed homotopy class. This completes the proof.

Next we state the formula for *the second derivative of energy* for a C^2 variation through smooth maps. Consider a C^2 map $F : M \times [0,1] \to N$, and set $u_t(x) = F(x,t)$ for $x \in M$. Set $V = F_*\left(\frac{\partial}{\partial t}\right) \in \Gamma(u_t^*(TN))$ and denote by ∇' the pullback connection from TN. Let $e_1, \ldots, e_n$ be an orthonormal basis for TM. Recall that

$$E(u_t) = \int_M \left(\sum h_{ij}(u_t(x)) g^{\alpha\beta}(x) \frac{\partial u_t^i}{\partial x^\alpha} \frac{\partial u_t^j}{\partial x^\beta} \right) \sqrt{g}\, dx \ .$$

Hence

$$\begin{aligned} \frac{d}{dt} E(u_t) &= \int_M \left[2h_{ij} g^{\alpha\beta} \frac{\partial V^i}{\partial x^\alpha} \frac{\partial u_t^j}{\partial x^\beta} + \partial h_{ij} \cdots \right] \sqrt{g}\, dx \\ &= 2 \int_M g^{\alpha\beta} \left\langle \nabla'_{\frac{\partial}{\partial x^\alpha}} V \ , \ F_*\left(\frac{\partial}{\partial x^\beta}\right) \right\rangle \sqrt{g}\, dx \\ &= 2 \int_M \sum_\alpha \left\langle \nabla'_{e_\alpha} v \ , \ F_*(e_\alpha) \right\rangle dv_M \ . \end{aligned}$$

Moreover, from [EL], we get

$$\begin{aligned} \frac{d^2}{dt^2} E(u_t) = 2 \int_M \Big[\sum_\alpha \| \nabla'_{e_\alpha} V \|^2 &- \sum_\alpha \langle R^N(V, F_*(e_\alpha))V, F_*(e_\alpha) \rangle \\ &+ \sum_\alpha \langle \nabla'_{e_\alpha} \left(\nabla'_{\frac{\partial}{\partial t}} V \right), F_*(e_\alpha) \rangle \Big] dv_M \ . \end{aligned}$$

The next lemma gives conditions under which the third term on the right can be ignored.

Lemma 3.2. *Suppose either of the following holds:*

(i) The map $u = u_0$ is harmonic, and $V(x,0)$ has compact support in the interior of M, or

(ii) For each $x \in M$, the curve $\sigma_x(t) = F(x,t)$ is a constant speed goedesic in N.

Then we have

$$\frac{d^2}{dt^2}E(u_t)\bigg|_{t=0} = 2\int_M \Big[\| \nabla' V\|^2 - \sum_\alpha \langle R^N(V, u_*(e_\alpha))V, u_*(e_\alpha)\rangle\Big] dv_M \ .$$

Proof: If u is harmonic, then the third term vanishes because it represents the first derivative of energy with respect to an initial variation given by $\nabla'_{\frac{\partial}{\partial t}} V$. On the other hand, if the curves $\sigma_x(t)$ are constant speed geodesics, then we have $\nabla'_{\frac{\partial}{\partial t}} V \equiv 0$ since this represents the acceleration vector of the curve σ_x. This proves the lemma.

Remark 3.3. *The integral expression above is the index form associated with the operator $\mathcal{L}V = \Delta' V + \sum_\alpha R^N(u_*(e_\alpha), V)u_*(e_\alpha)$ which is the Jacobi operator (linearized operator) associated with the second derivative of energy.*

Remark 3.4. *If N has nonpositive curvature, then the previous lemma tells us that the energy is a convex function along geodesic homotopies in analogy with the case in which $N = R^k$.*

We may apply this result to show that harmonic mappings minimize energy in their homotopy class if N has nonpositive curvature. The result can also be derived from the theorems of [ES], [Hm], and [Hr]. The following is a much more direct proof.

Theorem 3.5. *If u is a smooth harmonic map from M to N with $K_N \le 0$, then $E(u) \le E(\varphi)$ for any $\varphi : M \to N$ homotopic (rel ∂M) to u.*

Proof: Let $G : M \times [0,1] \to N$ be the geodesic homotopy from u to φ whose existence is shown above. Then, by applying the lemma, we get

$$\frac{d^2}{dt^2}E(u_t) = 2\int_M (\| \nabla' V\|^2 - \sum_\alpha \langle R^N(V, G_*(e_\alpha))V \ , \ G_*(e_\alpha)\rangle) dv_M \ ,$$

where $V = G_*\left(\frac{\partial}{\partial t}\right)$, $u_t(x) = G(x,t)$. In particular, since N has non-positive curvature, we see that the function $t \mapsto E(u_t)$ is a convex function on $[0,1]$. Since $u = u_0$ is harmonic we have $\frac{d}{dt}E(u_t) = 0$ at $t = 0$. Therefore $E(u_t)$ is nondecreasing in t and hence $E(u) \leq E(\varphi)$ as required.

The same argument gives a simple proof of the uniqueness theorem of Hartman [Hr].

Theorem 3.6. *If $\partial M \neq \phi$, $u_0 : M \to N$ is harmonic, and $K_N \leq 0$, then any harmonic map $u_1 : M \to N$ homotopic to u_0 (rel. ∂M) must coincide with u_0.*

If $\partial M = \phi$, u_0, u_1 are homotopic harmonic maps, and $K_N \leq 0$, then u_t are all harmonic, $\nabla' V \equiv 0$, and $\langle R^N(V, u_{t*}(X))V, u_{t*}(X)\rangle = 0$ for any $X \in T_xM$. In particular, if $K_N < 0$ and any u_t has rank greater than 1 at some point of M, then $u_0 \equiv u_1$ on M.

Proof: Let u_t, $t \in [0,1]$ be the geodesic homotopy constructed above. Since the function $t \mapsto E(u_t)$ is convex and has slope zero at 0 and 1 (u_0, u_1 are harmonic), we must have $\frac{d^2}{dt^2}E(u_t) \equiv 0$ for $0 \leq t \leq 1$. Therefore we have

$$\nabla' V \equiv 0 \quad \text{and} \quad \sum_\alpha \langle R^N(V, G_*(e_*))V \ , \ G_*(e_\alpha)\rangle \equiv 0 \qquad (*)$$

on $M \times [0,1]$. Now we have

$$e_\alpha \|V\|^2 = 2\langle V, \nabla'_{e_\alpha} V\rangle = 0 \ , \quad \alpha = 1, \dots, n$$

and hence $\|V\|$ is constant in x for each t. If $\partial M \neq \phi$, then $V = 0$ on $\partial M \times [0,1]$ and hence $V \equiv 0$ on $M \times [0,1]$. Thus $u_0 \equiv u_1$ on M. If $\partial M = \phi$ and $K_N < 0$, then from $(*)$ we see that V is parallel to $G_*(e_\alpha)$ for $\alpha = 1, \dots, n$. If V is not identically zero at $t = t_0$ then $V(x, t_0)$

is nowhere zero since $\|V\|$ is constant in x. Then it follows that u_{t_0} has rank 1 everywhere on M, a contradiction. Therefore $V(x, t_0) \equiv 0$, and hence $V(x,t) \equiv 0$ for all $(x,t) \in M \times [0,1]$ because the speed of geodesic $t \mapsto G(x,t)$ is constant in t. This proves the theorem.

4. *A Priori* Estimates for Harmonic Maps

Here we consider estimates which can be derived for smooth harmonic maps. These estimates can be exploited to prove existence theorems for maps into manifolds N of nonpositive curvature. The Bochner formula for harmonic maps (see Chapter I for a derivation) says

$$\begin{aligned}\frac{1}{2}\triangle|du|^2 = \| \nabla' \, du\|^2 - \sum_{\alpha,\beta} \langle R^N(u_*e_\alpha, u_*e_\beta)u_*e_\alpha, u_*e_\beta\rangle \\ + \sum_i \mathrm{Ric}^M(u^*\theta_i, u^*\theta_i)\end{aligned}$$

where $e_1, \dots, e_n$ is an orthonormal basis for TM, $\theta_1, \dots, \theta_k$ is an orthonormal basis for T^*N, and ∇' is the pullback connection from TN. We see from this formula that if M is compact (with possibly empty boundary) and N has nonpositive sectional curvature, then by throwing away first two terms we get

$$\triangle|du|^2 \geq -C|du|^2 \tag{4.1}$$

with C depending only on M. A standard result in PDE (see [M3, 5.3.1]) then implies the estimate

$$\sup_{\Omega} |du|^2 \leq C \int_M |du|^2 dv_M \tag{4.2}$$

for any open Ω compactly contained in M, where C depends only on M, n, and $\mathrm{dist}(\Omega, \partial M)$. More precisely, we have the estimate

$$\sup_{B_{\frac{r}{2}}(x)} |du|^2 \leq Cr^{-n} \int_{B_r(x)} |du|^2 dv_M \tag{4.3}$$

with $B_r(x) \subset\subset \Omega$.

The estimate (4.2) was first proved for the heat equation associated to the harmonic map problem in [ES]. The estimate (4.2) implies the following compactness result.

Proposition 4.1. *If M is compact with (possibly empty) boundary, and N is compact with nonpositive sectional curvatures, then the family of maps, for $\wedge > 0$*

$$\mathcal{F}_\wedge = \{u \in C^\infty(M,N) : u \textit{ is harmonic}, \quad E(u) \leq \wedge\}$$

is compact in the topology of uniform C^k convergence on compact subsets of M for any integer $k \geq 0$.

The proof of this proposition involves deriving estimates on all derivatives of maps $u \in \mathcal{F}_\wedge$ locally on M, and applying Ascoli's theorem on equicontinuity. From (4.2) one can obtain a Hölder estimate for $|du|^2$ in Ω uniformly in $\mathcal{F}_\wedge$. Since $\Delta u^i = \Gamma(du, du)$, where $\Gamma(du, du)$ is quadratic in du, it is a standard application of linear elliptic theory to obtain a uniform bound for the C^k norm in Ω for all $u \in \mathcal{F}_\wedge$. The inequality (4.1) relies heavily on the hypothesis that N has nonpositive sectional curvature. Conformal maps of S^2, or Example 2 of Section 2 shows that the above proposition does not hold for harmonic maps into compact manifolds without a curvature assumption.

The following a priori estimate shows that (4.3) does in fact hold for harmonic maps with small scaled total energy.

Theorem 4.2. *Suppose $u \in C^2(B_r, N)$ is harmonic with respect to a metric g on $B_r = \{x \in \mathbb{R}^n : |x| < r\}$ which satisfies*

$$\Lambda^{-1}(\delta_{ij}) \leq (g_{ij}) \leq \Lambda(\delta_{ij}), \quad |\partial_k g_{ij}| \leq \Lambda r^{-1}.$$

There exists $\epsilon > 0$ depending only on Λ, n, N such that if

$$r^{2-n} \int_{B_r} e(u) \leq \epsilon,$$

then u satisfies the inequality

$$\sup_{B_{r/2}} e(u) \le Cr^{-n} \int_{B_r} e(u),$$

where C depends only on Λ, n, N.

Remark 4.3. The proof we will give actually only requires that the g_{ij} be Hölder continuous on B_r.

Proof: The monotonicity inequality (1.4) implies

$$\sigma^{2-n} \int_{B_\sigma(x)} e(u) \le C\rho^{2-n} \int_{B_\rho(x)} e(u) \tag{4.4}$$

provided $x \in B_r$, and $0 < \sigma < \rho < r - |x|$. If we set $r_1 = \frac{3}{4}r$, then inequality (4.4) implies

$$\sigma^{2-n} \int_{B_\sigma(x)} e(u) \le \left(\frac{3}{4}\right)^{n-2} Cr^{2-n} \int_{B_r} e(u) \tag{4.5}$$

for any $x \in B_{r_1}$, $0 < \sigma \le r_1 - |x|$. In (4.4) and (4.5) the constant C depends only on n, Λ.

We will exploit (4.5) for σ small. First observe that there exists $\sigma_0 \in (0, r_1)$ such that

$$(r_1 - \sigma_0)^2 \sup_{B_{\sigma_0}} e(u) = \max_{0<\sigma\le r_1} (r_1 - \sigma)^2 \sup_{B_\sigma} e(u).$$

Moreover, there exists a point $x_0 \in \overline{B}_{\sigma_0}$ such that

$$e_0 = e(u)(x_0) = \sup_{B_{\sigma_0}} e(u).$$

Set $\rho_0 = \frac{1}{2}(r_1 - \sigma_0)$, and observe that from the choice of σ_0, x_0

$$\sup_{B_{\rho_0}(x_0)} e(u) \le \sup_{B_{\sigma_0+\rho_0}} e(u) \le 4e_0 \tag{4.6}$$

Now define a map $v \in C^2(B_{r_0}, N)$, $r_0 = (e_0)^{1/2}\rho_0$, by

$$v(y) = u\left(\frac{y - x_0}{(e_0)^{1/2}}\right).$$

Thus v is a rescaled version of u chosen so that $e(v)(0) = 1$. From (4.6) we get

$$\sup_{B_{r_0}} e(v) \leq 4, \quad e(v)(0) = 1.$$

Therefore we see from (4.1) that $\Delta e(v) \geq -Ce(v)$ on B_{r_0}. Now if $r_0 \geq 1$, then we can apply the mean value inequality to get

$$1 = e(v)(0) \leq C \int_{B_1} e(v).$$

However, we have from (4.5)

$$\begin{aligned} \textstyle\int_{B_1} e(v) &= \left((e_0)^{1/2}\right)^{n-2} \textstyle\int_{B_{e_0^{-1/2}}(x_0)} e(u) \\ &\leq Cr^{2-n} \textstyle\int_{B_r} e(u) \\ &\leq C\epsilon. \end{aligned}$$

Thus if ϵ is chosen small enough, these two inequalities are in contradiction. Therefore we may assume $r_0 \leq 1$. It then follows from [M3, 5.3.1] that the mean value inequality holds for $e(v)$ on B_{r_0}, and we have

$$1 = e(v)(0) \leq Cr_0^{-n} \int_{B_{r_0}} e(v) = Cr_0^{-2} \rho_0^{2-n} \int_{B_{\rho_0}(x_0)} e(u).$$

Combining this with (2.5) we get

$$\rho_0^2 e_0 = r_0^2 \leq Cr^{2-n} \int_{B_r} e(u).$$

From the choice of σ_0 this implies

$$\max_{0<\sigma\leq r_1} (r_1 - \sigma)^2 \sup_{B_\sigma} e(u) \leq 4Cr^{2-n} \int_{B_r} e(u).$$

The conclusion of Theorem 4.2 now follows by taking $\sigma = \frac{1}{2}r$ and dividing by r^2.

The previous a priori estimate can be used to prove a partial regularity result for maps which are weak limits of smooth harmonic maps with uniformly bounded energy. It shows that the sequence converges in C^∞ topology on the complement of a closed set with finite Hausdorff measure in dimension $n-2$. Recently F. H. Lin[Li] has shown that this set is actually $n-2$ rectifiable.

Corollary 4.4. *Let M be compact with possibly empty boundary, and let N be compact. For $\Lambda > 0$, set*

$$\mathcal{F}_\Lambda = \{u \in C^\infty(M, N) \ : \ u \, is \ harmonic, \ E(u) \leq \Lambda\}.$$

Any map u in the weak L_1^2 closure of $\mathcal{F}_\Lambda$ is smooth and harmonic outside a closed singular set of locally finite Hausdorff $(n-2)$-dimensional measure.

Proof: Let $\{u_i\} \subset \mathcal{F}_\Lambda$ be a sequence converging weakly to u. Let $\overset{o}{M}$ denote the interior of M, and define a subset $\mathcal{S} \subset \overset{o}{M}$ by

$$\mathcal{S} = \bigcap_{r>0} \left\{ x \in \overset{o}{M} : \ \liminf_{i\to\infty} r^{2-n} \int_{B_r(x)} e(u_i) \geq \epsilon_0 \right\}.$$

Here ϵ_0 will denote a fixed positive number determined by Theorem 4.2. It follows from the montonicity of the quantity $r^{2-n} \int_{B_r(x)} e(u_i)$ that $\mathcal{S}$ is a relatively closed subset of $\overset{o}{M}$, for suppose that $x_j \in \mathcal{S}$ and $\lim x_j = x \in \overset{o}{M}$. For any $r > 0$ and all j, it follows from (1.4) that

$$\liminf_{i\to\infty} r^{2-n} \int_{B_r(x_j)} e(u_i) \geq \epsilon_0.$$

This immediately implies $\liminf_{i\to\infty} r_1^{2-n} \int_{B_{r_1}(x)} e(u_i) \geq \epsilon_0$ for any $r_1 > r$. Since $r > 0$ was arbitrary we have $x \in \mathcal{S}$.

For any $\delta > 0$, and any $\Omega \subset\subset \overset{o}{M}$ we can cover $\mathcal{S} \bigcap \overline{\Omega}$ by a finite collection of balls $\{B_{r_j}(x_j)\}$ so that the collection $\{B_{\frac{1}{2}r_j}(x_j)\}$ is disjoint, and so that $x_j \in \mathcal{S}$. For i sufficiently large we then have

$$\left(\frac{1}{2} r_j\right)^{2-n} \int_{B_{\frac{1}{2}r_j}(x)} e(u_i) \geq \epsilon_0 \quad \text{for all} \, j.$$

Summing on j we then get

$$\sum_j r_j^{n-2} \leq CE(u_i) \leq C\Lambda.$$

It follows that $\mathcal{H}^{n-2}(\mathcal{S} \cap \overline{\Omega}) \leq C\Lambda$.

We show that a subsequence of $\{u_i\}$ converges uniformly in C^k norm to u on compact subsets of $\overset{o}{M} \setminus \mathcal{S}$. Let $x \in \overset{o}{M} \setminus \mathcal{S}$, and observe that for some r we have

$$r^{2-n}\int_{B_r(x)} e(u_i) < \epsilon_0$$

for infinitely many i. By Theorem 4.2 we have uniform C^1, and hence C^k, estimates on u_i in $B_{\frac{1}{2}r}(x)$. Thus a subsequence of $\{u_i\}$ converges uniformly in C^k norm in $B_{\frac{1}{2}r}(x)$. By uniqueness of weak limits, the limit is u. Therefore u is regular outside $\mathcal{S}$, and by a diagonal subsequence argument a subsequence of $\{u_i\}$ converges uniformly in C^k norm to u on compact subsets of $\overset{o}{M} \setminus \mathcal{S}$. This completes the proof of Corollary 4.4.

Theorem 4.2 requires the smallness of energy as a hypothesis. While this hypothesis is not generally satisfied at every point for maps into arbitrary manifold (e.g. conformal maps of $\mathbb{S}^2$), under certain assumptions on the target, it can be verified. In particular, if there is a smooth bounded strictly convex function g defined on N, we can verify the hypothesis. The following argument appears in [GH]. To see this, observe that $g \circ u$ is subharmonic, in fact we may assume

$$\Delta(g \circ u) \geq \epsilon_1 e(u), \quad 0 \leq g \circ u \leq C. \tag{4.7}$$

Given $x \in \overset{o}{M}$, $B_r(x) \subset \overset{o}{M}$, we can apply Green's formula to $\varphi(x) g \circ u$ where φ has compact support in $B_r(x)$ with $\varphi(x) = 1$

$$g \circ u(x) = \int_{B_r(x)} G(x,y)\Delta\left(\varphi g \circ u(y)\right) dv(y).$$

Choosing φ to be equal to 1 on $B_{\frac{1}{2}r}(x)$ with compact support, we can easily see from (4.7)

$$\int_{B_{\frac{1}{2}r}(x)} G(x,y) e(u)(y) dv(y) \leq C.$$

In particular, this implies (assume $n > 2$, for one handles the case $n = 2$ by a simple modification of these arguments.)

$$\int_{B_{\frac{1}{2}r}(x)} d(x,y)^{2-n} e(u)(y) dv(y) \leq C.$$

On the other hand, if we set $R(\sigma) = \sigma^{2-n} \int_{B_\sigma(x)} e(u)$, we have

$$\begin{aligned}\int_0^{\frac{1}{2}r} \frac{R(\sigma)}{\sigma} \mathrm{d}\sigma &= (2-n)^{-1} \int_0^{\frac{1}{2}r} \frac{\mathrm{d}}{\mathrm{d}\sigma} (\sigma^{2-n}) \left(\int_{B_\sigma(x)} e(u) \right) \mathrm{d}\sigma \\ &\leq (n-2)^{-1} \int_{B_{\frac{1}{2}r}(x)} d(x,y)^{2-n} e(u)(y) \mathrm{d}v(y)\end{aligned}$$

Since $R(\sigma)$ is a monotone function of σ, we have for any $\sigma_0 \in (0, \frac{1}{2}r)$

$$R(\sigma_0) \log \left(\sigma_0^{-1} \frac{1}{2} r \right) \leq C.$$

This shows that if σ_0 is chosen small, then $R(\sigma_0)$ is small, and Theorem 4.2 applies in $B_{\sigma_0}(x)$. We have therefore shown

Corollary 4.5. *If M is as above, and $u \in C^\infty(M,N)$ is harmonic such that a bounded strictly convex function g exists on $u(M)$, then for $\Omega \subset\subset M$ we have*

$$\|u\|_{C^k(\Omega)} \leq C$$

where C depends only on k, n, M, N, Ω, and $E(u)$. In particular, Proposition 4.1 *holds under this convex function hypothesis.*

Remark 4.6. Existence, regularity, and a priori estimates have been given in [HKW, HW] under the assumption that the image of u lies in a strictly convex ball. In fact, the results of [HW] also apply to the more difficult case when the metric of M is assumed only bounded, measurable, and (locally) uniformly Euclidean.

5. A Local Existence Theorem

Under the assumption that N is a complete, simply connected manifold with nonpositive curvature, we will derive a useful replacement for inequality (4.3). Let $\overline{u}$ be a chosen point in N, and define $\rho(u(x)) = d(u(x), \overline{u})$. Recall that in $\mathbb{R}^n$, $\left(\frac{1}{2}r^2\right)_{ij} = \delta_{ij}$, where r is a distance function and $\left(\frac{1}{2}r^2\right)_{ij}$ is the Hessian of $\frac{1}{2}r^2$. The Hessian comparison theorem implies that in N with $K_N \leq 0$ the distance function is more convex that in the flat case, i.e.

$$\left(\frac{1}{2}\rho^2\right)_{ij} \geq h_{ij} \ .$$

Viewing ρ as a composite function and applying the chain rule, we get

$$\begin{aligned}\frac{1}{2}\Delta d(u(x),\overline{u})^2 &= \sum_{i,\alpha}\left(\left(\frac{1}{2}\rho^2\right)_i \frac{\partial u^i}{\partial x^\alpha}\right)_\alpha \\ &= \sum_{i,j,\alpha}\left(\frac{1}{2}\rho^2\right)_{ij}\frac{\partial u^i}{\partial x^\alpha}\frac{\partial u^j}{\partial x^\alpha} + \sum_i \left(\frac{1}{2}\rho^2\right)_i \Delta u^i \ .\end{aligned}$$

If we assume that u is harmonic and $u^1, \ldots, u^k$ are normal coordinates in N, then the second term vanishes. Thus we obtain the formula:

$$\frac{1}{2}\Delta d(u(x),\overline{u})^2 \geq |du|^2 \ . \tag{5.1}$$

Now define

$$\operatorname*{OSC}_{\Omega_1}(u) \equiv \sup\{d(u(x),u(y)) : x,y \in \Omega_1\} \ , \ \Omega_1 \subset \Omega \ .$$

Thus $\operatorname*{OSC}_{\Omega_1}(u)$ is the diameter of the image $u(\Omega_1)$.

Proposition 5.1. *There exists a constant c depending only on (Ω, g) such that*

$$\sup_{B_{r/2}(x)} |du| \leq \frac{c}{r}\operatorname*{OSC}_{B_r(x)} u \ , \quad B_r(x) \subset\subset \Omega \ .$$

Proof: Let $\varphi \in C_c^\infty(B_r)$ be such that $\varphi \equiv 1$ on $B_{\frac{r}{2}}$ and $|\nabla \varphi| \leq \frac{c}{r}$. Then by (5.1) and integration by parts

$$\begin{aligned}&\int_{B_r}\varphi^2|du|^2 \leq \frac{1}{2}\int_{B_r}\varphi^2\Delta d(u(x),\overline{u})^2 \\ &= -\frac{1}{2}\int_{B_r}\nabla\varphi^2\cdot\nabla d(u(x),\overline{u})^2 \leq 2\int_{B_r}|\varphi|\cdot|d|\cdot|\nabla\varphi|\cdot|\nabla d|\,dv_M \ .\end{aligned}$$

Since

$$\frac{\partial}{\partial x^\alpha}d(u(x),\overline{u}) = \frac{\partial\rho}{\partial u^i}(u(x))\cdot\frac{\partial u^i}{\partial x^\alpha} \ ,$$

we see that $|\nabla d| \leq |du|$. Hence

$$\int_{B_r} \varphi^2 |du|^2 \leq 2 \int_{B_r} |\varphi| \cdot |d| \cdot |\nabla \varphi| \cdot |du|$$
$$\leq \frac{1}{2} \int_{B_r} \varphi^2 |du|^2 + 4 \int_{B_r} d(u(x), \overline{u})^2 |\nabla \varphi|^2 ,$$

and so

$$\int_{B_r} \varphi^2 |du|^2 \leq 8 \int_{B_r} d(u(x), \overline{u})^2 |\nabla \varphi|^2 \leq c r^{n-2} (\underset{B_r}{\mathrm{OSC}}(u))^2 .$$

Then the desired result follows from the definiton of φ and (4.3).

From Proposition 5.1 and the harmonic map equation, we know that $|\Delta u^i|$ is bounded. Then by potential theory we get a $C^{1,\alpha}$ estimate on u. Hence by Schauder theory we can obtain the following result.

Corollary 5.2. *The following a priori estimate holds for harmonic maps to simply connected manifolds of nonpositive curvature*

$$r^{2+\alpha}[\nabla \nabla u]_{\alpha, B_{\frac{r}{2}}} + r^2 |\nabla \nabla u|_{B_{\frac{r}{2}}} + r |\nabla u|_{B_{\frac{r}{2}}} \leq c \underset{B_r}{\mathrm{OSC}}\, u ,$$

where c depends on (Ω, g), $(u(\Omega), h)$ and

$$[f]_{\alpha, \Omega_1} \equiv \sup \left\{ \frac{|f(x) - f(y)|}{d(x,y)^\alpha} : x \neq y \ , \ x, y \in \Omega_1 \right\} , \quad \alpha \in (0,1) .$$

We now treat the existence theorems for maps into manifolds of nonpositive curvature. An application of the Leray-Schauder degree theory based on the *a priori* estimates above (and the boundary estimate below) can be used to prove the following.

Theorem 5.3 (Local Existence). *Suppose $\Omega \subset M$ is compact with boundary $\partial\Omega$. Given $\varphi \in C^{0,\beta}(\partial\Omega, N)$, where N is complete, simply connected with nonpositive curvature, there is a harmonic map $u \in C^\infty(\Omega, N) \cap C^{0,\beta}(\overline{\Omega}, N)$ with $u = \varphi$ on $\partial\Omega$.*

In order to prove this theorem we recall that one can extend $\varphi \in C^\infty(\Omega) \cap C^{0,\beta}(\overline{\Omega})$ such that $\Delta\varphi = 0$. Secondly, we introduce a weighted

Hölder space $C^{k,\alpha}_{(-\gamma)}(\Omega)$: Define

$$\|u\|_{k,\alpha,-\gamma} = \sup_{\delta\in[0,\delta_0]} \delta^{-\gamma+k+\alpha}[\nabla^k u]_{\alpha,\Omega_\delta} + \sum_{|\sigma|\leq k} \sup_{x\in\Omega} \delta(x)^{-\gamma+|\sigma|}| \nabla^\sigma u|(x) ,$$

where

$$\Omega_\delta = \{x \in \Omega : d(x, \partial\Omega) > \delta\} , \quad \delta \leq \delta_0 ,$$

and $\delta(x) = d(x, \partial\Omega)$. We then define

$$C^{k,\alpha}_{(-\gamma)}(\Omega) = \{u \in C^{k,\alpha}(\Omega) : \|u\|_{k,\alpha,-\gamma} < \infty\} .$$

$C^{k,\alpha}_{(-\gamma)}$ is obviously a Banach space, and Δ is a bounded linear operator from $C^{2,\alpha}_{(-\beta)}(\Omega)$ into $C^{0,\alpha}_{(2-\beta)}(\Omega)$. Moreover Δ has bounded inverse [GT, Section 6.5].

Finally, we need the following lemma.

Lemma 5.4 (Boundary Estimate). *Assume that Ω satisfies an exterior cone condition. Suppose that $\beta \in (0,1)$, u is harmonic, and $u - \varphi \in C^{2,\alpha}_{(-\beta)}(\Omega)$. Then $\sup_{x\in\Omega} \delta(x)^{-\beta}|u - \varphi|(x) \leq c$, where $c = c(\Omega, \varphi)$ is independent of u.*
*(**Note:** This lemma is not true if $\beta = 1$.)*

Proof: Since $d^2(u, \bar{u})$ is subharmonic (because of (5.1)), it follows from the maximum principle that $\sup_\Omega d(u, \bar{u}) = \sup_{\partial\Omega} d(\varphi, \bar{u}) \leq C$. Let $v^1, \dots, v^n$ be Riemannian normal coordinates in N centered at $\varphi(x_0)$ and suppose $x \in \Omega$ is close enough to x_0 such that x_0 is the unique nearest point of $\partial\Omega$ to x. Then $|v(x)| = d(u(x), \varphi(x_0))$ and $|v(x_0)| = 0$.

Now we claim that $|v|$ is subharmonic. To see this note that by (5.1), $\Delta|v|^2 \geq 2|dv|^2$. Since $\Delta|v|^2 = 2|v|\Delta|v| + 2\left|\nabla|v|\right|^2$ and $2\left|\nabla|v|\right|^2 \leq 2|dv|^2$, we have $\Delta|v| \geq 0$, as claimed.

By hypothesis there exists, for every point $y \in \partial\Omega$, a finite right circular cone K with vertex y, satisfying $\overline{K} \cap \overline{\Omega} = y$. At each point $y \in \partial\Omega$ taken as origin, a suitable local barrier can be chosen in the form

$w(x) = r^\beta w_0(\xi)$ where $0 < \beta < 1$ and $w_0(\xi)$ is a first eigen-function of $\triangle$ on $(M \sim K) \cap \partial B_1(y)$. Therefore

$$w(x) > 0 \quad \text{on} \quad \Omega \cap B_r(x_0) \sim \{x_0\}\ , \quad \Delta w \le 0\ ,$$

and

$$c^{-1}|x - x_0|^\beta \le w(x) \le c|x - x_0|^\beta\ .$$

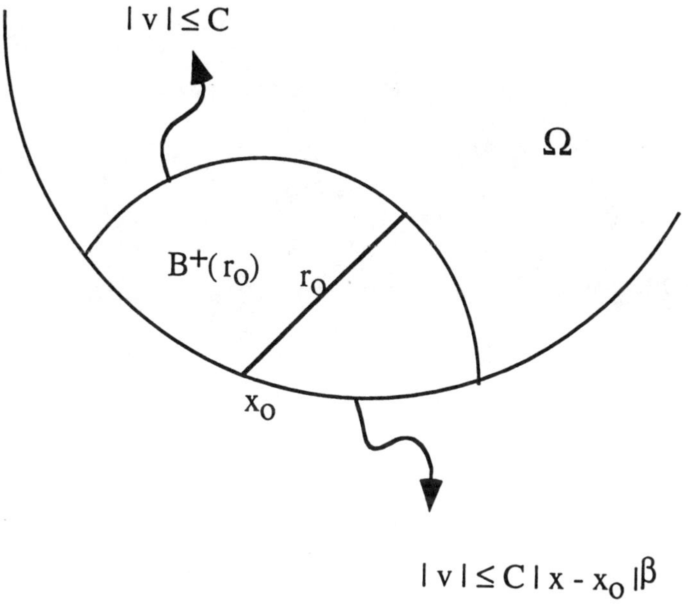

Let $B^+_{r_0} = \Omega \cap B_{r_0}(x_0)$. Since $|v| \le c$ on $\Omega \cap \partial B_{r_0}(x_0)$ and $|v| \le c|x - x_0|^\beta$ on $\partial\Omega \cap B_{r_0}(x_0)$, it follows that $|v| \le \tilde{c}w$ on $\partial B^+_{r_0}$. Hence by maximum principle, $|v| \le \tilde{c}w$ in $B^+_{r_0}$. Note that if Ω is bounded then the manifold distance $|v(x)|$ is equivalent to the Euclidean distance $|u(x) - \varphi(x_0)|$. This completes the proof of Lemma 5.4.

Remark 5.5. *By Lemma* 5.4 *we see that if u is harmonic and $u - \varphi \in C^{2,\alpha}_{(-\beta)}(\Omega)$, then $\operatorname{OSC}_{B^+_{2\delta(x)}} u \le c\delta(x)^\beta$. Hence by Corollary* 5.2 *we have*

$$\|u - \varphi\|_{2,\alpha,-\beta} \le C_0$$

where C_0 depends on φ, but not u. We are now in a position to prove the local existence theorem.

Proof of Theorem 5.3: Let $v = u - \varphi$. Let $N(u)$ be the nonlinear term of harmonic map equation written in a fixed normal coordinates u; that is, $\Delta u + N(u) = 0$. Let us now consider any map u with $\|u - \varphi\|_{2,\alpha,-\beta} \leq C_0$. We assume that we have extended φ so that $\Delta\varphi = 0$, so that if $\Delta u + N(u) = f$, then $\Delta v + N(v + \varphi) = f$, and hence $v + \Delta^{-1}N(v + \varphi) = \Delta^{-1}f$. Define

$$T = I + K \quad \text{where} \quad K(v) = \Delta^{-1}N(v + \varphi) \ .$$

Then the operator T maps $C^{2,\alpha}_{(-\beta)}(\Omega)$ into $C^{2,\alpha}_{(-\beta)}(\Omega)$.

Claim: K is a compact operator on $C^{2,\alpha}_{(-\beta)}(\Omega)$. Let $v \in C^{2,\alpha}_{(-\beta)}$. Since

$$N(u) = g^{\alpha\beta}\Gamma^i_{jk}(u(x))\frac{\partial u^j}{\partial x^\alpha}\frac{\partial u^k}{\partial x^\beta} \ ,$$

we see that $|N(v + \varphi)| \leq C(|\nabla v|^2 + |\nabla \varphi|^2)$. Therefore we have for all x

$$\delta(x)^{2-2\beta}|N(v+\varphi)|(x) \leq C((\delta(x)^{1-\beta}|\nabla v|(x))^2+1) \leq C((\|v\|_{2,\alpha,-\beta})^2+1).$$

On the other hand, we have

$$\delta(x)^{3-2\beta}|\nabla N(v+\varphi)|(x) \leq C[|\nabla v|^3+|\nabla v|\cdot|\nabla\nabla v|+|\nabla v|^2+1]\delta(x)^{3-2\beta} \ .$$

Since

$$\begin{aligned}
|\nabla v|^2\delta(x)^{3-2\beta} &= (\delta(x)^{1-\frac{2}{3}\beta}|\nabla v|(x))^3 \\
&\leq c\delta(x)^{1-\beta}|\nabla v|(x) \leq c\|v\|_{2,\alpha,-\beta} \ , \\
(\delta(x)^{1-\beta}|\nabla v|)\cdot(\delta(x)^{2-\beta}|\nabla\nabla v|) &\leq (\|v\|_{2,\alpha,-\beta})^2 \ , \quad \text{and} \\
|\nabla v|^2\delta(x)^{3-2\beta} &\leq c(\|v\|_{2,\alpha,-\beta})^2 \ ,
\end{aligned}$$

we obtain $\delta(x)^{3-2\beta}|\nabla N(v + \varphi)|(x) \leq C(\|v\|_{2,\alpha,-\beta} + 1)$ for any x. It follows that $\|N(v + \varphi)\|_{1,\alpha,2-2\beta} \leq c\|v\|_{2,\alpha,-\beta}$. Since Δ^{-1} is bounded, $\|K(v)\|_{3,\alpha,-2\beta} \leq c\|v\|_{2,\alpha,-\beta}$. We now use the fact that $C^{3,\alpha}_{(-2\beta)}(\Omega)$ is compactly embedded in $C^{2,\alpha}_{(-\beta)}(\Omega)$. Thus K is a compact operator on $C^{2,\alpha}_{(-\beta)}(\Omega)$. Therefore we may apply the Leray-Schauder degree theory (see [LS]): If $T_t = I + tK$ and $\mathcal{B}_R(0) = \{v \in C^{2,\alpha}_{(-\beta)}(\Omega) : \|v\|_{2,\alpha,-\beta} \leq R\}$, then $\deg(T_t|_{\mathcal{B}_R(0)}, 0)$ is independent of t provided $0 \notin T_t(\partial\mathcal{B}_R(0))$ for $0 \le t \le 1$.

To check that $0 \notin T_t(\partial\mathcal{B}_R(0))$ for $0 \le t \le 1$, we can observe that if $T_t(v) = 0$, then we have the equation $\Delta u + tN(u) = 0$ where $u = v + \varphi$. If we equip N with the metric h_t defined by $h_t(u) = h(tu)$, then this equation expresses the condition that u is harmonic with respect h_t. (Recall that u is a globally defined normal coordinate.) Note that $h_0(u)$ is the Euclidean metric and $K_{h_t} \le t$ for $0 \le t \le 1$. Thus we see that all t the *a priori* estimates obtained in Remark 5.5 are valid. Thus if we choose $R \ge 2C_0$, then we can guarantee that $0 \notin T_t\ (\partial\mathcal{B}_R(0))$. Therefore $\deg(T_1|_{\mathcal{B}_R(0)}, 0) = \deg(T_0|_{\mathcal{B}_R(0)}, 0) = 1$ by the linear existence theory for h_0. This implies the existence of the required harmonic map u and completes the proof of Theorem 5.3.

Theorem 5.6 (Boundary Regularity). *Suppose u is harmonic, $u = \varphi$ on $\partial\Omega$, $\varphi \in C^{k,\alpha}(\partial\Omega)$, and $u \in C^\infty(\Omega) \cap C^{0,\beta}(\overline{\Omega})$, where β is near 1. Then $u \in C^{k,\alpha}(\overline{\Omega})$.*

Proof. By Hölder continuity of u, we get $\underset{B_{\delta(x)}}{\mathrm{OSC}}\, u \le c\delta(x)^\beta$. Hence, from Proposition 5.1 we have

$$|du|(x) \le c\delta(x)^{-1}\delta(x)^\beta = c\delta(x)^{\beta-1}\ .$$

Therefore $\int_\Omega |du|^{2p} dx \le c$ if $(\beta - 1)\cdot 2p > -1$, or $1 > (1-\beta)\cdot 2p$. From the harmonic map equation, $|N(u)| = |\Delta u| \le c|du|^2$. Hence $\int_\Omega |\Delta u|^p dv_M \le c$ if $1 > (1-\beta)2p$. Recall the basic L^p elliptic estimate:

$$\|u\|_{2,p,\Omega} \le c\|\Delta u\|_{0,p,\Omega} + c\|u\|_{0,p,\Omega}\ , \quad 1 < p < \infty\ .$$

It follows from the Sobolev embedding theorem that if p is large enough, then $|u|_{C^{1,\alpha}(\Omega)} \le c$, where $\alpha = 1 - \dfrac{n}{p}$. (Note that $p \to \infty \Rightarrow \alpha \to 1$). Hence $|N(u)|_{C^{0,\alpha}(\Omega)} \le c$. Therefore by Schauder theory we have $\|u\|_{2,\alpha,\Omega} \le c$. Applying the method of difference quotients and induction on k, we can show that $u \in C^{k,\alpha}(\overline{\Omega})$.

6. Homotopy Dirichlet Problem

We now give a proof of the theorems of Eells-Sampson [ES] and Hamilton [H] based on the local existence theorem and the *a priori* estimates. We first need an extended version of the boundary estimate.

Lemma 6.1. *Assume $\varphi \in C^{0,\alpha}(\Omega, N)$ where N has nonpositive curvature and Ω has smooth boundary. Assume $u \in C^\infty(\Omega, N) \cap C^{0,\alpha}(\bar{\Omega}, N)$ is a harmonic map from Ω to N with $u = \varphi$ on $\partial\Omega$. We then have*

$$\|u\|_{0,\alpha,\Omega\cap B_r(x_0)} \leq C(\|\varphi\|_{0,\alpha}, E(u))$$

where $x_0 \in \partial\Omega$ and r is sufficiently small that $\Omega \cap B_r(x_0)$ is near a Euclidean half-ball.

Proof We define $g(x) = d(u(x), \varphi(x_0))$, then $0 \leq g(x) \leq c_1$ for all $x \in (\partial\Omega) \cap B_r(x_0)$. To derive a bound on g in $\Omega \cap B_r(x_0)$ we note that $|\bigtriangledown g| \leq |du|$, and hence we have

$$\int_{\Omega\cap B_r(x_0)} |\bigtriangledown g|^2 \leq C\,.$$

Let $(g - c_1)_+$ be the positive part of $g - c_1$. Then $(g - c_1)_+ \equiv 0$ on $(\partial\Omega) \cap B_r(x_0)$. By the Poincaré inequality,

$$\int_{\Omega\cap B_r(x_0)} (g - c_1)_+^2 \leq C \int_{\Omega\cap B_r(x_0)} |\bigtriangledown g|^2 \leq C\,.$$

Hence $\int_{\Omega\cap B_r(x_0)} g^2 \leq C$. From subharmonicity of g^2 it follows that $g \leq C$ on $\Omega \cap B_{\frac{r}{2}}(x_0)$. We can then bound $|u|_{0,\alpha}$ as in the proof of Lemma 5.4. This completes the proof of Lemma 6.1.

We now prove the main theorem of this section.

Theorem 6.2 (Eells-Sampson, Hamilton, Hartman). *Suppose M is a compact manifold with (possibly empty) boundary, and suppose N is compact with $K_N \leq 0$. Given $\varphi \in C^{0,\beta}(\overline{M}, N)$, $\beta \in \left(\frac{1}{2}, 1\right)$, there exists a harmonic map $u \in C^\infty(M, N) \cap C^{0,\beta}(\overline{M}, N)$ with $u = \varphi$ on ∂M and $u \sim \varphi$ (rel. ∂M). If φ is regular ($C^{k,\alpha}$) on $\partial\Omega$, then u is regular ($C^{k,\alpha}$) up to $\partial\Omega$.*

Proof First we may assume that $E(\varphi) < \infty$. Cover $\overline{M}$ with small balls $\{B_{r_i}(x_i)\}_i$, and replace φ in each $B_{r_i}(x_i)$ by a map $\hat{\varphi}_i$ with finite energy such that

$$|\hat{\varphi}_i|_{C^{0,\beta}(\overline{B}_{r_i}(x_i))} \leq C|\varphi|_{C^{0,\beta}(\overline{B}_{r_i}(x_i))} \quad \text{and} \quad |d\hat{\varphi}_i|(x) \leq C\delta(x)^{\beta-1} .$$

By hypothesis $\left(\beta > \dfrac{1}{2}\right)$, we have $\int_{B_{r_i}(x_i)} |d\hat{\varphi}_i|^2 \leq C$. (This explains why we need to assume $\beta \in \left(\dfrac{1}{2}, 1\right)$.)

Let $E_0 = \inf\{E(v) : v \in \Im_\varphi\}$ where $\Im_\varphi = \{v : v \in C^\infty(M) \cap C^{0,\beta}(\overline{M}),$ $v = \varphi$ on ∂M, and $v \sim \varphi\}$. Let $\{v_j\}$ be a minimizing sequence in $\Im_\varphi$. By the weak compactness of bounded sets in $L_1^2(M, N)$ (see Remark 2.2) a subsequence of $\{v_j\}$, again denoted $\{v_j\}$, converges weakly to a limit $u \in L_1^2(M, N)$. We show that u is a smooth harmonic map. First assume $B_{r_i}(x_i)$ is contained in M^0, the interior of M. By the local existence theorem we can find a harmonic map $h_j \in C^\infty(B_{r_i}(x_i), N) \cap C^{0,\beta}(\overline{B}_{r_i}(x_i), N)$ with $h_j = v_j$ on $\partial B_{r_i}(x_i)$. By Proposition 4.1 a subsequence of $\{h_j\}$ converges to a smooth limit h on compact subsets of $B_{r_i}(x_i)$. We will prove that $h = u$ a.e. in a neighborhood of x_i. To see this, let g_s be the geodesic homotopy from h_j to v_j (see Proposition 3.1). We have for $s \in [0, 1]$

$$\frac{d^2}{ds^2} E(g_s) \geq 2 \int_{B_{r_i}(x_i)} \| \nabla' V \|^2$$

where $V = G_*\left(\dfrac{\partial}{\partial s}\right)$, $G(x, s) = g_s(x)$ (see Lemma 3.2).

We now claim the following: Suppose $g_0, g_1 \in \Im_\varphi$, $g_0 \equiv g_1$ on $M \sim \Omega$, and g_s is a unique geodesic homotopy from g_0 to g_1. Given $\varepsilon > 0$, there exists $\delta > 0$ such that if $E(g_0)$, $E(g_1) < E_0 + \delta$ then $\int_M |g_0 - g_1|^2 < \varepsilon$. Since $E(g_s)$ is convex in s, $E(g_s)$ is nearly constant, so that for some $s_0 \in [0, 1]$ and some $\delta > 0$ we must have

$$\frac{d^2}{ds^2} E(g_s) \leq \delta \quad \text{at} \quad s = s_0 .$$

Notice that by the compatibility of ∇' with the inner product we have

$$\| \nabla' V\|^2 \geq \left| \nabla \|V\| \right|^2 .$$

Also we have $V = 0$ on $\partial M \times [0,1]$, and $\|V\|(x,s)$ is independent of s since $\|V\|$ represents the length of the tangent vector to a constant speed geodesic. It follows from the Poincaré inequality

$$\int_{B_{r_i}(x_i)} \left| \nabla \|V\| \right|^2 \geq c \int_{B_{r_i}(x_i)} \|V\|^2 .$$

Thus we have

$$\begin{aligned} \delta &\geq \frac{d^2}{ds^2} E(g_s) \geq 2 \int_{B_{r_i}(x_i)} \| \nabla' V\|^2 \geq 2 \int_{B_{r_i}(x_i)} \left| \nabla \|V\| \right|^2 \\ &\geq c \int_{B_{r_i}(x_i)} \|V\|^2 . \end{aligned}$$

Now recall that $N \subset \mathbb{R}^K$ isometrically, so we have

$$|v_j - h_j| \leq d(v_j, h_j)$$

where $d(\cdot,\cdot)$ denotes geodesic distance on N. Also, note that $\|V\|(x)$ is the length of a particular geodesic from $h_j(x)$ to $v_j(x)$ and hence dominates the geodesic distance. Therefore we have

$$\int_{B_{r_i}(x_i)} |v_j - h_j|^2 \leq c\delta .$$

Since $\varepsilon \equiv c\delta \to 0$, and h_j converges to h while v_j converges weakly to u it follows that $u = h$ a.e. and hence u is smooth in $B_{r_i}(x_i)$. A similar argument using Lemma 6.1 works also on $\partial\Omega$. Thus we have $u \in C^\infty(M,N) \cap C^{0,\beta}(\overline{M},N)$. Finally we must show that u is homotopic to φ. One way to do this is to show that two smooth maps which have bounded energy and are L^2 close are actually homotopic (maps into a manifold of nonpositive curvature). A proof which is more in the present spirit is to modify v_j by harmonic replacement on balls. Let $B_1, \ldots, B_p$ be a covering of $\overline{M}$ by balls. The argument given above shows that we can replace v_j by a harmonic map $h_{j,1}$ in B_1, and the result, denoted $v_{j,1}$, becomes uniformly close to u in the interior of B_1

and homotopic to v_j. We then repeat the argument replacing $v_{j,1}$ on B_2 to get $v_{j,2}$ which by the above argument will be uniformly close to u in the interior of $B_1 \cup B_2$ and homotopic to v_j. Finally we get a map $v_{j,p}$ which is homotopic to v_j and uniformly close to u. The maps $v_{j,p}$ and u are then homotopic and it follows that $u \sim \varphi$. This concludes the proof of Theorem 6.2

7. Existence and Regularity of Weak Solutions

In this section we discuss existence questions for minima, and regularity results. We first prove a "higher regularity" result which says that Hölder continuous weak solutions are regular. This result is known to experts, but only a special case [BG] appears explicitly in the literature. Note that a weak solution which has small oscillation has been shown to be Hölder continuous in [HKW], while a Lipschitz weak solution can easily be shown regular by standard linear elliptic theory. In our first lemma we bridge the gap between C^α and Lipschitz.

Lemma 7.1. *Let $u \in C^\alpha(M,N) \bigcap L_1^2(M,N)$ for some $\alpha \in (0,1)$ be a weakly harmonic map. Then u is a smooth map.*

Proof : We first show that u is locally Lipschitz. Let $x_0 \in M$, and $r_0 > 0$ such that the image of $B_{r_0}(x_0)$ under u lies in a uniformly Euclidean normal coordinate chart. Throughout this argument we work in this chart, so that u is thought of as a vector-valued function. We also choose normal coordinates y^β in $B_{r_0}(x_0) \subset M$. Thus, from the Hölder continuity of u we have

$$(7.1) \qquad \begin{aligned} g_{\beta\gamma}(y) &= \delta_{\beta\gamma} + O(\sigma^2) \\ h_{ij}(u(y)) &= \delta_{ij} + O(\sigma^{2\alpha}) \end{aligned}$$

for $|y| \le \sigma$. The fact that u is weakly harmonic implies for any $\sigma \in (0, r_0)$

$$(7.2) \qquad \left| \int_{B_\sigma(x_0)} \langle \nabla\eta, \nabla u \rangle \right| \le C\big(\sup_{B_\sigma(x_0)} |\eta| \big) \int_{B_\sigma(x_0)} e(u) dv_M$$

for any $\eta \in C^0(\bar{B}_\sigma, \mathbb{R}^k) \bigcap L_1^2(B_\sigma, \mathbb{R}^k)$ with $\eta \equiv 0$ on ∂B_σ. Let v be the solution of the linear Laplace equation

$$\sum_\gamma \partial_\gamma \partial_\gamma v = 0 \,\text{in}\, B_\sigma$$

$$v = u \,\text{on}\, \partial B_\sigma$$

where ∂_γ denotes the partial derivative with respect to y^γ. Since $|\partial v|^2 = \sum_{i,\gamma}\left(\dfrac{\partial v^i}{\partial y^\gamma}\right)^2$ is a (Euclidean) subharmonic function, we have the mean value inequality

$$\frac{\mathrm{d}}{\mathrm{d}r}\left(\fint_{B_r} |\partial v|^2\,\mathrm{d}y\right) \geq 0 \,\text{for}\, 0 \leq r \leq \sigma \tag{7.3}$$

where $\fint_{B_r} f$ denotes the Euclidean average value of f. Because v is harmonic we have

$$\int_{B_\sigma} |\partial(u-v)|^2\,\mathrm{d}y = \int_{B_\sigma} (\partial(u-v)\cdot\partial u)\mathrm{d}y.$$

From (7.1) we have

$$\left|\int_{B_\sigma} (\partial(u-v)\cdot\partial u)\mathrm{d}y - \int_{B_\sigma} \langle\nabla(u-v), \nabla u\rangle \mathrm{d}v_M\right|$$
$$\leq C\sigma^\alpha \int_{B_\sigma} (|\partial u|^2 + |\partial v|^2)$$

Using the energy minimizing property of v and (7.1) in combination with the above inequalities we have

$$\int_{B_\sigma} |\partial(u-v)|^2\,\mathrm{d}y \leq \int_{B_\sigma} \langle\nabla(u-v), \nabla u\rangle \mathrm{d}v_M + C\sigma^\alpha \int_{B_\sigma} e(u)\mathrm{d}v_M.$$

Applying (7.2) with $\eta = u - v$ we get

$$\int_{B_\sigma} \langle\nabla(u-v), \nabla u\rangle \mathrm{d}v_M \leq C \sup_{B_\sigma} |u-v| \int_{B_\sigma} e(u)\mathrm{d}v_M.$$

Combining this with (7.1) and the previous inequality we have

$$\int_{B_\sigma} |\partial(u-v)|^2\,\mathrm{d}y \leq C\sigma^\alpha \int_{B_\sigma} |\partial u|^2 \mathrm{d}y.$$

We note the elementary inequality $|\partial u|^2 \leq |\partial(u-v)|(|\partial u|+|\partial v|)+|\partial v|^2$, so that we have from the Schwartz inequality

$$A_{\frac{1}{2}\sigma}(u) \leq (\fint_{B_{\frac{1}{2}\sigma}} |\partial(u-v)|^2 dy)^{\frac{1}{2}}[(A_{\frac{1}{2}\sigma}(u))^{\frac{1}{2}} + (A_{\frac{1}{2}\sigma}(v))^{\frac{1}{2}}] + A_{\frac{1}{2}\sigma}(v)$$

where $A_\rho(\cdot) = \fint_{B_\rho} |\partial \cdot|^2 dy$. Applying (7.3) and the previous inequality we get

$$A_{\frac{1}{2}\sigma}(u) \leq C\sigma^\beta A_\sigma(u)^{\frac{1}{2}}[A_{\frac{1}{2}\sigma}(u)^{\frac{1}{2}} + A_\sigma(v)^{\frac{1}{2}}] + A_\sigma(v)$$

where $\beta = \alpha/2$. Using the energy minimizing property of v and elementary inequalities we get

$$A_{\frac{1}{2}\sigma}(u) \leq (1 + C\sigma^\beta)A_\sigma(u).$$

Setting $\sigma_i = r_0 2^{-i}$ we get

$$A_{\sigma_{i+1}}(u) \leq \left(1 + C\sigma_i^\beta\right) A_{\sigma_i}(u).$$

Upon iteration we get

$$\fint_{B_{2^{-i}r_0}} |\partial u|^2 \, dy \leq \bar{C} \fint_{B_{r_0}} |\partial u|^2 \, dy \leq \bar{\bar{C}}$$

where $\bar{C} = \prod_{j=0}^\infty \left(1 + Cr_0^\beta 2^{-i\beta}\right) < \infty$. Thus we see that for any $\sigma \in (0, r_0)$ we have

$$\fint_{B_\sigma(x_0)} |\partial u|^2 \, dy \leq C.$$

Since x_0 was an arbitrary point we conclude that $|\partial u|^2$ is an L^∞ function. Therefore it follows that u is locally Lipschitz. The equation for u then shows that Δu is L^∞ locally on M. Hence by linear elliptic regularity theory (see [M3, 6.2.5]) we see that u is locally L_2^p for $p < \infty$. Therefore we get $\Delta u \in L_{1,\text{loc}}^p$ and hence $u \in L_{3,\text{loc}}^p$. Repeating the argument shows that u is smooth. This proves Lemma 7.1.

Remark 7.2. The smoothness assumptions necessary for the previous argument are that $g_{\beta\gamma}$ be Hölder continuous and that h_{ij} be $C^{1,\alpha}$. One can then show that u is $C^{1,\alpha}$.

Remark 7.3. The proof of Lemma 7.1 can be modified in a straightforward way to prove boundary regularity assuming that u satisfies an appropriate Dirichlet boundary condition.

There are various regularity questions which arise in the harmonic map theory. For example, one can ask about regularity of weakly harmonic maps, stationary points, or energy minimizing maps. The regularity of weakly harmonic maps from a surface has been shown by F. Hélein [H]. This regularity theorem for harmonic maps with isolated singularities was shown earlier in [SaU]. Our next theorem proves regularity of stationary maps defined on a surface. The special case when the map is weakly conformal was treated by M. Gruter [Gr]. Examples of weak solutions with three dimensional domains which are everywhere singular have been constructed by T. Riviére [ ], and partial regularity of stationary maps in all dimensions have been proven by L. C. Evans [E] and F. Bethuel [B2].

Theorem 7.4. *Suppose* $n = \dim M = 2$, *and* $u \in L_1^2(M, N)$ *is a stationary point for energy. Then* u *is a smooth harmonic map on the interior of* M.

Proof: If suffices to prove that a stationary map defined on a unit disk D is smooth. From Lemma 1.1 we know that the function

$$\Phi(z) = \left(\|u_*\left(\frac{\partial}{\partial x}\right)\|^2 - \|u_*\left(\frac{\partial}{\partial y}\right)\|^2\right) - 2i\langle u_*\left(\frac{\partial}{\partial x}\right), u_*\left(\frac{\partial}{\partial y}\right)\rangle$$

is holomorphic in D. If Φ is identically zero, then u is weakly conformal and our theorem follows from [Gr]. Otherwise Φ has at most a finite number of zeroes $z_1, \ldots, z_s$ in $D_{1/2} = \{|z| < 1/2\}$. If we can prove that u is smooth on $D_{1/2} \setminus \{z_1, \ldots, z_s\}$, then our theorem follows from [SaU]. In this way we are reduced to the case that Φ is nowhere zero in D. We deal with this case as follows. Let $f(z)$ be an analytic function in D satisfying $f^2 = -\Phi$ in D. Let $v(z)$ be the real harmonic function given by

$$v(z) = \frac{1}{2}\mathrm{Re}\int_0^z f(\zeta)d\zeta.$$

Thus v satisfies $\dfrac{\partial v}{\partial z} = \dfrac{1}{2}f$, $v(0) = 0$. Define a harmonic map $\tilde{u} : D \to N \times \mathbb{R}$ by $\tilde{u} = (u(z), v(z))$, and observe that the Hopf differential of $\tilde{u}$ is given by

$$\tilde{\Phi}(z) = \Phi(z) + 4\left(\frac{\partial v}{\partial z}\right)^2(z) \equiv 0.$$

Thus $\tilde{u}$ is a weakly conformal harmonic map, and hence by the theorem of [Gr] is smooth. Therefore u is smooth and we have proved Theorem 7.4.

Remark 7.5. Notice that Corollary 4.4 asserts partial regularity for maps in any dimension which are locally (weakly) approximable by smooth harmonic maps. It is not at all clear that an arbitrary stationary map can be so approximated, nor is it clear that a map which can be so approximated is stationary.

We now consider existence questions for minimizing maps. We first recall that the Dirichlet problem was solved by the direct method to produce minimizing maps in Proposition 2.4. If in addition to specifying the boundary values of a minimizing map, we attempt to specify its homotopy class, we encounter the difficulty that the homotopy class is not generally a weakly closed condition in L_1^2. We suppose for simplicity that $\partial M = \emptyset$, and let $\varphi \in C^\infty(M, N)$ be a given map. Let E_φ denote the infimum of energy taken over all smooth maps v which are homotopic to φ. An old observation of Morrey is that if $M = N = \mathbb{S}^n$, $n > 2$, and φ is the identity map, then $E_\varphi = 0$. The following is a general result of B. White [W].

Proposition 7.6. $E_\varphi = 0$ *if and only if the homomorphism induced by* φ *on both* π_1 *and* π_2 *is zero.*

Several authors ([L], [SaU], [ScY2]) have proven existence theorems for minimizing maps of surfaces with specified behavior on π_1. Precisely, we consider the following definition.

Definition. *Two continuous maps v, φ are π_1-equivalent if the homomorphism $v_* : \pi_1(M, x_0) \to \pi_1(N, v(x_0))$ and $\varphi_* : \pi_1(M, x_0) \to \pi_1(N, \varphi(x_0))$ are conjugate along some path σ from $v(x_0)$ to $\varphi(x_0)$; i.e. $v_* = \sigma^{-1}\varphi_*\sigma$.*

It is easy to see that continuous maps which are L_1^2 close (even weakly) are π_1-equivalent. Thus we see that if the continuous maps were dense in $L_1^2(M, N)$ we could associate with any map $u \in L_1^2(M, N)$ a π_1-equivalence class of continuous maps (maps which approximate u sufficiently well). We could then hope to preserve this π_1-equivalence class under weak convergence, and hence construct a map of least energy among all maps which are π_1-equivalent to a given map φ. Simple examples show (see [SU2]) that the continuous maps are not generally dense in $L_1^2(M, N)$ (see [B1] for a more detailed discussion of this issue); however, if $\dim M = 2$ it is shown in [SU2] that $C^\infty(M, N)$ is dense in $L_1^2(M, N)$. This can be used to prove the following result of [ScY2].

Theorem 7.7. *Suppose* $\dim M = 2$, *and* $\varphi \in C^\infty(M, N)$. *There exists a harmonic map $u \in C^\infty(M, N)$ which is π_1-equivalent to φ, and which has least energy over all L_1^2 maps v which are π_1-equivalent to φ.*

Proof: Let $\mathcal{F}_\varphi = \{v \in L_1^2(M, N) : v \text{ is } \pi_1\text{-equivalent to } \varphi\}$ where the meaning of the π_1-equivalence class of $v \in L_1^2(M, N)$ is the equivalence class of any smooth approximating map. We then see that bounded subsets of $\mathcal{F}_\varphi$ are weakly closed (hence weakly compact). Thus by the direct method we can construct $u \in \mathcal{F}_\varphi$ of least energy. The smoothness of u then follows (as in [ScY2]) by adapting a theorem of Morrey [M2]. This completes a sketch of the proof of Theorem 7.7.

The generalization of the above result to higher dimensions is more complicated. The idea is to prove that a map $u \in L_1^2(M,N)$ can be approximated by a map $\hat{u}$ which is continuous on a suitable chosen 2-skeleton Z^2 of M. Since $\pi_1(Z^2)$ is isomorphic under the inclusion map to $\pi_1(M)$, the map $\hat{u}$ defines a π_1-equivalence class. Two such approximations can be shown to be π_1-equivalent. It can then be shown that the energy can be minimized in a π_1-equivalence class. The following result is proven in [W].

Theorem 7.8. *Let $\varphi \in L_1^2(M,N)$, and denote by $\mathcal{F}_\varphi$ the set of $L_1^2(M,N)$ maps v which are π_1-equivalent to φ (as described above). There exists $u \in \mathcal{F}_\varphi$ of least energy over all maps in $\mathcal{F}_\varphi$. The regularity properties of u are described below.*

We now describe the regularity properties of energy minimizing maps, in particular those which are constructed in Propostion 2.4 and Theorem 7.8. These results are proved in [SU1], [SU2]. First note that any smooth harmonic map $u : \mathbb{S}^n \to N$ for $n \geq 2$ gives rise to a weakly harmonic map $\bar{u} : \mathbb{R}^{n+1} \to N$ with an isolated singularity at 0 by setting

$$\bar{u}(x) = u\left(\frac{x}{|x|}\right) \quad \text{for } x \neq 0.$$

Such a map $\bar{u}$ is called a <u>tangent map</u>. If $\bar{u}$ minimizes energy on compact subsets of $\mathbb{R}^{n+1}$, then $\bar{u}$ is a <u>minimizing tangent map</u>. Thus a minimizing tangent map has the property that for any bounded domain $\Omega \subset \mathbb{R}^{n+1}$ and any map $v \in L_1^2(\Omega, N)$ with $v = \bar{u}$ on $\partial\Omega$, we have $E_\Omega(\bar{u}) \leq E_\Omega(v)$. (Actually, it suffices to take Ω to be the unit ball in $\mathbb{R}^{n+1}$ because of the homogeneity of $\bar{u}$.) Certain manifolds admit no nontrivial tangent maps; e.g. any manifold N whose universal cover carries a strictly convex function. On the other hand, a theorem of [SaU] shows that any compact manifold N whose universal cover is not contractible contains a harmonic $\mathbb{S}^2$, i.e. a nonconstant smooth harmonic $u : \mathbb{S}^2 \to N$. Thus such manifolds always contain tangent maps. The condition that a tangent map $\bar{u}$ be minimizing is clearly much stronger, and one expects such $\bar{u}$ to be rare (see [SU3] for some evidence). The tangent maps indicate that harmonic maps will in general have singularities, so in order to develop a regularity theory one must allow for this possibility. We say that a point x in the interior of M is a regular point for a map u if u is C^1 in a neighborhood of x. The set $\mathcal{R}$ of regular points is then a (possibly empty) open subset of $\text{Int}(M)$. The complement of $\mathcal{R}$ is denoted $\mathcal{S}$, and is the singular set of u. That $\mathcal{S}$ may be nonempty for a harmonic map is shown by the tangent maps $\bar{u}$. The Hausdorff dimension, $\dim \mathcal{S}$, of a set $\mathcal{S}$ is the greatest lower bound of real numbers $t \geq 0$ such that $\mathcal{S}$ can be covered by balls $\{B_{r_i}(x_i)\}$ with $\sum_i r_i^t < \epsilon$ for any preassigned positive number ϵ. It is clear that for a set $\mathcal{S} \subset M^n$ we have $0 \leq \dim \mathcal{S} \leq n$. The interior regularity theorem of [SU1] is as follows.

Theorem 7.9. *Suppose* $u \in L_1^2(M, N)$ *is a minimizing map. Then* $\dim \mathcal{S} \leq n-3$. *In particular* $\mathcal{S} = \emptyset$ *if* $n = 2$. *If* $n = 3$ *then* $\mathcal{S}$ *is a discrete set of points. Morover, if every minimizing tangent map from* $\mathbb{R}^{p+1}$ *into* N *is constant for* $2 \leq p \leq q$, *then* $\dim \leq n - q - 2$ $(\mathcal{S} = \emptyset$ *if* $n < q+2)$. *If* $n = q + 2$, *then* $\mathcal{S}$ *is discrete.*

Remark 7.10. The case $n = 2$ of the Theorem 7.9 was proved by Morrey [M2]. A partial regularity result for systems of functions minimzing general variational integrals was obtained by Giaquinta-Guisti [GG1], [GG2]. The case $n = 2$ of their theorem had been done in an earlier paper of Morrey [M1].

Remark 7.11. In [SU2], the question of boundary regularity has been considered. It was shown that minimizing maps are regular in a neighborhood of ∂M provided a Dirichlet boundary condition is imposed on the map. The ideas of [GG1] have been extended to the boundary by Jost-Meier [JM] to assert boundary regularity for elliptic systems minimizing energy-type integrals.

8. THE HEAT EQUATION METHOD AND NONCOMPACT TARGETS

Eells and Sampson [ES] pioneered the use of the heat equation method to prove existence of harmonic mappings between Riemannian manifolds. They considered the case in which M is compact (without boundary), and N has nonpositive sectional curvatures (and is compact or complete with an embedding restriction). Later, Hamilton [Ha] successfully used the heat equation method in obtaining harmonic maps on manifolds M with boundary, having prescribed boundary behavior (Dirichlet or Neumann). In this section we will study the heat equation method for the Dirichlet boundary value problem from compact manifolds with nonempty boundary to arbitrary complete manifolds of nonpositive curvature. We first record some of Hamilton's results, and refer to his book [Ha] for their proof.

Let $p \in M$, and let $\{e_\alpha : \alpha = 1, \dots, n\}$ be a smooth orthonormal frame for the tangent space of M defined in a neighborhood of p. Let $\{f_i : i = 1, \dots, k\}$ be a locally defined orthonormal frame for N in a neighborhood of $u(p)$. Let $\{\theta_\alpha : \alpha = 1, \dots, n\}$ and $\{\omega_i : i = 1, \dots, k\}$ be the dual coframes of $\{e_\alpha\}$ and $\{f_i\}$ respectively. We recall the first structural equations for M and N

$$d\theta_\alpha = \sum_{\beta=1}^{n} \theta_{\alpha\beta} \wedge \theta_\beta, \quad \theta_{\alpha\beta} = -\theta_{\beta\alpha}$$

$$d\omega_i = \sum_{j=1}^{k} \omega_{ij} \wedge \theta_j, \quad \omega_{ij} = -\omega_{ji}$$

where $\{\theta_{\alpha\beta}\}$ and $\{\omega_{ij}\}$ are the connection one-forms determined by these equations. Also recall the curvature tensors $K_{\alpha\beta\gamma\delta}$ and R_{ijlm} of M and N given by the formulae

$$d\theta_{\alpha\beta} = \sum_{\gamma=1}^{n} \theta_{\alpha\gamma} \wedge \theta_{\gamma\beta} - \frac{1}{2} \sum_{\gamma,\delta=1}^{n} K_{\alpha\beta\gamma\delta}\theta_\gamma \wedge \theta_\delta$$

$$d\omega_{ij} = \sum_{m=1}^{k} \omega_{im} \wedge \omega_{mj} - \frac{1}{2} \sum_{l,m=1}^{k} R_{ijlm}\omega_l \wedge \omega_m.$$

Then $K_{\alpha\beta\gamma\delta}$, R_{ijlm} satisfty the usual symmetries.

Now suppose $u : (a, b) \times M \to N$ is a smooth map. We write

$$u^*\omega_i = u_{i\alpha}\theta_\alpha + u_{it}\mathrm{d}t$$

where θ_α and ω_i for $1 \le \alpha \le n$, $1 \le i \le k$ are locally defined orthonormal frames of one forms on M and N respectively. We consider the initial value problem for the harmonic map heat equation. Let u_0 be a given map from M to N. We seek a map $u : [0, \infty) \times M \to N$

$$\begin{array}{lll} u_{it} = \sum_{\alpha=1}^{n} u_{i\alpha\alpha} & \text{for} & 1 \le i \le k \\ u(0,p) = u_0(p) & \text{for} & p \in M \\ u(t,p) = u_0(p) & \text{for} & t \in [0,\infty)\ \ p \in \partial M. \end{array} \tag{8.1}$$

The right hand side of the heat equation above is the tension field of the map u written with respect to orthonormal frames on M and N.

The following proposition is a consequence of Hamilton's work.

Proposition 8.1. *Suppose M is a compact Riemannian manifold with boundary, and N is complete with nonpositive sectional curvatures. There exists $T \le \infty$ such that (8.1) has a solution $u : [0, T) \times M \to N$ which is continuous on $[0, T) \times M$ and C^∞ except on $\{0\} \times \partial M$. Moreover, if the image of u stays inside a compact subset of N, then T may be increased. If the compact subset is independent of T, then we can take $T = \infty$, and there exists a C^∞ harmonic mapping $u_\infty : M \to N$ such that*

$$\begin{array}{l} u_\infty(p) = u_0(p) \quad \textit{for}\, p \in \partial M \\ u_\infty \textit{ is homotopic to}\, u_0 \textit{ relative to}\, \partial M \textit{ and} \\ \lim_{t\to\infty} u(t,\cdot) = u_\infty(\cdot) \quad \textit{in } C^\infty \textit{ topology.} \end{array}$$

The first assertion of Proposition 8.1 follows from [Ha, Part IV, Section 11]. The second and third assertions follow from [Ha, p. 157-158], where we replace the condition of convex boundary in the image by the hypothesis that the image of u remains in a compact subset of N. The fact that a choice of subsequence of $t \to \infty$ is unnecessary follows from [Ha, Part V, Section 13].

We will prove that if M is compact with nonempty boundary, and N is complete with nonpositive sectional curvatures, then the solution of the heat equation has image contained in a fixed compact subset of N. By Proposition 8.1, this will then establish existence of a harmonic map with prescribed boundary values and prescribed homotopy class.

Suppose $u : [0,T) \times M \to N$ is a solution of the heat equation (8.1). Let $\{\theta_\alpha\}$ and $\{\omega_i\}$ be as above, and define $u_{i\alpha}$ and u_{it} by

$$u^*\omega_i = \sum_{\alpha=1}^{n} u_{i\alpha}\theta_\alpha + u_{it}\mathrm{d}t \quad \text{for } 1 \le i \le k. \tag{8.2}$$

We let $u_t : M \to N$ be the map given by

$$u_t(p) = u(t,p) \quad \text{for } p \in M.$$

The function

$$\kappa(u) = \sum_{i=1}^{k} u_{it}^2$$

is well-defined function on $[0,T) \times M$, and we compute its Laplacian as a function on M for fixed t. We have

$$\Delta\kappa(u) = 2\left(\sum_{i=1}^{k} u_{it}u_{it\alpha}\right)_\alpha = 2\sum_{i,\alpha} u_{it\alpha}^2 + 2\sum_{i,\alpha} u_{it}u_{it\alpha\alpha} \tag{8.3}$$

where $u_{it\alpha}$ and $u_{it\alpha\beta}$ are defined by

$$\mathrm{d}_M u_{it} + \sum_{j=1}^{k} u_{jt}u_t^*\omega_{ji} = \sum_{\alpha=1}^{n} u_{it\alpha}\theta_\alpha \tag{8.4}$$

$$\mathrm{d}_M u_{it\alpha} + \sum_{j=1}^{k} u_{jt\alpha}u_t^*\omega_{ji} + \sum_{\beta=1}^{n} u_{it\beta}\theta_{\beta\alpha} = \sum_{\beta=1}^{n} u_{it\alpha\beta}\theta_\beta. \tag{8.5}$$

We have written $\mathrm{d} = \mathrm{d}_M + \frac{\partial}{\partial t}\mathrm{d}t$. Define a_{ij} by

$$u^*\omega_{ij} = u_t^*\omega_{ij} + a_{ij}\mathrm{d}t \quad \text{for } 1 \le i,j \le k. \tag{8.6}$$

Now let $u_{i\alpha t}$, $u_{i\alpha t\beta}$, $u_{i\alpha\beta t}$ be defined by

$$\frac{\partial u_{i\alpha}}{\partial t} + \sum_{j=1}^{k} u_{j\alpha}a_{ji} = u_{i\alpha t} \tag{8.7}$$

$$\text{(8.8)} \qquad \mathrm{d}_M u_{i\alpha t} + \sum_{j=1}^{k} u_{j\alpha t} u_t^* \omega_{ji} + \sum_{\beta=1}^{n} u_{i\beta t} \theta_{\beta\alpha} = \sum_{\beta=1}^{n} u_{i\alpha t\beta} \theta_\beta$$

$$\text{(8.9)} \qquad \frac{\partial u_{i\alpha\beta}}{\partial t} + \sum_{j=1}^{k} u_{j\alpha\beta} a_{ji} = u_{i\alpha\beta t}.$$

We wish to re-express $u_{it\alpha\alpha}$ by commuting indices, so we exterior differentiate (8.2) and collect terms involving $\mathrm{d}t$

$$u_t^* \omega_{ij} \wedge u_{jt} \mathrm{d}t + a_{ij} \mathrm{d}t \wedge u_{j\alpha} \theta_\alpha = \frac{\partial u_{i\alpha}}{\partial t} \mathrm{d}t \wedge \theta_\alpha + \mathrm{d}_M u_{it} \wedge \mathrm{d}t.$$

Rearranging, and applying (8.4) and (8.7) we have

$$\text{(8.10)} \qquad u_{it\alpha} = u_{i\alpha t} \quad \text{for } 1 \leq i \leq k,\ 1 \leq \alpha \leq n.$$

Exterior differentiating the equation

$$\mathrm{d}u_{i\alpha} + \sum_{j=1}^{k} u_{j\alpha} u^* \omega_{ji} + \sum_{\beta=1}^{n} u_{i\beta} \theta_{\beta\alpha} = \sum_{\beta=1}^{n} u_{i\alpha\beta} \theta_\beta + u_{i\alpha t} \mathrm{d}t$$

collecting terms involving $\mathrm{d}t$, rearranging, and using (8.7), (8.8), (8.9) we obtain the formula

$$\text{(8.11)} \qquad u_{i\alpha t\beta} - u_{i\alpha\beta t} = -R_{ijlm} u_{j\alpha} u_{l\beta} u_{mt}.$$

Combining (8.1), (8.10), and (8.11) we have

$$\sum_{i,\alpha} u_{it} u_{it\alpha\alpha} = \sum_{i,\alpha} u_{it} u_{i\alpha t\alpha} = - \sum_{i,j,l,m,\alpha} R_{ijlm} u_{i\alpha} u_{jt} u_{l\alpha} u_{mt} + u_{it} u_{itt}.$$

Thus (8.3) becomes

$$\text{(8.12)} \quad \Delta\kappa(u) = \frac{\partial\kappa(u)}{\partial t} + 2\sum_{i,\alpha} u_{it\alpha}^2 - 2 \sum_{i,j,k,l,\alpha} R_{ijlm} u_{i\alpha} u_{jt} u_{l\alpha} u_{mt}.$$

We will need the following standard Sobolev inequality for M.

Lemma 8.2. *There exists a constant c depending only on M such that the inequality (assume $n > 2$)*

$$\left(\int_M |h|^{\frac{2n}{n-2}} dv_M \right)^{\frac{n-2}{n}} \leq c \int_M |\nabla h|^2 dv_M$$

holds for any smooth function h with $h = 0$ on ∂M. If $n = 2$, then for any $p > 2$ we have

$$\left(\int_M |h|^{2p} dv_M\right)^{\frac{1}{p}} \le c(p) \int_M |\nabla h|^2 dv_M$$

We now derive exponential decay estimates on the function $\kappa(u)$. The first result gives us an integral decay estimate.

Lemma 8.3. *Suppose M is compact with boundary and N is complete with nonpositive sectional curvatures. Let $u : [0,T) \times M \to N$ be a solution of the heat equation (8.1). Let $\mu \in (0,1) \cap (0, T/2)$. The following inequality holds*

$$\int_M \kappa(u)(t,x) dv_M(x) \le ce^{-\epsilon t} \quad \textit{for } t \in (\mu, T)$$

where c, $\epsilon > 0$ are constants independent of t.

Proof: Since the sectional curvatures of N are nonpositive, formula (8.12) implies

$$\text{(8.13)} \qquad \Delta\kappa(u) \ge \frac{\partial \kappa(u)}{\partial t} + 2\sum_{i,\alpha} u_{it\alpha}^2.$$

Now $|\nabla\kappa(u)|^2 = 4\sum_{\alpha=1}^n \left(\sum_{i=1}^k u_{it}u_{it\alpha}\right) \le 4\kappa(u)\sum_{i,\alpha} u_{it\alpha}^2$, so (8.13) implies

$$\text{(8.14)} \qquad \Delta\kappa(u) \ge \frac{\partial \kappa(u)}{\partial t} + 2|\nabla\sqrt{\kappa(u)}|^2.$$

For $t > 0$, $\kappa(u) = \nabla\kappa(u) = 0$ on ∂M, so integrating (8.14) we obtain

$$\text{(8.15)} \qquad \frac{\mathrm{d}}{\mathrm{d}t}\int_M \kappa(u)\mathrm{d}v_M + 2\int_M |\nabla\sqrt{\kappa(u)}|^2 \mathrm{d}v_M \le 0.$$

Using the Poincaré inequality we have

$$\epsilon\int_M \kappa(u)\mathrm{d}v_M \le 2\int_M |\nabla\sqrt{\kappa(u)}|^2$$

where ϵ depends only on M. Combined with (8.15) this gives

$$\frac{\mathrm{d}}{\mathrm{d}t}\left(e^{\epsilon t}\int_M \kappa(u)\mathrm{d}v_M\right) \le 0.$$

Integrating this over $[\mu, t]$ gives the conclusion of Lemma 8.3.

Lemma 8.4. *Let M, N, T, and μ be as in Lemma 8.3. The following inequality holds*

$$\max_{x\in M}\kappa(u)(t,x)\le ce^{-\epsilon t}\quad \textit{for } t\in(2\mu,T)$$

where c, $\epsilon>0$ are constants independent of t.

Proof: We multiply (8.14) by $\kappa(u)^{p-1}$ and integrate over M to obtain

$$\int_M \kappa(u)^{p-1}\frac{\partial\kappa(u)}{\partial t}\mathrm{d}v_M+2\int_M\kappa(u)^{p-1}|\nabla\sqrt{\kappa(u)}|^2\mathrm{d}v_M\le 0$$

where we have thrown away a negative term from integration by parts. This can be rewritten

$$\frac{\mathrm{d}}{\mathrm{d}t}\int_M\kappa(u)^p\mathrm{d}v_M+\frac{2}{p}\int_M|\nabla\kappa(u)^{p/2}|^2\mathrm{d}v_M\le 0. \tag{8.16}$$

Applying the Sobolev inequality of Lemma 8.2 gives

$$\frac{\mathrm{d}}{\mathrm{d}t}\int_M\kappa(u)^p\mathrm{d}v_M+\frac{2}{pc}\left(\int_M\kappa(u)^{p\lambda}\mathrm{d}v_M\right)^{1/\lambda}\le 0 \tag{8.17}$$

where we set $\lambda=\frac{n}{n-2}$. (In case $n=2$, we can take λ to be some fixed number greater than one.) Inequality (8.16) implies that $\int_M\kappa(u)^p\mathrm{d}\nu_M$ is a nonincreasing function of t for any $p>1$, so we integrate (8.17) over the interval $[t-\delta,t]$ and use the fact that $\int_M\kappa(u)^{p\lambda}\mathrm{d}v_M$ is nonincreasing to obtain

$$\left(\int_M\kappa(u)^{p\lambda}(t,x)\mathrm{d}v_M(x)\right)^{1/\lambda}\le\frac{pc}{2\delta}\int_M\kappa(u)^p(t-\delta,x)\mathrm{d}v_M(x).$$

This inequality holds for any $\delta\in(0,\mu)$, $t\in(\mu,T)$, $p\ge 1$. Let i be a nonnegative integer, and choose $p=\lambda^i$, $\delta=\mu 2^{-i-2}$, and replace t by $t-\mu+\sum_{j=0}^{i+1}\mu 2^{-j-1}$. We then have

$$(J_{i+1})^{\frac{1}{\lambda^{i+1}}}\le\left(\frac{c\lambda^i}{2^{-i-1}\mu}\right)^{\frac{1}{\lambda^i}}(J_i)^{\frac{1}{\lambda^i}} \tag{8.18}$$

where we have introduced the notation

$$J_i=\int_M\kappa(u)^{\frac{1}{\lambda^i}}(t-\mu+\sum_{j=0}^{i}\mu 2^{-j-1},x)\mathrm{d}v_M(x).$$

Iterating (8.18) for $i = 0, 1, \ldots, l$ we have

$$(J_{l+1})^{\frac{1}{\lambda^{l+1}}} \le \left[\prod_{i=0}^{l} (c\lambda^i \mu^{-1})^{\frac{1}{\lambda^i}} 2^{\frac{-i}{\lambda^i}}\right] J_0. \tag{8.19}$$

It can be checked directly by taking logarithms that

$$\prod_{i=0}^{l} (c\lambda^i \mu^{-1})^{\frac{1}{\lambda^i}} 2^{\frac{-i}{\lambda^i}} \le C \quad \text{for all}\, l.$$

Now, using the fact that $\int_M \kappa(u)^{\lambda^{l+1}} dv_M$ is a nonincreasing function of t, we see that

$$\int_M \kappa(u)^{\lambda^{l+1}}(t,x) dv_M(x) \le J_{l+1}.$$

Thus (8.19) implies

$$\left(\int_M \kappa(u)^{\lambda^{k+1}}(t,x) dv_M(x)\right)^{\frac{1}{\lambda^{k+1}}} \le c \int_M \kappa(u)(t-\mu, x) dv_M(x)$$

For $t \in (2\mu, T)$. Letting $k \to \infty$, we have

$$\max_{x \in M} \kappa(u)(t,x) \le c \int_M \kappa' u)(t-\mu, x) dv_M(x).$$

If $t \in (2\mu, T)$, we apply Lemma 8.3 to obtain the required inequality.

We can now state the existence theorem.

Theorem 8.5. *Suppose M is compact with (nontrivial) boundary, and N is complete with nonpositive sectional curvatures. Let $u_0 : M \to N$ be a C^∞ mapping. There exists a C^∞ harmonic mapping $u : M \to N$ satisfying $u = u_0$ on ∂M and u homotopic (relative to ∂M) to u_0.*

Proof : For a fixed $x \in M$, we consider the t parameter curve $\{u(t,x) : 2\mu \le t < T\}$, and compute its length, l

$$l = \int_{2\mu}^{l} \left|\frac{\partial u}{\partial t}\right| (t,x) dt = \int_{2\mu}^{T} \sqrt{\kappa(u)}(t,x) dt.$$

By lemma 8.4, we then have

$$l \le \frac{2}{\epsilon} c^{1/2} e^{-\epsilon\mu}$$

a bound independent of x and T. This implies that the image of u stays in a compact subset of N, so Proposition 8.1 applies to give the required existence result.

Remark 8.6. *It is interesting to note that existence is false in general if N is noncompact with nonpositive (even strictly negative) sectional curvatures, and M is compact without boundary. A simple counter-example is given [ES, p.155] where N is a surface of revolution in 3-space and M is S^1. This example shows that in case M has no boundary, the image of u need not stay in a compact subset of N.*

References

[B1] F. Bethuel, *A characterization of maps in $H^1(B^3, S^2)$ which can be approximated by smooth maps*, Ann Inst Poincaré Anal Non Lin **7** (1990), 269-186.

[B2] F. Bethuel, *On the singular set of stationary harmonic maps*, Manuscripta Math **78**, (1993), 417-443.

[BG] H. J. Borchers and W. J. Garber, *Analyticity of solutions of the $O(N)$ non-linear σ-model*, Comm. Math. Phys. **71** (1980), 299-309.

[BCL] H. Brezis, J.-M. Coron, and E. Lieb, *Harmonic maps with defects*, Comm Math Phys **107** (1986), 649-705.

[CG] S. S. Chern and S. Goldberg, *On the volume-decreasing property of a class of real harmonic mappings*, Amer. J. Math. **97** (1975), 133-147.

[ES] J. Eells and J. Sampson, *Harmonic mappings of Riemannian manifolds*, Amer. J. Math. **86** (1964), 109-160.

[EL] J. Eells and L. Lemaire, *A report on harmonic maps*, Bull. London Math. Soc. **10** (1978), 1-68.

[E] L. C. Evans, *Partial regularity for stationary harmonic maps into spheres*, Arch Rat Mech Anal **116** (1991), 101-113.

[G] M. Giaquinta, *Multiple integrals in the calculus of variations and nonlinear elliptic systems*, Annals of Math. Studies **105**, 1983.

[GG1] M. Giaquinta and E. Giusti, *On the regularity of the minima of variational integrals*, Acta. Math. **148** (1982), 31-46.

[GG2] M. Giaquinta and E. Giusti, *The singular set of the minima of certain quadratic functionals*, Ann. Sc. Norm. Sup. Pisa. **11** (1984), 45-55.

[GH] M. Giaquinta and S. Hildebrandt, *A priori estimates for harmonic mappings*, J. Reine Angew. Math. **336** (1982), 124-164.

[Gr] M. Grüter, *Regularity of weak H-surfaces*, J Reine Angew Math **329** (1981), 1-15.

[Hm] R. Hamilton, *Harmonic maps of manifolds with boundary*, Lecture notes **471**, Springer 1975.

[Hr] P. Hartman, *On homotopic harmonic maps*, Can. J. Math. **19** (1967), 673-687.

[H] F. Hélein, *Regularité des applications faiblements harmoniques entre une surface et une variete Riemannienne*, CR Acad Sci Paris Ser I Math **312** (1991), 591-596.

[HKW] S. Hildebrandt, H. Kaul, and K. O. Widman, *An existence theorem for harmonic mappings of Riemannian manifolds*, Acta Math. **138** (1977), 1-16.

[HW] S. Hildebrandt and K. O. Widman, *On the Hölder continuity of weak solutions of quasilinear elliptic systems of second order*, Ann. Sc. Norm. Sup. Pisa IV (1977), 145-178.

[JM] J. Jost and M. Meier, *Boundary regularity for minima of certain quadratic functionals*, Math. Ann. **262** (1983), 549-561.

[L] L. Lemaire, *Applications harmoniques de surfaces Reimanniennes*, J. Diff. Geom. **13** (1978), 51-78.

[LS] J. Leray and J. Schauder, *Topologie et équations fonctionelles*, Ann Sci École Norm Sup **51** (1934), 45-78.

[Li] F.-H. Lin, *Gradient estimate and blow-up analysis for stationary harmonic maps*, In preparation, 1996.

[M1] C. B. Morrey, *On the solutions of quasilinear elliptic partial differential equations*, Trans. A. M. S. **43** (1938), 126-166.

[M2] C. B. Morrey, *The problem of Plateau on a Riemannian manifold*, Ann. of Math. **49** (1948), 807-851.

[M3] C. B. Morrey, *Multiple integrals in the calculus of variations*, Springer-Verlag, New York, 1966.

[P] P. Price, *A monotonicity formula for Yang-Mills fields*, Manuscripta Math. **43** (1983), 131-166.

[R] T. Riviére, *Everywhere discontinuous harmonic maps into spheres*, Acta Math **175** (1995), 197-226.

[Sc] R. Schoen, *Analytic aspects of the harmonic map problem*, MSRI Publ **2** (1984), Springer, New York-Berlin, 321-358.

[ScY1] R. Schoen and S. T. Yau, *Compact group actions and the topology of manifolds with non-positive curvature*, Topology **18** (1979), 361-380.

[ScY2] R. Schoen and S. T. Yau, *Existence of incompressible minimal surfaces and the topology of three dimensional manifolds with non-negative scalar curvature*, Ann. of Math. **110** (1979), 127-142.

[SaU] J. Sacks and K. Uhlenbeck, *The existence of minimal immersions of 2-spheres*, Ann. of Math. **113** (1981), 1-24.

[SU1] R. Schoen and K. Uhlenbeck, *A regularity theory for harmonic maps*, J. Diff. Geom. **17** (1982), 307-335.

[SU2] R. Schoen and K. Uhlenbeck, *Boundary regularity and the Dirichlet problem for harmonic maps*, J. Diff. Geom. **18** (1983), 253-268.

[SU3] R. Schoen and K. Uhlenbeck, *Regularity of minimizing harmonic maps into the sphere*, Invent Math **78** (1984), 89-100.

[W] B. White, *Infima of energy functionals in homotopy classes*, J Diff Geom **23** (1986), 127-142.

Chapter X

Sobolev Spaces and Harmonic Maps for Metric Space Targets

Introduction

In this chapter we present the work of [KS] which develops the theory of Sobolev spaces and much of the basic harmonic mapping theory from a purely intrinsic point of view, and without using the smoothness of the target space. Jost [J] has given an independent development of aspects of the same theory which allows the possibility that the domain also be singular. For smooth domains his results are less detailed than those presented here. In particular he doesn't derive the pointwise properties of Sobolev maps nor of harmonic maps which we obtain.

When one studies variational problems for maps between Riemannian manifolds one must consider spaces which we denote $W^{1,p}(\Omega, X)$. Here Ω is a compact domain in a Riemannian manifold, X is a second Riemannian manifold, $p \in [1, \infty)$, and $W^{1,p}$ indicates that the first derivatives of the map are $L^p(\Omega)$. For $p > n$ such maps will be continuous, and the corresponding space $W^{1,p}(\Omega, X)$ can be given the structure of a smooth Banach manifold. This is because, for $p > n$, any map which is close in $W^{1,p}$ distance to a map u_0 can be described as a pointwise small deformation of u_0. This linear space of $W^{1,p}$ deformations is then a Banach space on which one can locally model $W^{1,p}(\Omega, X)$. For $p \leq n$ this is no longer possible, and the definition of the space $W^{1,p}(\Omega, X)$ becomes much less clear. This problem was first encountered by C.B. Morrey [Mo] in case $n = \dim \Omega = 2$ and $p = 2$. A great deal of effort was spent by Morrey to give a definition of this space. In more recent times people have exploited the embedding theorem of J. Nash, and considered X to be a smooth submanifold of a Euclidean space $\mathbb{R}^K$. If we define $W^{1,p}(\Omega, X)$ to be the subset of the Banach space $W^{1,p}(\Omega, \mathbb{R}^K)$ consisting of those maps with image essentially in X, it turns out that this gives a workable definition for many purposes. This is the approach we have taken in the previous chapter. An aesthetic drawback of this definition is that the space $W^{1,p}(\Omega, X)$ should depend only on the metric of X and not on the embedding of X into $\mathbb{R}^K$. A much more serious difficulty arises if one attempts to consider maps to spaces X which are not smooth Riemannian manifolds. These could include Riemannian spaces with singularities, smooth Finsler manifolds, or infinite-dimensional spaces. In the first chapter of this paper we develop a direct intrinsic approach to this problem, and define $W^{1,p}(\Omega, X)$ for (X, d) any complete metric space. This is shown to reduce to the usual space $W^{1,p}(\Omega)$ for $X = \mathbb{R}$, and reduces to the space described above for X a smooth compact Riemannian manifold. We also derive the results which are essential for doing variational theory. These include lower semicontinuity of the $p-$energy, Rellich-type compactness results, and an L^p-trace theory for restrictions of maps to hypersurfaces.

To illustrate the idea of the Sobolev space theory, assume for simplicity of notation that $\Omega \subset \mathbb{R}^n$ is a Euclidean domain. If $u : \Omega \to X$ is a map, $x \in \Omega$, and $V \in \mathbb{R}^n$, we can give a formal definition of the norm of the directional derivative of u in the direction V at x by

$$|u_*(V)| = \lim_{\varepsilon \to 0} \left(\frac{d(u(x), u(x + \varepsilon V))}{\varepsilon} \right).$$

To define the $p-$Sobolev energy we can raise the distance quotient to the pth power, integrate over the unit vectors $V \in S^{n-1}$ and let

$$e_\varepsilon(x) = \int_{S^{n-1}} \left(\frac{d(u(x), u(x + \varepsilon V)}{\varepsilon} \right)^p d\sigma(V).$$

The $p-$energy density $e(x)$ is then expected to be the limit as $\varepsilon \to 0$ of $e_\varepsilon(x)$. One of the main results of §1 is that the measures $e_\varepsilon(x)dx$ converge (weakly) in an almost monotone fashion as $\varepsilon \to 0$ provided that their total masses are uniformly bounded. Further it is shown that for $p > 1$ the limiting measure is absolutely continuous with respect to Lebesgue measure, and hence may be written as $e(x)dx$ for an L^1 function $e(x)$. This convergence result may be considered as an analogue to the definition of the length of a continuous curve $\gamma : [0,1] \to X$ where X is a metric space. The monotone property of the $e_\varepsilon(x)dx$ is analogous to the property that the approximation to the length given by

$$\sum_{i=1}^{m} d(\gamma(x_i), \gamma(x_{i-1}))$$

is increased when the partition $\{x_0, \dots, x_m\}$ is refined. The proof of the monotone property relies only on the triangle inequality and judicious changes of variable. The convergence of the measures $e_\varepsilon(x)dx$ seems to be essential in deriving any reasonable definition of the space $W^{1,p}(\Omega, X)$. If we were forced to choose a subsequence $\varepsilon_i \to 0$, we would not be able to show for example that a map which minimizes energy on Ω is also a minimizer on any subdomain of Ω. The absolute continuity of the energy is also important in many further results. Both of these properties are used heavily in the analysis of energy minimizing maps.

In §2 we construct least energy maps ($p = 2$) with prescribed values on $\partial\Omega$, into Alexandrov spaces (X, d) of non-positive curvature. (The curvature condition is given in terms of triangle comparisons, see (2.1) for the precise definition.) Here an Alexandrov space is a complete metric space for which any pair of points can be joined by a curve whose length is equal to the distance between the points. The idea of using triangle (or quadrilateral) comparisons to define curvature bounds in metric spaces seems to be due to an Austrian mathematician A. Wald in the 1930's [Wa]. It has been developed by a Russian school of mathematicians centered around A.D. Alexandrov, starting in the late 1940's.

We are able to construct (unique) least-energy maps for the Dirichlet problem because the energy satisfies a strong convexity property under the non-positive curvature assumption. We do not require the space X to be locally compact. We show that the minimizing map is Lipschitz in the interior of Ω, with the Lipschitz constant locally bounded in terms of the total energy and the distance to $\partial\Omega$. The proof of continuity given here involves a rough form of the Eells-Sampson Bochner formula [ES]. The boundary continuity of a minimizing map to a metric space of non-positive curvature has been obtained recently by T. Serbinowski [Se]. His result says that the minimizing map is C^α up to the boundary for any $\alpha < 1$, provided that the boundary map is also C^α.

An important property which is needed to further advance the theory of harmonic maps is derived in §2.3. This is the result that any finite energy map into a space of non-positive curvature has an induced distance function which yields an infinitesimally Riemannian metric. Thus one can write the usual (trace) formula for energy density that holds for maps to smooth Riemannian manifolds. Note that for a general metric space target X the induced metric will only be Finsler.

In §2.5 we develop some general averaging methods for maps into non-positively curved spaces. We quantify the general principle that averaging decreases energy. We then apply these results to the study of homotopy and equivariant mapping problems. Precisely, we consider the existence problem for equivariant maps defined on the universal covering $\tilde{M}$ of a complete Riemannian manifold M, into a non-positively curved metric space X. These maps are assumed to be equivariant for a given homomorphism $\rho : \Gamma \to Isom(X)$, where $\Gamma = \pi_1(M)$. Assuming Γ is finitely generated we construct a locally Lipschitz equivariant map with local Lipschitz constant bounded in terms of the infimum of the "translation function" $\delta : X \to \mathbb{R}^+$ (see 2.6iii). If M is compact, this produces a finite energy equivariant map with a sharp bound on its total energy. We then use the local Dirichlet problem together with a delicate averaging argument to construct a uniformly locally Lipschitz minimizing sequence. We prove convergence of minimizing sequences in case the domain has a nonempty boundary, or when the maps are lifts of maps between compact spaces.

The theory of harmonic maps into smooth manifolds of non-positive curvature begins with the work of J.Eells and H. Sampson [ES], and for manifolds with boundary with R. Hamilton [Ha]. These theories were developed using the heat equation method. A variational approach to these results exploiting the energy convexity was given by the second author [Sch]. This was extended to the study of maps into locally compact polyhedral spaces of non-positive curvature in the paper of Gromov and Schoen[GS], along with certain applications to rigidity questions for discrete groups. Our work here is a strong generalization of these results.

1. Sobolev Space Theory for Maps to Metric Spaces

In this section we construct the spaces $W^{1,p}(\Omega, X)$ (for $p > 1$) and the space $BV(\Omega, X)$ (for $p = 1$), in case (Ω^n, g) is a Riemannian domain and (X, d) is a complete metric space. We only use the triangle inequality in the target to define Sobolev maps, much as one does when studying rectifiable curves into metric spaces. In fact, our approach represents a higher-dimensional (and higher p) generalization of curve theory.

We recall the definition of $L^p(\Omega, X)$ in §1.1. For fixed $u \in L^p(\Omega, X)$ we construct an approximate energy-density function $e_\varepsilon(x)$, by using the distance function d to measure average displacements of u in $\varepsilon-$neighborhoods about x. We average symmetrically, so that if $X = \mathbb{R}$ and u is smooth, then $e_\varepsilon(x) \to c_{n,p}|\nabla u(x)|^p$ as $\varepsilon \to 0$. For technical reasons it is convenient to work with various symmetric averages: In (1.2ii) we first define $e_\varepsilon(x)$ essentially as in the introduction, taking it to be an average of $d^p(u(x), u(y))$, where y is in the $\varepsilon-$sphere $S(x, \varepsilon)$ about x. Then for suitable measures ν on the interval $(0, 2)$ we define ${}_\nu e_\varepsilon(x)$ to be averages (with respect to $d\nu(\lambda)$) of the spherical averages

$e_{\lambda\varepsilon}(x)$. Although this process is technically useful, it adds a layer of complication to our arguments. In order to focus on the main idea used to construct the Sobolev energy measure, we will restrict ourselves in this overview to a Euclidean domain and to the choice of ν (1.2vii) which leads to ball averages:

$$\tilde{e}_\varepsilon(x) = (n+p) \int_{B(x,\varepsilon)} \frac{d^p(u(x),u(y))}{\varepsilon^p} \frac{dy}{\varepsilon^n}. \tag{1.0i}$$

The $\tilde{e}_\varepsilon(x)$ are bounded continuous functions (away from $\partial\Omega$), and the corresponding measures $e_\varepsilon(x)dx$ define linear functionals $\tilde{E}_\varepsilon(f)$ for $f \in C_c(\Omega)$, via integration. We say that $u \in L^p(\Omega, X)$ has finite energy E whenever

$$\sup_{\substack{0\le f\le 1\\ f\in C_c(\Omega)}} \limsup_{\varepsilon\to 0} \tilde{E}_\varepsilon(f) \equiv E < \infty. \tag{1.0ii}$$

In this case we write $u \in W^{1,p}(\Omega, X)$ if $p > 1$, or $u \in BV(\Omega, X)$ if $p = 1$. (See 1.3).) For such u we show that

$$\lim_{\varepsilon\to 0} \tilde{E}_\varepsilon(f) \equiv E(f)$$

exists for each $f \in C_c(\Omega)$. The key idea in this step is a "refinement lemma" (Lemma 1.3.1) which generalizes the fact that approximate lengths of curves increase when a partition is refined. Since the functional E is linear and bounded (1.0ii) it follows from the Riesz representation theorem that there is an energy-density measure de for the map u, with $e_\varepsilon d\mu_g \rightharpoonup de$ weakly, and with $e(\Omega) = E$. We sketch the refinement lemma and its consequences here, for the particular approximate energy functions (1.0i) and a Euclidean domain. For $f \in C_c(\Omega)$ with $f \ge 0$, and $\varepsilon > 0$ we define the slightly larger function

$$f_\varepsilon(x) = f(x) + \omega(f,\varepsilon)(x).$$

Here $\omega(f,\varepsilon)(x)$ is the oscillation of f in $B(x,\varepsilon)$ (1.3iii). For small $\varepsilon > 0$, $f_\varepsilon \in C_c(\Omega)$. Now, let any finite partition of the interval $[0,1]$ be given, into subintervals of lengths λ_i, $i = 1, \dots, m$. The refinement lemma is the inequality

$$\tilde{E}_\varepsilon(f)^{1/p} \le \sum_i \lambda_i (\tilde{E}_{\lambda_i \varepsilon}(f_\varepsilon))^{1/p}. \tag{1.0iii}$$

This inequality is a direct consequence of the iterated (X, d) and L^p triangle inequalities, as the following argument shows. We may write

$$\tilde{E}_\varepsilon(f) = (n+p) \iint_{|x-y|<\varepsilon} f(x) \frac{d^p(u(x), u(y))}{\varepsilon^p} \frac{dydx}{\varepsilon^n}. \tag{1.0iv}$$

For small ε, x in the support of f and $|y - x| < \varepsilon$ we may partition the segment xy into pieces of length $\lambda_i |x - y|$, and call the corresponding partition

$$x = x_0, x_1, \dots x_m = y.$$

The (X, d) triangle inequality implies

$$d(u(x), u(y)) \le \sum_{i=1}^m d(u(x_{i-1}), u(x_i)).$$

Thus from the iterated L^p triangle inequality we have

$$\tilde{E}_\varepsilon(f)^{1/p} \le \sum_i \left((n+p) \iint_{|x-y|<\varepsilon} f(x) \frac{d^p(u(x_{i-1}), u(x_i))}{\varepsilon^p} \frac{dydx}{\varepsilon^n} \right)^{1/p}. \tag{1.0v}$$

For each $i = 1, \dots, m$ we change variables in the corresponding integral, from (x, y) to (x_{i-1}, x_i). Write

$$\mu_i = \sum_{j=1}^i \lambda_j.$$

Then we have the estimates

$$dxdy = \frac{dxdx_i}{\mu_i^n} = (\frac{\mu_i}{\lambda_i})^n dx_{i-1} \frac{dx_i}{\mu_i^n} = \frac{dx_{i-1}dx_i}{\lambda_i^n}$$
$$f(x_{i-1}) \le f(x) + \omega(f, \varepsilon)(x) = f_\varepsilon(x)$$
$$|x_{i-1} - x_i| < \lambda_i \varepsilon.$$

Applying these estimates (and also multiplying and dividing terms by λ_i), we see that (1.0v) implies (1.0iii).

The refinement estimate implies the existence of a limit measure because it gives a quantitative estimate of the sense in which the approximate energy functionals increase as ε decreases. In particular, for fixed ε we let ε' be small, and take each $\lambda_i = \frac{1}{[\varepsilon/\varepsilon']}$ (where [] denotes the greatest integer function) to deduce

$$\tilde{E}_{[\varepsilon/\varepsilon']\varepsilon'}(f) \leq \tilde{E}_{\varepsilon'}(f_\varepsilon)$$

as a special case of (1.0iii). We let $\varepsilon' \to 0$ and note that the numbers $\tilde{E}_\varepsilon(f)$ vary continuously in ε (because we used ball-average approximate energy densities), to conclude

$$\tilde{E}_\varepsilon(f) \leq \liminf_{\varepsilon' \to 0} \tilde{E}_{\varepsilon'}(f_\varepsilon).$$

Noting that

$$\tilde{E}_{\varepsilon'}(f_\varepsilon) = \tilde{E}_{\varepsilon'}(f) + \tilde{E}_{\varepsilon'}((\omega(f,\varepsilon)(x))$$

and using the finite-energy hypothesis we see that

$$\limsup_{\varepsilon \to 0} \tilde{E}_\varepsilon(f) \leq \liminf_{\varepsilon' \to 0} \tilde{E}_{\varepsilon'}(f).$$

This shows that the limit functional is well defined on non-negative functions $f \in C_c(\Omega)$. It is then an easy matter to show the limit functional is defined on all of $C_c(\Omega)$, and hence to deduce the existence of the limit Sobolev energy measure.

We carry out the general version of the above argument in §1.2-1.5. The lemmas are separated in such a way so that we may appeal to them again in §1.7, where we construct directional energy measures for Sobolev maps. In this case one fixes a Lipschitz vector field Z and defines the ε-approximate energy density by

$${}^Z e_\varepsilon(x) = \frac{d^p(u(x), u(\overline{x}(x,\varepsilon)))}{\varepsilon^p},$$

where $\overline{x}(x,\varepsilon)$ is the point obtained by flowing time ε along Z, starting at x. The approximate directional-energy measures also converge to a limit measure, and in §1.8 we derive some useful estimates and properties of the directional energies, in addition to proving that the Sobolev energy is an average of the directional energies.

In §1.6 we prove the lower semicontinuity of Sobolev energies: If

$$\{u_i\} \subset W^{1,p}(\Omega, X) \quad (\text{or } \{u_i\} \subset BV(\Omega, X)),$$

is a sequence with uniformly bounded energies, and if $u_i \to u$ in $L^p(\Omega, X)$, then u is a finite energy map and its energy measure de^u satisfies

$$de^u \leq \liminf_{i\to\infty} de^{u_i}.$$

In this section we also verify that for $X = \mathbb{R}$ our construction gives the usual Sobolev (and BV) spaces, and that the energy densities correspond in the expected way.

In §1.9 we develop the differentiability theory of directional energies. By restricting to integral curves of vector fields, one is able to reduce to understanding finite-energy maps from intervals to X, and so one can mimic classical differentiation theory. The final results (Theorem 1.9.6) are that for $p > 1$ the directional energies are absolutely continuous with respect to Lebesgue measure, so can be written as

$$|u_*(Z)|^p d\mu_g(x)$$

for an L^p function $|u_*(Z)|$, and that for suitable choices of ν the ε-approximate energies converge almost everywhere to $|u_*(Z)|^p$. It is then an easy matter in §1.10 to show that the Sobolev energy measure is also given by an L^1 density function, in case $p > 1$.

In §1.11 we collect some estimates for the directional energy density functions which are needed in chapter 2. The L^p trace theory for Sobolev maps from Lipschitz domains is developed in §1.12. Our approach is to use auxiliary transverse vector fields and to show that u has well-defined limits along almost all of the corresponding integral curves, i.e. we follow the classical approach. In Theorem 1.12.2 we show that a sequence of bounded energy maps which converges in $L^p(\Omega, X)$ to a limit map ($p > 1$), has the property that the corresponding trace maps also converge, in $L^p(\partial\Omega, X)$. We also characterize maps u, v with equal traces as those for which the real-valued functions $d(u, v)$ are Sobolev functions with trace zero. These facts are useful for the study of energy minimizing maps, in Section 2. Theorem 1.12.3 states that if Ω can be decomposed into Lipschitz subdomains, then finite energy maps with equal traces on the boundaries define in the aggregate a finite energy map on Ω, with total energy given by the sum of the individual energies. This theorem will be used in the second chapter, where we use replacement techniques to study the equivariant harmonic map problem.

Finally, in §1.13, we include a natural precompactness theorem, which generalizes the fact that sequences in $W^{1,p}(\Omega, \mathbb{R})$ (or in $BV(\Omega, \mathbb{R})$) of uniformly bounded norm have convergent subsequences in $L^p(\Omega, \mathbb{R})$. We do not use this result in Section 2, but it is natural and follows quickly from the tools developed in here, so we include it.

1.1. Preliminary definitions. We will say that (Ω, g) is a Riemannian domain if it is a connected, open subset of a Riemannian manifold (M, g), having the property that its metric completion $\overline{\Omega}$ is a compact subset of M. For $x, y \in \Omega$ we will denote the distance between x and y (on (M, g)) by $|x - y|$.

Define

$$\Omega_\varepsilon = \{x \in \Omega | \text{dist}(x, \partial\Omega) > \varepsilon\}.$$

For $x \in \Omega$, $v \in T_x\Omega$, let $\exp(x, v)$ denote the (exponential) tangent map, i.e. $\exp(x, v) = \gamma(1)$, where γ is the constant-speed geodesic satisfying $\gamma(0) = x$, $\gamma'(0) = v$.

If Z is a Lipschitz vector field on $\overline{\Omega}$, we will write $Z \in \Gamma(T\overline{\Omega})$. In analogy to the exponential map, write $\overline{x}(x, t)$ for the flow induced by

Z, i.e. $\overline{x}(x,t) = \gamma(t)$ where γ solves

$$\begin{cases} \frac{d}{dt}\gamma = Z(\gamma(t)) \\ \gamma(0) = x. \end{cases}$$

Write

$${}^Z\Omega_\varepsilon \equiv \{x \in \Omega \mid \operatorname{dist}(x, \partial\Omega) > \varepsilon |Z|_\infty\}.$$

If (Ω, g) is a Riemannian domain, if (X, d) is a complete metric space, and if $1 \leq p < \infty$, then there is a natural definition of the space $L^p(\Omega, X)$. It is the set of Borel-measurable functions $u : \Omega \to X$ having separable range for which

$$\int_\Omega d^p(u(x), Q) d\mu_g(x) < \infty$$

for some $Q \in X$.

If u and v are two Borel-measurable functions with separable range in X, then the pair $(u(x), v(x))$ is a measurable function to $X \times X$. Thus $d^p(u(x), v(x)) : \Omega \to X \times X \to \mathbb{R}$ is a measurable function. Thus integrals of $d^p(u(x), v(x))$ are well-defined, of which the one above is an example ([F] 2.3.2). It is straightforward to show that $L^p(\Omega, X)$ is a complete metric space, with distance function D defined by

$$D^p(u, v) = \int_\Omega d^p(u(x), v(x))\, d\mu_g(x).$$

The proof of this fact follows from the considerations in ([F] 2.4.12), but we quickly sketch the ideas. The triangle inequality for d, combined with that for real-valued L^p functions implies that if u is L^p, then

$$\int_\Omega d^p(u(x), P)\, d\mu_g(x) < \infty$$

for any $P \in X$. Another application of these two triangle inequalities shows that $D^p(u, v)$ is finite whenever $u, v \in L^p(\Omega, X)$. The triangle inequality for D follows by the same argument. The proof that $L^p(\Omega, X)$ is complete mimics the usual proof for real-valued functions: Given a Cauchy sequence one finds a subsequence which converges *a.e.*, shows the limit function is in $L^p(\Omega, X)$, and then that the sequence converges to the limit function in the D metric.

1.2. Approximate energies. Fix $1 \le p < \infty$ and $u \in L^p(\Omega, X)$. Let $V \in \Gamma(T\overline{\Omega})$ be a smooth vector field on $\overline{\Omega}$. Then for $\varepsilon > 0$ small the map $y = \exp(x, \varepsilon V)$ is a diffeomorphism between ${}^V\Omega_\varepsilon$ and its image, approaching the identity map as $\varepsilon \to 0$. Therefore the map $x \to u(\exp(x, \varepsilon V)$ is in $L^p({}^V\Omega_\varepsilon, X)$ and we have the estimate

(1.2i)
$$\int_{{}^V\Omega_\varepsilon} d^p(u(x), u(x), \exp(x, \varepsilon V)))d\mu_g(x) \le C,$$

independently of ε.

Now, for $(x, y) \in \Omega \times \Omega$ define

$$e_\varepsilon(x, y) = \frac{d^p(u(x), u(y))}{\varepsilon^p}.$$

For $x \in \Omega_\varepsilon$ define

$$S(x, \varepsilon) = \{y \text{ s.t. } |y - x| = \varepsilon\}$$

$$d\sigma_{x,\varepsilon}(y) = (n-1) - \text{dimensional surface measure on } S(x, \varepsilon).$$

Finally, define the (spherically averaged) ε-approximate energy density function for $x \in \Omega_\varepsilon$, by

(1.2ii)
$$e_\varepsilon(x) = \int_{S(x,\varepsilon)} e_\varepsilon(x, y)\frac{d\sigma_{x,\varepsilon}(y)}{\varepsilon^{n-1}}.$$

(Define $e_\varepsilon(x)$ to be zero otherwise .) We claim that e_ε is a real-valued L^1-function, with

(1.2iii)
$$\int_{\Omega_\varepsilon} e_\varepsilon(x) d\mu(x) \le C\varepsilon^{-p}.$$

To see this we reduce (by standard partition of unity arguments) to the case where $\overline{\Omega}$ has a global orthonormal frame $\{e_1, \dots e_n\}$. Identify $\omega = \omega^i\partial_i \in S^{n-1}(0, 1) \subset \mathbb{R}^n$ with $\omega^i e_i \in S(0, 1)_x \in T\Omega_x$. Then the map

$$(x, \omega) \longrightarrow \exp(x, \varepsilon\omega) \longrightarrow u(\exp(x, \varepsilon\omega))$$

is measurable and is easily seen to be in $L^p(\Omega_\varepsilon \times S(0,1))$. In fact, by Tonelli's and Fubini's Theorems, and by applying the estimate (1.2i) we have

$$\int_{\Omega_\varepsilon} \int_{S(0,1)} e_\varepsilon(x, \exp(x, \varepsilon\omega)) d\sigma(\omega) d\mu(x) \leq C\varepsilon^{-p}$$

(for some other generic constant C). For $y = \exp(x, \varepsilon\omega)$ we note that

$$\frac{d\sigma_{x,\varepsilon}(y)}{\varepsilon^{n-1} d\sigma(\omega)}$$

is a continuous bounded function of (x, ω), independently of ε. Hence we may multiply the integrand above by this factor and maintain measurability. Furthermore we may deduce

$$\int_{\Omega_\varepsilon} \int_{S(x,\varepsilon)} e_\varepsilon(x, y) \frac{d\sigma_{x,\varepsilon}(y)}{\varepsilon^{n-1}} d\mu(x) \leq C\varepsilon^{-p}.$$

This verifies (1.2iii).

It is convenient to work with a variety of average energies. Let ν be any Borel measure on the interval $(0, 2)$ satisfying

$$\text{(1.2iv)} \qquad \nu \geq 0, \qquad \nu((0,2)) = 1, \qquad \int_0^2 \lambda^{-p} d\nu(\lambda) < \infty.$$

Define the approximate energy density function ${}_\nu e_\varepsilon(x)$ by averaging the spherical averages $e_\varepsilon(x)$:

$$\text{(1.2v)} \qquad {}_\nu e_\varepsilon(x) = \int_0^2 e_{\lambda\varepsilon}(x) d\nu(\lambda),$$

for $x \in \Omega_{2\varepsilon}$ (and ${}_\nu e_\varepsilon(x) = 0$ otherwise). It is easy to see that ${}_\nu e_\varepsilon(x)$ is measurable, and from (1.2iii) and the integrability requirement in (1.2iv) we can estimate

$$\text{(1.2vi)} \qquad \int_{\Omega_{2\varepsilon}} {}_\nu e_\varepsilon(x) d\mu(x) \leq C\varepsilon^{-p}.$$

A particular choice of ν which we will have occasion to use (besides the choice $\nu = \delta(1)$ which corresponds to our original energy density approximant) is the one which leads to uniform ball averaging, namely

$$d\nu(\lambda) = (n+p)\lambda^{n+p-1}d\lambda \qquad 0<\lambda<1. \tag{1.2vii}$$

For $u:\Omega\to\mathbb{R}$ smooth it is easy to see from our definitions that

$$\begin{aligned}\lim_{\varepsilon\to 0} {}_\nu e_\varepsilon(x) &= c_{n,p}\,|\nabla u(x)|^p \\ c_{n,p} &= \int_{S^{n-1}} |x^1|^p\,d\sigma(x),\end{aligned} \tag{1.2viii}$$

(where $x=(x^1,\dots x^n)\in\mathbb{R}^n$ and $S^{n-1}=\{|x|=1\}$). In particular, $c_{n,2}=\omega_n$. If $u:\Omega\to N^k$ is a smooth map between Riemannian manifolds, then one can check that for $p=2$, $e_\varepsilon(x)\to\omega_n|\nabla u(x)|^2$ as well. For $p\neq 2$, however, e_ε does not converge to a multiple of $|\nabla u(x)|^p\equiv(|\nabla u(x)|^2)^{p/2}$, when $k>1$.

1.3. The functionals ${}_\nu E_\varepsilon$. Let $1\le p<\infty$, $u\in L^P(\Omega,X)$, and ν as in (1.2iv). Then for $\varepsilon>0$ and $f\in C_c(\Omega)$ define

$${}_\nu E_\varepsilon(f) = \int_\Omega f(x)\,{}_\nu e_\varepsilon(x)\,d\mu(x). \tag{1.3i}$$

(In case $\nu=\delta(1)$ we suppress the ν in our notation.) We say that u has finite energy (and write $u\in W^{1,p}(\Omega,X)$ for $p>1$ and $u\in BV(\Omega,X)$ for $p=1$), if for some ν as above

$$\sup_{\substack{f\in C_c(\Omega)\\ 0\le f\le 1}}\left(\limsup_{\varepsilon\to 0}\ {}_\nu E_\varepsilon(f)\right) \equiv\ {}_\nu E\ <\infty. \tag{1.3ii}$$

(We show in Lemma 1.4.1 that the expression above is finite or infinite independently of ν.) Let Ω be a locally compact metric space. For $f\in C_c(\Omega)$ define

$$\begin{aligned}|f| &= \max_{x\in\Omega}|f(x)| \\ \omega(f,\varepsilon)(x) &= \max_{|y-x|\le\varepsilon}|f(y)-f(x)| \\ \omega(f,\varepsilon) &= \max_x\,\omega(f,\varepsilon)(x).\end{aligned} \tag{1.3iii}$$

For $C > 0$ define

$$f_\varepsilon^C(x) = (1 + C\varepsilon)\,(f(x) + \omega(f, 2\varepsilon)(x)). \tag{1.3iv}$$

We now prove a basic "subpartitioning lemma" which will essentially guarantee the existence of energy-density measures for finite energy maps. It is the integral analog of the (trivial) lemma (for curves into metric spaces) that the approximate length of a curve with respect to a partition increases if the partition is refined. This monotone property is the essential ingredient in showing that rectifiable curves into metric spaces have well-defined length measures, and the lemma below will play the same role here.

Lemma 1.3.1. *Let $1 \le p < \infty$, $u \in L^p(\Omega, X)$, ν as in (1.2iv). Then for $f \in C_c(\Omega)$, $f \ge 0$, there exists a constant $C > 0$ (depending only on Ricci curvature control of the metric g), so that the following "subpartition estimate" holds for all sufficiently small $\varepsilon > 0$:*

$$ {}_\nu E_\varepsilon(f)^{1/p} \le \sum_i \lambda_i \left({}_\nu E_{\lambda_i \varepsilon}(f_\varepsilon^C)\right)^{1/p}. \tag{1.3v}$$

Here

(1.3vi) $\sum \lambda_i = 1$, *each* $\lambda_i > 0$, *and the sum is a finite one.*

Proof. We first consider the case $\nu = \delta(1)$, i.e. e_ε in (1.2ii). For $\varepsilon > 0$ small we may write $E_\varepsilon(f)$ more symmetrically as an integral over $\Omega \times \Omega$:

$$E_\varepsilon(f) = \iint\limits_{|x-y|=\varepsilon} f(x) e_\varepsilon(x,y)\, d\sigma_\varepsilon(x,y), \tag{1.3vii}$$

where $d\sigma_\varepsilon(x,y)$ is normalized $(2n-1)$-dimensional surface measure on $\{|x-y| = \varepsilon\}$, which can be expressed in terms of $d\sigma_{x,\varepsilon}(y)$ from (1.2ii):

$$d\sigma_\varepsilon(x,y) \equiv \frac{d\sigma_{x,\varepsilon}(y)}{\varepsilon^{n-1}}\, d\mu(x) = \frac{d\sigma_{y,\varepsilon}(x)}{\varepsilon^{n-1}}\, d\mu(y). \tag{1.3viii}$$

Let $\{\lambda_1 \dots, \lambda_\ell\}$ satisfy (1.3vi). For any x in the support of f, and for ε sufficiently small, there is a unique geodesic from x to any y with

$|x - y| = \varepsilon$. Let $\varphi : [0,1] \to \Omega$ be the (constant-speed) geodesic path from x to y. Define the subdivision

$$x_0 = x$$
$$x_i = \varphi\left(\sum_{k=1}^{i} \lambda_k\right), \qquad i = 1, \dots, \ell.$$

Then

$$|x_i - x_{i-1}| = \lambda_i \varepsilon,$$

and the X-triangle inequality implies

$$d(u(x),\, u(y)) \le \sum_{i=1}^{n} d(u(x_{i-1}), u(x_i)).$$

The standard (iterated) L^p-triangle inequality (applied to $(fe_\varepsilon)^{1/p}$ in (1.3vi)) yields

(1.3ix)

$$E_\varepsilon(f)^{1/p} \le \sum_{i=1}^{l} \lambda_i \left(\iint\limits_{|x_i - x_{i-1}| = \lambda_i \varepsilon} f(x)\; e_{\lambda_i \varepsilon}(x_{i-1}, x_i)\; d\sigma_\varepsilon(x,y) \right)^{1/p}.$$

For each fixed i we may change variables from (x,y) to (x_{i-1}, x_i) in the corresponding term of the above inequality. It is easy to show that

$$d\sigma_\varepsilon(x,y) \le (1 + C\varepsilon)\, d\sigma_{\lambda_i \varepsilon}(x_{i-1}, x_i), \tag{1.3x}$$

where the constant C depends on Ricci curvature control of g. (If Ω is Euclidean, the last inequality is an equality with $C = 0$.) We dominate $f(x)$ by $f(x_{i-1}) + \omega(f,\varepsilon)(x_{i-1})$ in (1.3ix) and also use (1.3x) to deduce the sub-partition estimate (1.4i) in this case.

The proof for general ν is essentially the same. We may write

$${}_\nu E_\varepsilon(f) = \int_0^2 \iint\limits_{|x-y|<2\varepsilon} f(x) e_\varepsilon(x,y)\, d\sigma_{\lambda\varepsilon}(x,y)\, \lambda^{-p} d\nu(\lambda).$$

One derives the general version of (1.3ix) by repeating the geodesic subdivision detailed above, this time for arbitrary $x \in \operatorname{supp}(f)$ with

$|x-y| < 2\varepsilon$. The result is

$$\text{(1.3xi)} \quad {}_{\nu}E_{\varepsilon}(f)^{1/p} \leq$$

$$\sum \lambda_i \left(\int_0^2 \iint_{|x_i - x_{i-1}| < 2\varepsilon\lambda_i} f(x)\, e_{\lambda_i \varepsilon}(x_{i-1}, x_i)\, d\sigma_{\lambda\varepsilon}(x,y)\, \lambda^{-p} d\nu(\lambda) \right)^{1/p} .$$

Noting that

$$d\sigma_{\lambda\varepsilon}(x,y) \leq (1 + C\varepsilon) d\sigma_{\lambda_i \lambda\varepsilon}(x_{i-1}, x_i),$$

we may proceed as above to conclude (1.4i) in this general case. □

1.4. Functional analysis lemmas. We prove two lemmas which enable us to conclude the existence of the energy-density measures for finite-energy maps. We will use these lemmas again when we discuss directional energies.

Lemma 1.4.1. *Let Ω be a locally compact metric space. Let $\{\mathcal{L}_\varepsilon\}_{0<\varepsilon\leq\varepsilon_0}$ be a family of positive linear functionals on $C_c(\Omega)$. Let $1 \leq p < \infty$. Suppose there exists a $C > 0$ so that for $f \in C_c(\Omega), f \geq 0$, for the function f_ε^C defined in (1.3iv), and for any $\{\lambda_i\}$ satisfying (1.3vi) the subpartition inequality below holds when $\varepsilon > 0$ is small:*

$$\text{(1.4i)} \qquad (\mathcal{L}_\varepsilon(f))^{1/p} \leq \sum_i \lambda_i \left(\mathcal{L}_{\lambda_i \varepsilon}(f_\varepsilon^C)\right)^{1/p} .$$

Let ν be a non-negative Borel measure satisfying (1.2iv). Assume

$$\text{(1.4ii)} \qquad {}_{\nu}\mathcal{L}_\varepsilon(f) \equiv \int_0^2 \mathcal{L}_{\rho\varepsilon}(f) d\nu(\rho)$$

is well-defined (i.e. $\mathcal{L}_\varepsilon(f)$ is Borel measurable in ε for any $f \in C_c(\Omega)$). Assume

$$\sup_{\substack{f \in C_c(\Omega) \\ 0 \leq f \leq 1}} \left(\limsup_{\varepsilon \to 0} \ {}_{\nu}\mathcal{L}_\varepsilon(f) \right) \equiv \ {}_{\nu}L < \infty.$$

Then also,

$$\text{(1.4iii)} \qquad \sup_{\substack{f \in C_c(\Omega) \\ 0 \leq f \leq 1}} \left(\limsup_{\varepsilon \to 0} \ \mathcal{L}_\varepsilon(f) \right) \equiv L < \infty.$$

Proof. By hypothesis, if $f \geq 0$, $f \in C_c(\Omega)$ then for $\varepsilon > 0$ small

$$_\nu\mathcal{L}_\varepsilon(f) = \int_0^2 \mathcal{L}_{\rho\varepsilon}(f)d\nu(\rho) \leq C|f|,$$

where C is independent of f. Hence

$$\int_0^1 \int_0^2 \mathcal{L}_{\mu\rho\varepsilon}(f)d\nu(\rho)d\mu \leq C|f|,$$

i.e.

$$\int_0^2 \left(\int_0^1 \mathcal{L}_{\mu\rho\varepsilon}(f)d\mu \right) d\nu(\rho) \leq C|f|.$$

Now, there is a $\delta > 0$ so that $\nu((\delta, 2)) > 1/2$. Thus there is a $\rho \in (\delta, 2)$, with

$$\int_0^1 \mathcal{L}_{\mu\rho\varepsilon}(f)d\mu \leq 2C|f|.$$

Writing $\varepsilon' = \varepsilon\delta$, we have

$$\int_0^1 \mathcal{L}_{\mu(\rho/\delta)\varepsilon'}(f)d\mu \leq 2C|f|.$$

Changing variables to $\mu' = \mu\,(\rho/\delta)$, noting $1 \leq \rho/\delta \leq {}^2/\delta$, yields

$$\text{(1.4iv)} \qquad \int_0^1 \mathcal{L}_{\mu'\varepsilon'}(f)d\mu' \leq \frac{4C}{\delta}\,|f|,$$

for $f \in C_c(\Omega)$ and ε' small (depending on $\text{supp}(f)$).

Now fix $f \in C_c(\Omega)$, $0 \leq f \leq 1$. We have from (1.4i)

$$\mathcal{L}_\varepsilon(f) \leq 2^p\mathcal{L}_{\lambda\varepsilon}(f_\varepsilon^C) + 2^p\mathcal{L}_{(1-\lambda)\varepsilon}\,(f_\varepsilon^C),$$

for small ε. In particular,

$$\text{(1.4v)} \qquad \mathcal{L}_\varepsilon(f) \leq 2^{p+1} \int_0^1 \mathcal{L}_{\lambda\varepsilon}\,(f_\varepsilon^C)d\lambda.$$

For ε small the functions f_ε^C have uniformly-bounded supremum and uniform compact support. Hence we use (1.4iv) for the function f_ε^C to estimate the integral in (1.4v), obtaining

$$\mathcal{L}_\varepsilon(f) \leq C'. \quad \square$$

Lemma 1.4.2. *Let Ω be a locally compact metric space. Let $\{\mathcal{L}_\varepsilon\}_{0<\varepsilon\leq\varepsilon_0}$ be a family of positive linear functionals on $C_c(\Omega)$. Let $1 \leq p < \infty$. Suppose there exists a $C > 0$ so that for $f \in C_c(\Omega), f \geq 0$, for the function f_ε^C defined in (1.3iii), and for any $\{\lambda_i\}$ satisfying (1.3vi) the subpartition inequality (1.4i) holds when $\varepsilon > 0$ is small. Suppose also that the boundedness hypothesis (1.4iii) holds. Then*

(1.4vi) $$\lim_{\varepsilon\to 0} \mathcal{L}_\varepsilon(f) \equiv \mathcal{L}(f)$$

exists for any $f \in C_c(\Omega)$, and defines a positive linear functional $\mathcal{L}$ with $\|\mathcal{L}\| = L$. Furthermore, we have the inequality

(1.4vii) $$\mathcal{L}_\varepsilon(f) \leq \mathcal{L}(f_\varepsilon^C)$$

when $f \in C_c(\Omega), f \geq 0$, and $\varepsilon > 0$ is sufficiently small.

Proof. Let $\varepsilon' \ll \varepsilon$, and write [ ] for the greatest integer function. Let $\lambda_i = \left(\frac{\varepsilon'}{\varepsilon}\right)$ for $i = 1, \ldots k$, where $k = [\frac{\varepsilon}{\varepsilon'}]$. Let $\lambda_{k+1} = 1 - \left(\frac{\varepsilon'}{\varepsilon}\right) k$ if necessary. Applying (1.4i) and using (1.6i) for sufficiently small ε, we estimate

(1.4viii)
$$(\mathcal{L}_\varepsilon(f))^{1/p} \leq [\frac{\varepsilon}{\varepsilon'}] \left(\frac{\varepsilon'}{\varepsilon}\right) (\mathcal{L}_{\varepsilon'}(f_\varepsilon^C))^{1/p} + \left(\frac{\varepsilon'}{\varepsilon}\right) ((L+1)\,|f_\varepsilon^C|)^{1/p}.$$

Letting $\varepsilon' \to 0$ we see

(1.4ix) $$\mathcal{L}_\varepsilon(f) \leq \liminf_{\varepsilon'\to 0} \mathcal{L}_{\varepsilon'}(f_\varepsilon^C).$$

Thus by the boundedness hypothesis (1.4iii) we see that

$$\mathcal{L}_\varepsilon(f) \leq \liminf_{\varepsilon'\to 0} \mathcal{L}_{\varepsilon'}(f) + |L|\,(C\varepsilon|f| + (1 + C\varepsilon)\omega(f,\varepsilon)),$$

so that

$$\limsup_{\varepsilon\to 0} \mathcal{L}_\varepsilon(f) \leq \liminf_{\varepsilon'\to 0} \mathcal{L}'_\varepsilon(f).$$

Thus (1.4vi) holds for $f \geq 0$. The statement for arbitrary $f \in C_c(\Omega)$ follows. It is clear $\mathcal{L}$ is a positive linear functional with $\|\mathcal{L}\| = L$. The estimate (1.4vii) follows from (1.4viii), as $\varepsilon' \to 0$. □

Remark 1.4.3. *1.4.3. If the functionals $\mathcal{L}_\varepsilon(f)$ vary continuously in ε for any $f \in C_c(\Omega)$, then we may deduce (1.4ix) using only the subpartition estimate: (1.4i) immediately implies (for $\varepsilon' < \varepsilon$)*

$$\mathcal{L}_{[\frac{\varepsilon}{\varepsilon'}]\varepsilon'}(f) \leq \mathcal{L}_{\varepsilon'}(f_\varepsilon^C) \tag{1.4x}$$

which gives (1.4ix) as $\varepsilon' \to 0$. (Without the boundedness assumption (1.4iii), of course, both of these numbers may be infinite.) Note that if ν satisfies (1.2iv) and is absolutely continuous with respect to λ, then the functionals ${}_\nu E_\varepsilon$ (1.3i) satisfy the continuity hypothesis. We will use of this fact when discussing directional energies.

1.5. The energy-density measure.

Theorem 1.5.1. *Let $1 \leq p < \infty$, $u \in L^p(\Omega, X)$ have finite energy ${}_{\nu_1}E$ with respect to some measure ν_1 satisfying (1.2iv). Then it has finite energy with respect to all such ν, and each measure ${}_\nu e_\varepsilon(x)d\mu(x)$ converges weakly to the same "energy density" measure de, having total mass ${}_{\nu_1}E$. Furthermore, the estimate (1.4vii) holds for $\mathcal{L}_\varepsilon = {}_\nu E_\varepsilon$ and $\mathcal{L} = E$.*

Proof. By Lemma 1.4.1 u has finite energy with respect to the standard measure, $\nu = \delta(1)$. Hence its energy is finite with respect to any ν satisfying (1.2iv). From Lemmas 1.3.1,1.4.1,1.4.2, and the Riesz representation theorem for continuous linear functionals on $C_c(\Omega)$ we deduce that each ${}_\nu e_\varepsilon(x)d\mu(x)$ converges weakly to a limit measure. The definition (1.2v) of ${}_\nu E_\varepsilon$ shows that the limit measure is the one arising from $\nu = \delta(1)$, i.e the weak limit of $e_\varepsilon(x)d\mu(x)$. The fact that (1.4vii) holds as indicated is clear. □

Remark 1.5.2. *1.5.2. For $1 < p < \infty$, $u \in W^{1,p}(\Omega, X)$, it is easy to see that $u \in W^{1,p'}(\Omega, X)$, for any $1 \leq p' < p$, and that the p'-energy density measure $de_{p'}$ is absolutely continuous with respect to Lebesgue measure. (We will say that $u \in W^{1,1}(\Omega, X)$ if $u \in BV(\Omega, X)$ and its energy density de is absolutely continuous with respect to Lebesgue* *measure.) In fact, if E is the total (Sobolev) p-energy of u, then there is a constant C (depending only on dimension n) so that for any Borel measurable set $S \subset \Omega$,*

$$e_{p'}(S) \leq C(\mu_g(S))^{\frac{p-p'}{p}} E^{\frac{p'}{p}}.$$

To see why this is true, it suffices to consider the case $S \subset\subset \Omega$. In this case, for any $\delta > 0$ we may pick $f \in C_c(\Omega)$, $0 \leq f \leq 1$, with $f \equiv 1$ on S and $\mu_g(supp(f) \leq \mu_g(S) + \delta$. Then by Hölder

$$\begin{aligned} E_{\varepsilon,p'}(f) &= \int\limits_{\Omega} f\, e_{\varepsilon,p'}(x) d\mu(x) \\ &\leq C \int\limits_{\Omega} \left(f^{\frac{p}{p'}} e_{\varepsilon,p}(x) d\mu(x) \right)^{\frac{p'}{p}} (\mu(supp(f)))^{\frac{p-p'}{p}}. \end{aligned}$$

Here we have used $E_{\varepsilon,p}(f)$ and $e_{\varepsilon,p}(x)$ to denote the approximate energy functional and ε-approximate energy function, corresponding to the power p. As $\varepsilon \to 0$ we conclude

$$e_{p'}(S) \leq E_{p'}(f) \leq C E^{\frac{p'}{p}} (\mu(S) + \delta)^{\frac{p-p'}{p}},$$

which shows our claim as $\delta \to 0$.

1.6. Lower semicontinuity, and consistency when $X = \mathbb{R}$.

Theorem 1.6.1. *Let $1 \leq p < \infty, \{u_k\} \subset W^{1,p}(\Omega, X)$ if $p > 1$ (or $\{u_k\} \subset BV(\Omega, X)$ if $p = 1$). Let $u_k \to u$ in the L^p metric. Write e^k for the energy-density measure of u_k, $E^k \equiv e^k(\Omega)$. Assume there exists $E < \infty$ with each $E^k < E$. Then $u \in W^{1,p}(\Omega, X)$ if $p > 1$ (or $u \in BV(\Omega, X)$ if $p = 1$), and its energy-density measure e satisfies*

$$de \leq \liminf\, de^k$$

as measures.

Proof. For $d\nu(\lambda) = (n + p)\lambda^{n+p-1}\, d\lambda$ (ball averages (1.2vii)) and for fixed $f \in C_c(\Omega)$ we have as a consequence of Lemma 1.3.1, (1.4vii) of

Lemma 1.4.2, and Theorem 1.5.1, that

$$(1.6i) \qquad {}_\nu E^k_\varepsilon(f) \leq E^k(f^C_\varepsilon) = E^k(f) + E^k(f^C_\varepsilon - f).$$

Because we are using ball averages thc energy-density functions ${}_\nu e^k_\varepsilon(x)$ converge uniformly to ${}_\nu e_\varepsilon(x)$ as $k \to \infty$ (and $\varepsilon > 0$ is fixed). Thus

$$(1.6ii) \qquad \lim_{k\to\infty} {}_\nu E^k_\varepsilon(f) = {}_\nu E_\varepsilon(f).$$

Hence

$${}_\nu E_\varepsilon(f) \leq \liminf_{k\to\infty} E^k(f) + E(C\varepsilon|f| + \omega(f, 2\varepsilon)).$$

Thus u has finite energy and if we let $\varepsilon \to 0$ we also deduce that

$$E(f) \leq \liminf_{k\to\infty} E^k(f). \quad \square$$

Theorem 1.6.2.

$$W^{1,p}(\Omega, \mathbb{R}) = W^{1,p}(\Omega) \quad for \quad p > 1,$$
$$BV(\Omega, \mathbb{R}) = BV(\Omega).$$

Furthermore, the energy densities are constant multiples of each other in all cases.

Proof. We first assume (Ω, g) is Euclidean. If $u \in L^p(\Omega, \mathbb{R})$ is smooth, then $e_\varepsilon(x) \to c_{n,p}|\nabla u(x)|^p$ uniformly on compact subsets of Ω, so it is clear that the measure de equals $c_{n,p}|\nabla u(x)|^p d\mu_x$. If $u \in L^p(\Omega, \mathbb{R})$ then we mollify it with a C^∞ approximate identity:

$$\eta_t(x) = t^{-\eta}\eta(\frac{x}{t})$$
$$\eta(x) \in C_0^\infty(B(0,1)), \quad \eta \geq 0, \quad \int_{B(0,1)} \eta = 1$$
$$\eta(x) = \varphi(|x|).$$

We write $u_t = u * \eta_t$.

It is well-known that if $u \in W^{1,p}(\Omega)$ then $u_t \to u$ in $W^{1,p}_{loc}(\Omega)$, i.e. on any compact subset the convergence is in the $W^{1,p}$-norm [GT]. In particular, for $p > 1$ we have that $|\nabla u_t|^p d\mu \rightharpoonup |\nabla u|^p d\mu$ (weakly) as measures. By (the usual) Sobolev-Space semicontinuity, it is also true that $\int_{\Omega_t} |\nabla u_t|^p d\mu \to \infty$ in case $u \in L^p(\Omega)$ but $u \notin W^{1,p}(\Omega)$. In case $p = 1$ it is true that $u \in BV(\Omega)$ implies $|\nabla u_t| d\mu$ converges weakly to the BV-measure $|\nabla u|$, and that $u \notin BV(\Omega)$ implies $\int_{\Omega_t} |\nabla u_t| d\mu \to \infty$)[Gi]. Thus Theorem 1.6.2 follows if we can show the corresponding statements for our energy densities.

If $u \in L^p(\Omega, \mathbb{R})$ and u does not have finite energy (in our sense) then it must follow that $\int_{\Omega_t} |\nabla u_t|^p d\mu \to \infty$, since otherwise lower semicontinuity (1.6i) would be violated by choosing an appropriate subsequence. Thus we need only show that if u is a finite-energy map, then $c_{n,p}|\nabla u_t|^p d\mu = de^{u_t} \rightharpoonup de$. From our comments above, and from Theorem 1.6.1, we know that

$$de^u \leq \liminf_{t \to 0} de^{u_t}, \tag{1.6iii}$$

so we only need the reverse inequality. We use $d\nu(\lambda) = (n+p)\lambda^{n+p-1} d\lambda$, i.e. ball averages, for the approximate energies and we take $f \in C_c(\Omega)$, $f \geq 0$. Then

$$\begin{aligned} {}_\nu E_\varepsilon^{u_t}(f) &= \int_\Omega f(x) {}_\nu e_\varepsilon^{u_t}(x) dx \\ &= (n+p) \int_\Omega \int_{B(0,1)} f(x) \frac{|u^t(x) - u^t(x+\varepsilon v)|^p}{\varepsilon^p} dv dx. \end{aligned}$$

By properties of convolution and Jensen's inequality

$$\begin{aligned} |u^t(x) - u^t(x+\varepsilon v)|^p &\leq |u(\cdot) - u(\cdot + \varepsilon v)|^p * \eta_t \\ &= \int_{B(0,1)} |u(x - tw) - u(x - tw + \varepsilon v)|^p \eta(w)\, dw. \end{aligned}$$

Hence

$$\begin{aligned} &{}_\nu E_\varepsilon^{u_t}(f) \leq \\ &\frac{n+p}{\varepsilon^p} \int_{B(0,1)} \int_{B(0,1)} \int_\Omega f(x) \eta(w) |u(x - tw) - u(x - tw + \varepsilon v)|^p \, dx dv dw. \end{aligned}$$

Letting $z = x - tw$, estimating $|f(x) - f(z)| \leq \omega(f,t)$, yields

$$ {}_\nu E^{u_t}_\varepsilon(f) \leq \frac{n+p}{\varepsilon^p} \iiint f(z)\eta(w)|u(z) - u(z+\varepsilon v)|^p \, dzdvdw + C\omega(f,t), $$

where C depends on the energy of u (for ε small). That is

$$ {}_\nu E^{u_t}_\varepsilon(f) \leq {}_\nu E^{u}_\varepsilon(f) + C\omega(f,t), $$

so $(\varepsilon \to 0)$,

$$ E^{u_t}(f) \leq E^u(f) + C\omega(f,t). $$

Hence

$$ \limsup_{t\to 0} \; E^{u_t}(f) \leq E^u(f), $$

i.e.

(1.6iv) $$ \limsup_{t\to 0} \; de^{u_t} \leq de^u. $$

In case (Ω, g) is a general Riemannian domain, we proceed as above and eventually are left to verify (1.6iv), which we do as follows. Express Ω as a finite union of open subsets, each of which have approximately Euclidean local coordinate charts. Use a subordinate partition of unity $\{\eta_i\}$ to express $E(f)$ as a sum of $E(\eta_i(f))$. One may then work as above to estimate ${}_\nu E^{u_t}_\varepsilon(\eta_i f)$, introducing a small error term into (1.6iv) from the fact that the metric isn't exactly Euclidean. This error term approaches zero as one takes subsets with diameter approaching zero. Thus one can verify (1.6iv) in the general case as well. □

Corollary 1.6.3. *Let $1 \leq p < \infty$, $u, v \in W^{1,p}(\Omega, X)$ (or $u, v \in BV(\Omega, X)$ if $p = 1$). Define the p-distance on $X \times X$, by*

$$ d^p((x_1,x_2),\,(y_1,y_2)) = d^p(x_1,y_1) + d^p(x_2,y_2). $$

Let $f : X \times X \to \mathbb{R}$ be Lipschitz continuous, with Lipschitz constant L (with respect to D). Then the map $h(x) = f(u(x), v(x))$ satisfies $h \in W^{1,p}(\Omega)$ (or $h \in BV(\Omega)$ if $p = 1$), and

(1.6v) $$ c_{n,p}|\nabla h|^p d\mu \leq L^p(de^u + de^v) $$

as measures.

Proof. The ε-energy density for h is an average (with respect to y) of

$$\frac{|h(x)-h(y)|^p}{\varepsilon^p} = \frac{|f(u(x),v(x)) - f(u(y),v(y))|^p}{\varepsilon^p}$$
$$\leq L^p \left(\frac{d^p(u(x),u(y)) + d^p(v(x),v(y))}{\varepsilon^p} \right).$$

The result follows immediately. A special case (which is easy to check directly) is $f(p,q) = d(p,q)$, whence if $u,v \in W^{1,p}(\Omega,X)$ (*resp.* $BV(\Omega,X)$) we have $d(u,v) \in W^{1,p}(\Omega,\mathbb{R})$ (*resp.* $BV(\Omega)$), and

(1.6vi) $$c_{n,p}|\nabla d(u,v)|^p d\mu \leq 2^p(de^u + de^v).$$

If $v \equiv P$, we can argue directly as in (1.6v) and remove the factor of 2^p to get

(1.6vii) $$c_{n,p}|\nabla d(u,P)|^p d\mu \leq de^u.$$

□

1.7. Directional energies. It is natural to define directional energies for maps $u \in L^p(\Omega,X)$, that is, measures of their rate of change in directions of smooth vector fields. For finite-energy maps ($u \in W^{1,p}(\Omega)$ or $u \in BV(\Omega)$) it will turn out that these directional energy densities are always well-defined measures, and the resulting calculus will be of use not only in this chapter, but in the later sections of the paper as well. Let $1 \leq p < \infty$, $u \in L^p(\Omega,X)$. Let Z be a Lipschitz vector field on $\overline{\Omega}$, $Z \in \Gamma(T\overline{\Omega})$, and recall the definitions $\overline{x}(x,t)$ and ${}^Z\Omega_\varepsilon$ from 1.1. For $x \in {}^Z\Omega_\varepsilon$ define the ε energy-density function (of u in the direction Z) by

(1.7i) $${}^Ze_\varepsilon(x) = \frac{d^p(u(x),u(\overline{x}(x,\varepsilon)))}{\varepsilon^p}.$$

(For $x \notin {}^Z\Omega_\varepsilon$ define ${}^Ze_\varepsilon(x) = 0$.) As in 1.1 we deduce ${}^Ze_\varepsilon \in L^1(\Omega,\mathbb{R})$, with

$$\int_\Omega {}^Ze_\varepsilon(x)d\mu(x) \leq C\varepsilon^{-p}.$$

For a non-negative Borel measure ν satisfying (1.2iv) we write (suppressing the Z-dependence here and below)

$$ {}_{\nu}e_{\varepsilon}(x) = \int\limits_{(0,2)} e_{\rho\varepsilon}(x)d\nu(\rho), \tag{1.7ii} $$

for $x \in {}^{Z}\Omega_{2\varepsilon}$ (${}_{\nu}e_{\varepsilon}(x) = 0$ otherwise). Clearly ${}_{\nu}e_{\varepsilon} \in L^{1}(\Omega, \mathbb{R})$ as well. For $f \in C_c(\Omega)$ write

$$ \begin{aligned} {}_{\nu}E_{\varepsilon}(f) &= \int_{\Omega} f(x)\, {}_{\nu}e_{\varepsilon}(x)d\mu(x) = \int_{(0,2)} {}_{\nu}E_{\rho\varepsilon}(f)d\nu(\rho), \\ {}_{\nu}E &= \sup_{\substack{0\leq f\leq 1 \\ f\in C_c(\Omega)}} \limsup_{\varepsilon\to 0} {}_{\nu}E_{\varepsilon}(f). \end{aligned} \tag{1.7iii} $$

In case some ${}_{\nu}E < \infty$ we say u has finite ($p-$) energy (in the direction Z). The analog of Theorem 1.7 for Sobolev (or BV) energies is the following result for directional energies:

Theorem 1.7. *Let* $1 \leq p < \infty$, $u \in L^{p}(\Omega, X)$, Z *as above. If for some* ν_1 *as above we have* ${}_{\nu_1}E < \infty$, *then* ${}_{\nu}E < \infty$ *for all such* ν. *For any such* ν *the measures* ${}_{\nu}e_{\varepsilon}(x)d\mu(x)$ *converge weakly to an energy density* de *as* $\varepsilon \to 0$, *which is independent of* ν *and has total mass* ${}_{\nu_1}E$.

Proof. By scaling the domain metric we may assume $|Z| \leq 1$. If we can verify the "subpartition estimate" (1.4i) for $\mathcal{L}_{\varepsilon}(f) \equiv {}_{\nu}E_{\varepsilon}(f)$, then Lemmas 1.3.1, 1.4.1, 1.4.2, and the Riesz representation theorem immediately imply our result, just as they yielded Theorem 1.5.1 Let $\{\lambda_i\}$ satisfy (1.3vi). Given x, $\overline{x}(x,\varepsilon)$, define $\overline{x}_0(x,\varepsilon) = x$ and

$$ \overline{x}_i(x,\varepsilon) = \overline{x}\left(x, (\sum_{j=1}^{i} \lambda_j)\varepsilon\right), \qquad i \geq 1. $$

Now

$$ {}_{\nu}E_{\varepsilon}(f) = \int_{\Omega}\int_{0}^{2} f(x)\, \frac{d^{p}(u(x), u(\overline{x}(x,\varepsilon\rho))}{\varepsilon^{p}}\, \rho^{-p}d\nu(\rho)d\mu(x). $$

Using triangle inequalities as in (1.3ix), we estimate

$$({}_\nu E_\varepsilon(f))^{\frac{1}{p}} \le$$
$$\sum_i \lambda_i \left(\iint f(x) \frac{d^p(u(\overline{x}_{i-1}(x,\varepsilon\rho)),\, u(\overline{x}_i(x,\varepsilon\rho)))}{\varepsilon^p \lambda_i^p} \rho^{-p} d\nu(\rho) d\mu(x) \right)^{\frac{1}{p}}.$$

For fixed i (and ρ) we change variables from x to $w = \overline{x}_{i-1}(x,\varepsilon\rho)$. Note that $\overline{x}_i(x,\varepsilon\rho) = \overline{x}(w,\varepsilon\rho\lambda_i)$, and estimate $f(x)$ above by $f(w) + \omega(f,2\varepsilon)(w)$, $d\mu(x) \le (1+C\varepsilon)\, d\mu(w)$. (In this case C depends not only on the metric g but also on the Lipschitz constant of Z.) Hence

(1.7iv)
$$\begin{aligned}
&({}_\nu E_\varepsilon(f))^{\frac{1}{p}} \\
&\quad \le \sum \lambda_i \left(\iint f_\varepsilon^C(w) \frac{d^p(u(w), u(\overline{x}(w,\varepsilon\rho\lambda_i)))}{(\varepsilon\lambda_i)^p} \rho^{-p} d\nu(\rho) d\mu(x) \right)^{\frac{1}{p}} \\
&\quad = \sum \lambda_i \; ({}_\nu E_{\lambda_i \varepsilon}(f_\varepsilon^C))^{\frac{1}{p}}. \quad \square
\end{aligned}$$

1.8. The calculus of directional energies. We collect some useful properties of the directional-derivative measures. We fix $1 \le p < \infty$, $u \in W^{1,p}(\Omega, X)$ if $p > 1$ or $u \in BV(\Omega, X)$ if $p = 1$. We use $E(f)$ to denote the (Sobolev or BV) functional induced by u, writing E for its norm. For $Z \in \Gamma(T\overline{\Omega})$ we write ${}^Z E(f)$ and ${}^Z E$ for the directional functional and its norm.

Theorem 1.8.1. *If $Z \in \Gamma(T\overline{\Omega})$ then ${}^Z E < \infty$ and there is a constant C (depending only on the dimension n of Ω) so that we can estimate*

(1.8.i)
$$d({}^Z e) \le C|Z|_\infty^p \, de.$$

If $z(x,t)$ is any one-parameter family of $C^{1,1}$ diffeomorphisms, having Lipschitz velocity fields and satisfying

$$z(x,0) = x, \qquad \frac{\partial}{\partial t} z(x,t)|_{t=0} = Z(x) \;\; (x \in \Omega)$$

for $Z \in \Gamma(T\overline{\Omega})$, then

(1.8ii)
$$\frac{d^p(u(x), u(z(x,\varepsilon)))}{\varepsilon^p} \rightharpoonup d({}^Z e)$$

as $\varepsilon \to 0$.

For $p = 1$ *and* $W \in \Gamma(T\overline{\Omega})$ *we have the triangle inequality*

$$d({}^{Z+W}e) \le d({}^{Z}e) + d({}^{W}e). \tag{1.8iii}$$

More generally, for any $1 \le p < \infty$ *and* $f \in C_c(\Omega)$ *with* $f \ge 0$ *we have*

$$({}^{Z+W}E(f))^{\frac{1}{p}} \le ({}^{Z}E(f))^{\frac{1}{p}} + ({}^{W}E(f))^{\frac{1}{p}}. \tag{1.8iv}$$

If h *is a Lipschitz function on* $\overline{\Omega}$, *then*

$$d({}^{hZ}e) = |h|^p d({}^{Z}e). \tag{1.8v}$$

In case (Ω, g) *has an orthonormal frame* $\{e_1, \dots, e_n\}$ *we identify* $S^{n-1} \subset \mathbb{R}^n$ *with* $S_x^{n-1} \subset T\Omega_x$ *by*

$$\omega = (\omega^1, \dots, \omega^n) \longmapsto \omega^i e_i,$$

and denote $\omega^i e_i \in \Gamma(T\Omega)$ *by* ω *as well. Then we have*

$$E(f) = \int_{S^{n-1}} {}^{\omega}E(f)d\sigma(\omega). \tag{1.8.1}$$

Proof. These results are consequences of the following technical estimate:

Lemma 1.8.2. *Let* $\psi : \Omega \to \psi(\Omega) \subset M$ *be a bi-Lipschitz map from* Ω *to its image. We think of* ψ *being close to the identity map, and write* $|\psi - id|_\infty = \delta$. *Assume* δ *is small enough so that the geodesic subdivision techniques of 1.4 are applicable whenever* $x \in \Omega_{3\delta}$ *and* $|x - y| < 3\delta$. *Let* $f \in C_c(\Omega_{3\delta},)$, $f \ge 0$. *Then there exists a constant* C *depending only on* n, p, *on the curvature of* (Ω, g) *and on first-derivative bounds for* ψ *and* ψ^{-1}, *so that*

$$\int_\Omega f(x)d^p(u(x), u(\psi(x)))d\mu_g(x) \le C\delta^p E|f|_\infty. \tag{1.8vii}$$

Proof. For $x \in \Omega_{3\delta}$ we write $\tilde{x}$ for the geodesic midpoint of x and $\psi(x)$. Then from the triangle inequality and an average of integration over $B(\tilde{x}, \frac{\delta}{2})$ we have

$$\begin{aligned} d^p(u(x), u(\psi(x))) &\leq 2^p \left(\fint_{B(\tilde{x},\frac{\delta}{2})} d^p(u(x), u(y)) d\mu(y) \right. \\ &\qquad \left. + \fint_{B(\tilde{x},\frac{\delta}{2})} d^p(u(\psi(x)), u(y))) d\mu(y) \right) \\ &\leq C \; ({}_\nu e_\delta(x) + {}_\nu e_\delta(\psi(x)))\, \delta^p. \end{aligned}$$

Here we are using ν corresponding to ball averages (1.2vii), and C depends on allowed quantities. We apply this estimate to the integrand in (1.8vii), change variables from x to $\psi(x)$ in the second term, and deduce

(1.8viii)
$$\int_\Omega f(x) d^p(u(x), u(\psi(x))) d\mu(x) \leq C\delta^p \left({}_\nu E_\delta(f) + {}_\nu E_\delta(\tilde{f}) \right),$$

where

$$\tilde{f}(x) = f(x) + \omega(f, \delta)(x).$$

From (1.4vii) we see that this last estimate is bounded above by

$$C\delta^p (E(f_\delta^C) + E((\tilde{f})_\delta^C)) \leq C\delta^p E|f|_\infty.$$

This proves (1.8vii). □

Our first claim (1.8i) follows immediately from the reasoning above: We take $\psi(x) = \overline{x}(x, \varepsilon)$ and deduce from (1.8vii) that ${}^Z E < \infty$. Then we take $\delta = \varepsilon |Z|_\infty$ in (1.8viii), divide by ε^p and let $\varepsilon \to 0$, to conclude (1.8i).

We show (1.8ii). From triangle inequalities we may estimate

$$\Big|\Big(\int f(x)\frac{d^p(u(x),u(z(x,\varepsilon)))}{\varepsilon^p}d\mu(x)\Big)^{\frac{1}{p}} - \Big(\int f(x)\frac{d^p(u(x),u(\overline{x}(x,\varepsilon)))}{\varepsilon^p}d\mu(x)\Big)^{\frac{1}{p}}\Big| \le \Big(\int f(x)\frac{d^p(u(z(x,\varepsilon)),u(\overline{x}(x,\varepsilon)))}{\varepsilon^p}d\mu(x)\Big)^{\frac{1}{p}}.$$

For small $\varepsilon > 0$ we consider ψ which maps $z(x,\varepsilon)$ to $\overline{x}(x,\varepsilon)$. Since $|\psi - id|_\infty$ is $o(\varepsilon)$ we deduce from (1.8vii) that the right hand side of the above inequality approaches zero as $\varepsilon \to 0$. This proves (1.8ii).

We next show (1.8iv), which yields (1.8iii) as a special case. We may assume we are working in a local coordinate chart. In this case we write

$$\Big(\int f(x)\frac{d^p(u(x),u(x+\varepsilon(Z+W)(x)))}{\varepsilon^p}d\mu(x)\Big)^{\frac{1}{p}} \le \Big(\int f(x)\frac{d^p(u(x),u(x+\varepsilon Z(x)))}{\varepsilon^p}d\mu(x)\Big)^{\frac{1}{p}} + \Big(\int f(x)\frac{d^p(u(y),u(y+\varepsilon W(x)))}{\varepsilon^p}d\mu(x)\Big)^{\frac{1}{p}},$$

where we have written y for $x + \varepsilon Z(x)$ in the bottom integral. By (1.8ii) the top integral converges to ${}^{Z+W}E(f)$ as $\varepsilon \to 0$, and the middle integral converges to ${}^{Z}E(f)$. In fact, the bottom one converges to ${}^{W}E(f)$, as one can deduce by changing variables from x to y, dominating $f(x)$ by $f(y) + \omega(f, \varepsilon|W|_\infty)(y)$, and applying (1.8ii) again. This shows (1.8iv).

We next show (1.8v). We first consider the case of constant h: If $h > 0$ then we observe that $h^p({}^{Z}e_{\varepsilon h}(x)) = {}^{hZ}e_\varepsilon(x)$. This implies the result as $\varepsilon \to 0$. In case $h = -1$, we note that ${}^{-Z}e_\varepsilon(\overline{x}(x,\varepsilon)) = {}^{Z}e_\varepsilon(x)$, so by the usual change of variables arguments we obtain $d({}^{-Z}e) = d({}^{Z}e)$. Thus (1.8v) holds when h is any constant.

Next we assume that there is a constant h_0 and a small number $\delta > 0$ so that $|h - h_0| \le \delta$. We apply the triangle inequality (1.8iv) twice, with vector fields $hZ, (h_0 - h)Z$ the first time and $h_0 Z, (h - h_0)Z$ the second time. Taking p^{th}-powers, applying (1.8i), using the result for constant multiples which we just showed, and using standard inequalities, we see that for any $\varepsilon > 0$,

$$^{Z}E(h_0^p f) \leq (1+\varepsilon)^p ({}^{hZ}E(f)) + (\frac{1+\varepsilon}{\varepsilon})^p C\delta^p E(f)$$

$$^{hZ}E(f) \leq (1+\varepsilon)^p ({}^{Z}E(h_0^p f)) + (\frac{1+\varepsilon}{\varepsilon})^p C\delta^p E(f).$$

These inequalities imply the measure inequalities

(1.8ix) $|h_0|^p d({}^{Z}e) - c(\varepsilon,\delta) de \leq d({}^{hZ}e) \leq |h_0|^p d({}^{Z}e) + c(\varepsilon,\delta) de,$

where $c(\varepsilon,\delta)$ has the property that it can be made arbitrarily small by choosing ε (and then) δ small. We can now show the general case of (1.8v). Pick $c(\varepsilon,\delta)$ small. Cover Ω with a finite number of open subsets $\{U_i\}$ so that there are constants $\{h_i\}$ with $|h - h_i| < \delta$ on U_i. Pick a subordinate partition of unity $\{\eta_i\}$ to the $\{U_i\}$. Note that ${}^{hZ}E(f) = \sum_i {}^{hZ}E(\eta_i f)$. We apply the estimate (1.8ix) in each U_i and sum, obtaining

(1.8x)

$$^{Z}E(\sum_i |h_i|^p \eta_i f) - c(\varepsilon,\delta)E(f) \leq {}^{hZ}E(f)$$

$$\leq {}^{Z}E(\sum_i |h_i|^p \eta_i f) + c(\varepsilon,\delta)E(f).$$

This last estimate implies ${}^{hZ}E(f) = {}^{Z}E(|h|^p f)$, which is equivalent to (1.8v). (The reason for the implication is that we may choose $c(\varepsilon,\delta), \varepsilon, \delta \to 0$ and covers $\{U_i\}$ with diameters approaching zero. Then the sums $\sum \| h_i|^p \eta_i f$ converge uniformly to $|h|^p f$, and then continuity of the linear functional ${}^{Z}E$ gives the conclusion.) Finally, we study (1.8vi). Recall the definition of $e_\varepsilon(x)$ given in (1.2ii). Compare this to the approximant obtained using the exponential map:

$$\tilde{e}_\varepsilon(x) \equiv \int_{S_x^{n-1}} \frac{d^p(u(x), u(exp(x,\varepsilon\omega)))}{\varepsilon^p} d\sigma(\omega).$$

Because of the uniform closeness of the respective surface measures, we see that

$$(1 - o(1))e_\varepsilon(x) \leq \tilde{e}_\varepsilon(x) \leq (1 + o(1))e_\varepsilon(x),$$

where $o(1)$ is a term which approaches zero uniformly (in x), as $\varepsilon \to 0$. Hence

(1.8xi) $$\tilde{e}_\varepsilon(x)d\mu \rightharpoonup de.$$

We also note that for the fixed vector field ω (as defined above (1.8vi)), the two diffeomorphism families $\psi_1(x,t) = \overline{x}(x,t)$ and $\psi_2(x,t) = exp(x, t\omega)$ have the same velocity vector field ω when $t = 0$. In succession we apply (1.8xi), Fubini's Theorem, the Lebesgue Dominated Convergence Theorem (justified by (1.8vii)), and then (1.8ii) for the pair (ψ_1, ψ_2):

$$\begin{aligned} E(f) &= \lim_{\varepsilon\to 0} \int f(x)\tilde{e}_\varepsilon(x)d\mu(x) \\ &= \lim_{\varepsilon\to 0} \int \left(\int f(x) \frac{d^p(u(x), u(exp(x,\varepsilon\omega)))}{\varepsilon^p} d\mu(x) \right) d\sigma(\omega) \\ &= \int \lim_{\varepsilon\to 0} \left(\int f(x) \frac{d^p(u(x), u(exp(x,\varepsilon\omega)))}{\varepsilon^p} d\mu(x) \right) d\sigma(\omega) \\ &= \int {}^\omega E(f) d\sigma(\omega). \end{aligned}$$

This shows (1.8vi) and completes the proof of Theorem 1.8.1. □

1.9. Differentiability theory for directional energies. For $p > 1$, $u \in W^{1,p}(\Omega, X)$, and $Z \in \Gamma(T\overline{\Omega})$ we show that the directional -derivative energies $d({}^Z e)$ are absolutely continuous with respect to Lebesgue measure, i.e. $d({}^Z e) = {}^Z e(x)d\mu(x)$, for an L^1 energy-density function ${}^Z e(x)$. Furthermore, the ε-energy functions converge almost everywhere to ${}^Z e(x)$. When Z is non-zero one can always choose local coordinates so that $Z = \partial^1$. This motivates us to first consider the special configuration below:

(1.9i) DEFINITION. Let $\Omega \subset \mathbb{R}^n$ be a bounded domain, and let g be the Euclidean metric. Let $1 \leq p < \infty$, $u \in L^p(\Omega, X)$, $\omega = \partial^1$, and let

u have finite p-energy in the direction ω,

$$E \equiv {}^{\omega}E < \infty,$$

and write $E(f)$ for the corresponding linear functional. (In the following discussion ω will be fixed and we will frequently suppress it in our notation.) Assume that each line in the direction of ω intersects Ω in at most one interval. Write Π for the projection of Ω onto the $(n-1)$-plane $\{x^1 = 0\}$. Thus we may write

$$\Omega = \{x = (t, y),\ y \in \Pi,\ t \in I_y \subset \mathbb{R}\}$$
$$E^y = E(u|_{I_y}).$$

That is, E^y is the p-energy of the 1-variable map $u|_{I_y}$ from I_y to X. In case $E^y < \infty$ we write $E^y(f)$ for the corresponding linear functional on $C_c(I_y)$. For a function $f \in C_c(\Omega)$ we also write f for its restriction fo I^y.

Lemma 1.9.1. *For the configuration (1.9i) we have*

$$E = \int E^y\, dy, \quad \textit{and } E(f) = \int E^y(f)\, dy, \quad \textit{for any } f \in C_c(\Omega).$$

Conversely, if $u \in L^p(\Omega, X)$ is a map for which $\int E^y dy$ is finite, then u has finite energy ${}^{\omega}E$, and the above identities hold.

Proof. If the first part of the lemma is true, then the converse statement follows easily. Indeed, from Theorem 1.5.1 and (1.4vii) we have the following estimate for $f \geq 0, f \in C_c(\Omega)$:

$$E_\varepsilon(f) = \int E^y_\varepsilon(f) dy \leq \int E^y(f^0_\varepsilon) dy \leq (|f|_\infty + o(\varepsilon)) \int E^y dy.$$

Hence ${}^{\omega}E < \infty$, and the first part of the lemma applies.

We now prove the first part of the Lemma. We use an absolutely continuous measure ν satisfying (1.2iv) to define our ε-energy densities. (For example, the choice $d\nu(\lambda) = (p+1)\lambda^p$ for $0 < \lambda < 1$ would correspond to interval averaging.) For any $f \in C_c(\Omega)$ with $0 \leq f \leq 1$, the fact that $E < \infty$ implies that given $\delta > 0$ and $C \geq 0$, there exists $\varepsilon_1 > 0$ so that

$$\lim_{\varepsilon' \to 0} E_{\varepsilon'}(f^C_{\varepsilon_1}) \leq E + \delta. \tag{1.9ii}$$

Since $E_{\varepsilon'}(f^C_\varepsilon) = \int E^y_{\varepsilon'}(f^C_\varepsilon)\, dy$, (1.9ii) and Fatou's Lemma imply

$$\int \liminf_{\varepsilon' \to 0} E^y_{\varepsilon'}(f^C_{\varepsilon_1})\, dy \leq E + \delta \tag{1.9iii}$$

Fix δ, ε_1 and set $C = 0$. Denote the integrand above by $h(y, \varepsilon')$. By the absolute continuity of ν, by Remark 1.4.3 (and the fact that each I_y is Euclidean),

$$E^y_\varepsilon(f) \leq \liminf_{\varepsilon' \to 0} h(y, \varepsilon'), \quad \text{for any } \varepsilon < \varepsilon_1.$$

Hence (taking the lim sup of the right-hand side and noting that δ was arbitrary)

$$\int \limsup_{\varepsilon \to 0} E^y_\varepsilon(f)\, dy \leq E. \tag{1.9iv}$$

Pick any sequence $\{f_i\} \subset C_c(\Omega)$ which is monotone increasing in i and converges (uniformly on compact subsets) to 1, as $i \to \infty$. The monotone convergence theorem and (1.9iv) yield

$$\int \sup_{0 \leq f \leq 1} \left(\limsup_{\varepsilon \to 0} E^y_\varepsilon(f)\, dy \right) \leq E,$$

i.e.

$$\int E^y\, dy \leq E. \tag{1.9v}$$

Thus for any $f \in C_c(\Omega)$

$$E(f) = \lim_{\varepsilon \to 0} \int E^y_\varepsilon(f)\, dy = \int E^y(f)\, dy, \tag{1.9vi}$$

by the Lebesgue Dominated Convergence Theorem. Again using the sequence $\{f_i\}$ described above, this time in (1.9vi), shows that $E = \int E^y\, dy$. $\square$

Lemma 1.9.2. *For the configuration (1.9i) there is a representative of u having the property that for almost all $I_y = (a_y, b_y)$ with $y \in \Pi$,*

$$\text{(1.9vii)} \qquad \begin{aligned} \lim_{t \to a_y^+} u(t,y) &\equiv u(a_y, y) \\ \lim_{t \to b_y^-} u(t,y) &\equiv u(b_y, y) \end{aligned}$$

exist. In case $p > 1$ (and for the y above) $u|_{I_y}$ is a Hölder continuous mapping, with exponent $\alpha = \frac{p-1}{p}$. In case $p = 1$ the functions $u|_{I_y}$ are maps of bounded variation. In all cases we have the estimate

$$\text{(1.9viii)} \qquad \int_{\Pi} d^p(u(a_y, y), u(b_y, y))dy \le {}^{\omega}E \max_{y \in \Pi}(b_y - a_y)^{p-1}.$$

Proof. We first deal with the case $p > 1$ and then indicate how to modify the argument in case $p = 1$. By Lemma 1.9.1 almost all $u|_{I_y} \in W^{1,p}(I_y, X)$. For such y write $u|_{I_y} = v$, and assume (by rescaling) that $I_y = I = (0,1)$. For such a $v \in W^{1,p}(I, X)$ and $t \in I$, define $w(s)$ by

$$w(s) = d(v(t+s), v(t)) \qquad \text{for } s \in (-t, 1-t).$$

It follows from (1.6vii) of Corollary 1.6.3 that $w \in W^{1,p}((-t, 1-t), \mathbb{R})$, with $|\nabla_s w|^p ds \le de_p$ as measures. In particular (a representative of) $w \in C^{\alpha}((-t, 1-t))$ (for $\alpha = \frac{p-1}{p}$), so is absolutely continuous. Therefore we have the estimate

$$\text{(1.9ix)} \qquad w(\varepsilon) - w(0) \le \int_0^{\varepsilon} |\nabla_s w| ds \le \int_t^{t+\varepsilon} e_1(t')dt'.$$

We would like to know that $w(0) = 0$ and $w(\varepsilon) = d(u(t), u(t+\varepsilon))$ in the above estimate. This is formally true but actually requires some technical justification since a priori v is only defined a.e. So, for any fixed $0 < \mu < \frac{1}{2}$, we pick $f \in C_c(\Omega)$ with $0 \le f \le 1$ and with $f \equiv 1$ on the interval $(\mu, 1-\mu)$. From (1.4vii) and $p' = 1$ we have

$$\int_I f(t) \, {}_{\nu}e_{\delta}(t)dt \le \int_I e_1(t)dt < \infty$$

for small $\delta > 0$. Here we take $d\nu(\lambda) = 2\lambda$, corresponding to interval averages. Thus by Fatou's Lemma,

$$\int_{\mu}^{1-\mu} \liminf_{\delta \to 0} {}_{\nu}e_{\delta}(t)dt < \infty,$$

so that

$$\liminf_{\delta \to 0} \qquad 2\int_{0}^{\delta} \frac{d(v(t), v(t+s))}{\delta} \frac{ds}{\delta} < \infty \qquad \text{a.e.t.}$$

For such a t we conclude that $w(0) = 0$, since the numerator in the expression above is $w(s)$ and w is Hölder continuous. For such a t the estimate (1.9ix) implies

$$\text{(1.9x)} \quad d(v(t), v(t+\varepsilon)) \leq \int_{0}^{\varepsilon} |\nabla_s w| ds \leq \int_{t}^{t+\varepsilon} e_1(t')dt' \qquad \text{a.e.}\varepsilon.$$

From the first half of this inequality we see that v is equivalent to a C^{α} mapping, with $\alpha = \frac{p-1}{p}$: By raising both sides to the pth power, using Hölder, and using the fact that $w \in W^{1,p}(\Omega, \mathbb{R}))$, we have the estimate (at first a.e. t, ε, but then everywhere by redefinition)

$$d^p(v(t), v(t+\varepsilon)) \leq \left(\int_{t}^{t+\varepsilon} |\nabla_s w|^p \right) \varepsilon^{p-1}.$$

This implies that for the $y \in \Pi$ with $u|_{I_y} \in W^{1,p}(\Omega, X)$ the limits (1.9vii) exist, and furthermore that the estimate

$$\text{(1.9xi)} \qquad d^p(u(a_y, y), u(b_y, y)) \leq (b_y - a_y)^{p-1} E^y$$

holds. Upon integration this yields (1.9viii).

In the case $p = 1$ it is necessary to modify the argument slightly. Now the function $w(s)$ is equivalent to a function of bounded variation, so may be taken to be continuous from the left, with only a countable number of discontinuities, and so that all one-sided limits exist. We replace (1.9ix) with the estimate

$$\text{(1.9xii)} \qquad w(\varepsilon^-) - w(0^+) \leq \mathrm{Var}\,(0, \varepsilon)w \leq e_1((t, t+\varepsilon)),$$

where the $+,-$ superscripts refer to limits from the right and left respectively. For t satisfying the finite liminf condition above, $w(0^+) = 0$. For such t we conclude

(1.9xiii)
$$d(v(t), v(t+\varepsilon)) = w(\varepsilon) \le \mathrm{Var}(0,\varepsilon)w \le e_1((t, t+\varepsilon)) \qquad \text{a.e.}\varepsilon.$$

We redefine v to be the limit from the left of $v(t')$, where t' satisfies (1.9xiii). It is straightforward to verify that this leaves v unchanged at such t', and that it creates a function which is continuous from the left, and satisfies (1.9xiii) everywhere. In particular, this representative for v will have only a countable number of discontinuities, exactly at points where the measure e_1 has point masses. We deduce that the map v is a map of bounded variation in the classical sense. In particular, all one sided limits exist. Also, returning to the intervals I_y with $E^y < \infty$ we get the estimate

$$d(u(a_y, y), u(b_y, y)) \le E^y.$$

Upon integration this yields (1.9viii), and the proof of Lemma 1.9.2 is complete. □

Lemma 1.9.3. *Let* $1 < p < \infty$, $u \in W^{1,p}(I, X)$. *Then (its Hölder continuous representative satisfies)*

$$\lim_{\varepsilon \to 0} \frac{d(u(t),\, u(t+\varepsilon))}{\varepsilon} = e_1(t) \quad a.e.$$

Proof. We see immediately from the inequality between the first and last terms of (1.9x) (and the Lebesgue Differentiation Theorem applied to the L^1-function $e_1(t)$) that

$$\limsup_{\varepsilon \to 0} \frac{d(u(t),\, u(t+\varepsilon))}{\varepsilon} \le e_1(t) \quad \text{a.e.t.} \qquad \text{(1.9xiv)}$$

Now let $\delta > 0$. Define

$$\begin{aligned} S_\delta \;\equiv\; & \{t \in I \,\text{s.t. } t \text{ is a Lebesgue point for } e_1, \text{ and} \\ & \liminf_{\varepsilon \to 0} \frac{d(u(t), u(t+\varepsilon))}{\varepsilon} < e_1(t) - \delta\}. \end{aligned}$$

By definition, for any fixed $\mu > 0$ we may cover S_δ by intervals $(t - \varepsilon, t + \varepsilon)$ with $\varepsilon < \mu$ and for which $t \in S_\delta$,

$$\frac{d(u(t),\, u(t+\varepsilon))}{\varepsilon} < e_1(t) - \delta,$$

$$e_1(t) - \frac{1}{\varepsilon}\int_t^{t+\varepsilon} e_1(s)ds < \frac{\delta}{2}.$$

By a well-known covering lemma [Rudin, Lemma 8.4] we may pick a finite disjoint subset of this cover, with the finite sum of the interval lengths at least $\frac{1}{4}^{\text{th}}$ the measure of S_δ. Denote the corresponding half-intervals by

$$\{I_i\}_{i=1}^k, \quad I_i = [t_i, t_i + \varepsilon_i).$$

Complete this finite collection of intervals to a partition of [0,1], by adding intervals

$$\{J_{i'}\}_{i'=1}^{k'}, \quad J_{i'} = [t_{i'}, t_{i'} + \varepsilon_{i'}),$$

also with maximum length bounded by μ. We have from (1.9x) and our covering hypothesis that

$$\begin{aligned} &\sum_i d(u(t_i),\, u(t_i+\varepsilon_i)) + \sum_{i'} d(u(t_{i'}),\, u(t_{i'}+\varepsilon_{i'})) \\ &\leq \sum_i \left(\int_{I_i} e_1(t)dt - \frac{\delta}{2}\varepsilon_i\right) + \sum_{i'} \int_{J_{i'}} e_1(t)dt \\ &\leq \int_I e(t)dt - \frac{\delta}{8}\,|S_\delta|. \end{aligned} \tag{1.9xv}$$

If we can show

$$\int_I e_1(t)dt = \lim_{\|P\|\to 0} \sum d(u(t_{i+1}),\, u(t_i)) \tag{1.9xvi}$$

for arbitrary partitions P of I, we conclude that the measure of S_δ is 0 by letting $\mu \to 0$ in (1.9x). Picking a sequence S_{δ_i} with $\delta_i \to 0$ then implies Lemma 1.9.3. We verify (1.9xvi) as follows:

Because $de_1 = e_1(t)dt$ is absolutely continuous it is easy to see that the total $p = 1$ energy E_1 for the map u is given by

$$\int_I e_1(t)dt = \lim_{\varepsilon\to 0} \int_0^{1-\varepsilon} \frac{d(u(t),\, u(t+\varepsilon))}{\varepsilon} dt$$

$$= \lim_{\varepsilon\to 0} \int_0^{\varepsilon} \sum_{i=0}^{[\frac{1}{\varepsilon}]-2} \frac{d(u(t+i\varepsilon),\, u(t+(i+1)\varepsilon))}{\varepsilon} dt$$

$$\equiv \lim_{\varepsilon\to 0} \frac{1}{\varepsilon} \int_0^{\varepsilon} \sum\nolimits_{\varepsilon}(t)dt.$$

Hence for $\delta > 0$, $\varepsilon > 0$ small, there exists $t_\varepsilon \in (0,\varepsilon)$ with $\sum_\varepsilon(t_\varepsilon) \geq E_1 - \delta$. Now let

$$P : 0 = t_0 < t_1 < \ldots < t_k = 1$$

be an arbitrary partition of I, subject to $N\|P\| < \varepsilon$ for N large. We pick a subpartition $P' \subset P$ for which t_i is within $\frac{\varepsilon}{N}$ of $t_\varepsilon + i\varepsilon$, $i = 1, \ldots ([\frac{1}{\varepsilon}] - 2)$. Note that this implies

$$\sum_i |t_\varepsilon + i\varepsilon - t_i| \leq \frac{1}{N}.$$

Hence

$$\sum_P d(u(t_j),\, u(t_{j+1})) \geq \sum_{P'} d(u(t_i),\, u(t_{i+1}))$$

$$\geq \sum\nolimits_{\varepsilon}(t_\varepsilon) - 2\sum \left|\int_{t_i}^{t_\varepsilon + i\varepsilon} e_1(t)dt\right|$$

$$\geq \int_I e_1(t)dt - \delta - 2\left(\frac{1}{N}\right)^{\frac{p-1}{p}} (E_p)^{\frac{1}{p}},$$

(where we have used Remark 1.5.2 in the last step). As δ and N are arbitrary, and since a one-sided estimate suffices (see (1.9x)), we conclude (1.9xvi) and the proof of Lemma 1.9.3 is complete. □

Lemma 1.9.4. *Let (Ω, g), ω and u be as in (1.9i). Let $1 < p < \infty$. Then for any $1 \leq p' < p$ the energy-density function $e_{p'}(x)$ (for the direction ω) satisfies*

(1.9xvii) $$e_{p'}(x) = (e_1(x))^{p'} \quad a.e.$$

Furthermore, there exists a representative of u *so that*

$$\lim_{\varepsilon\to 0} \frac{d^{p'}(u(t,y),\, u(t+\varepsilon,y))}{\varepsilon^{p'}} = e_{p'}(t,y) \quad a.e. \tag{1.9xviii}$$

Proof. From Lemma 1.9.1 and Lemma 1.9.3 we know we may pick a representative of u so that

$$\lim_{\varepsilon\to 0} \frac{d^{p'}(u(t,y),\, u(t+\varepsilon,y))}{\varepsilon^{p'}} = (e_1(t,y))^{p'} \quad a.e.$$

so (1.9xviii) will follow from (1.9xvii). The claim (1.9xvii) is true because of the following measure theory lemma.

Lemma. *Let* $1 < p < \infty$, $\{g_\varepsilon\} \subset L^p_{loc}(\Omega, \mathbb{R})$, $g_\varepsilon \geq 0$. *Suppose that on any compact subset of* Ω *the* L^p*-norms of* g_ε *are uniformly bounded as* $\varepsilon \to 0$. *Let* $g_\varepsilon(x) \to g(x)$ *a.e. as* $\varepsilon \to 0$. *If, for some* $1 \leq p' < p$ *it is also true that* $g_\varepsilon^{p'} d\mu \rightharpoonup h d\mu$ *with* $h \in L^1(\Omega, \mathbb{R})$, *then in fact* $h = g^{p'}$ *a.e.*

Proof. Let $f \in C_c(\Omega)$. Write $\Omega = G \cup B$, $G \cap B = \varphi$, where $g_\varepsilon \to g$ uniformly on G. (So B can be chosen with arbitrarily small positive measure). Then

$$\lim_{\varepsilon\to 0} \int_G g_\varepsilon^{p'} f \, d\mu = \int_G f g^{p'} \, d\mu,$$

and

$$\limsup_{\varepsilon\to 0} \int_B g_\varepsilon^{p'} f \, d\mu \leq |f|_\infty C^{\frac{p'}{p}} (\mu(B))^{\frac{p-p'}{p}},$$

where C depends on the uniform L^p estimates for the g_ε as $\varepsilon \to 0$, on the compact subset $supp(f)$. For fixed f we may make this last term smaller than any given $\delta > 0$ by picking $\mu(B)$ small. Combining this observation with Fatou's Lemma gives

$$\int\limits_\Omega f g^{p'} \, d\mu \leq \int\limits_\Omega f h d\mu = \lim_{\varepsilon\to 0} \int\limits_\Omega f g_\varepsilon^{p'} \, d\mu \leq \int\limits_\Omega f g^{p'} \, d\mu + \delta.$$

This proves the Lemma.

By applying this result to the functions

$$g_\varepsilon(t,y) = \frac{d(u(t,y),\, u(t+\varepsilon,y))}{\varepsilon}$$
$$h(x) = e_{p'}(x),$$

we immediately conclude (1.9xvii) and Lemma 1.9.4.

Lemma 1.9.5. *Let (Ω, g), ω and u be as in (1.9i). Let $1 < p < \infty$. Then the energy density measure de (for the direction ω) is absolutely continuous with respect to Lebesgue measure, $de = e_p(x) d\mu(x)$, and*

$$e_p(x) = e_1(x)^p \quad a.e.$$

Furthermore, there is a representative of u so that

$$\lim_{\varepsilon \to 0} \frac{d^p(u(t,y)),\ u(t+\varepsilon, y)}{\varepsilon^p} = e_p(t,y) \qquad a.e.$$

For any representative of u, and any choice of ν in (1.2iv) *for which $\lambda^{-p} d\nu(\lambda)$ is a bounded multiple of $d\lambda$, we have*

$$\lim_{\varepsilon \to 0} {}_{\nu}e_{\varepsilon}(x) = e_p(x) \qquad a.e.$$

Proof. As in Lemma 1.9.4 it suffices to show the first claim only. Note that the third claim will follow from the second, because if we change representatives for u, almost all $u|_{I_y}$ will remain the same L^p function, and for such y the ${}_{\nu}e_{\varepsilon}(t,y)$ will be unchanged, because of the restriction on ν. We establish the first claim by verifying the two measure inequalities

$$e_1^p \, d\mu \le de,$$

$$de \le e_1^p \, d\mu.$$

The first of these follows immediately from Fatou's Lemma:
Fix $f \in C_c(\Omega)$, then

$$\begin{aligned} \int f e_1^p \, d\mu &= \int \liminf_{\varepsilon \to 0} f \frac{d^p(u(t,y),\ u(t+\varepsilon, y))}{\varepsilon^p} \, d\mu \\ &\le \int_{\Omega} f \, de. \end{aligned}$$

To show the second we again fix $f \in C_c(\Omega)$. Then

$$\int_{\Omega} f \, de = \lim_{\varepsilon \to 0} \int_{\Omega} f \frac{d^p}{\varepsilon^p} \, d\mu,$$

where we suppress the various arguments $x = (t, y)$, $u(t, y)$, $u(t + \varepsilon, y)$. Hence

$$E(f) = \int_\Omega f \frac{d^p}{\varepsilon^p}\, d\mu + \delta_1(\varepsilon)$$

where $\delta_1(\varepsilon) \to 0$ as $\varepsilon \to 0$. But for $\varepsilon > 0$ fixed,

$$\int_\Omega f \frac{d^p}{\varepsilon^p}\, d\mu = \int_\Omega f \frac{d^{p'}}{\varepsilon^{p'}}\, d\mu + \delta_2(p')$$

where $\delta_2(p') \to 0$ as $p' \to p$. Applying the estimate (1.4vii) to the integral on the right side, and also Lemma 1.9.4, yields

$$\begin{aligned} E(f) &= \int_\Omega f \frac{d^{p'}}{\varepsilon^{p'}}\, d\mu + \delta_1(\varepsilon) + \delta_2(p') \\ &\leq \int_\Omega f_\varepsilon^C(x)\, e_1(x)^{p'}\, d\mu(x) + \delta_1(\varepsilon) + \delta_2(p'). \end{aligned}$$

Using the already established inequality $(e_1)^p\, d\mu \leq de$ to justify Lebesgue's Dominated Convergence Theorem, we let $p' \to p$ and deduce

$$\int_\Omega f\, de \leq \int_\Omega f_\varepsilon^C\, e_1(x)^p\, d\mu + \delta_1(\varepsilon).$$

Letting $\varepsilon \to 0$ we conclude the desired second inequality $de \leq (e_1^p)\, d\mu$. Thus Lemma 1.9.5 is shown. □

Theorem 1.9.6. *Let (Ω, g) be a Riemannian domain (1.1). Let*

$$u \in W^{1,p}(\Omega, X) \quad \text{for some } 1 < p < \infty,$$

and let $Z \in \Gamma(T\overline{\Omega})$. Then the energy-density measure $d^Z e$ for each $1 \leq p' \leq p$ is absolutely continuous with respect to Lebesgue measure. In particular, if we denote the $p = 1$ energy-density function by $|u_*(Z)|(x)$, then the p'-energy measure is given by

$$|u_*(Z)|^{p'}\, d\mu_g(x)$$

for each $1 \leq p' \leq p$. *For any choice of* ν *in* (1.2iv) *for which* $\lambda^{-p} d\nu(\lambda)$ *is a bounded multiple of* $d\lambda$ *we have (for each* $1 \leq p' \leq p$*)*

$$\lim_{\varepsilon \to 0} {}^Z_\nu e_\varepsilon(x) = |u_*(Z)|^{p'}(x) \qquad a.e.$$

Proof. We pick a local coordinate chart and express $Z = Z^i \partial_i$ in terms of the local coordinate direction fields. It suffices to prove our claim in the case $p' = p$. From the triangle inequality (1.8iv) and the homogeneity property (1.8v) we may estimate

$$^Z E(f) \leq |Z|^p_\infty n^p \sum_{i=1}^n {}^{\partial_i} E(f).$$

Applying Lemma 1.9.5 this yields

$$d(^Z e) \leq |Z|^p_\infty n^p \sum_{i=1}^n |u_*(\partial_i)|^p(x) d\mu(x). \tag{1.9xix}$$

Since the Euclidean measure $d\mu(x)$ and the metric measure $d\mu_g(x)$ are uniformly equivalent, this last inequality proves the absolute continuity of $d^Z e$.

To prove the pointwise convergence of the ${}^Z_\nu e_\varepsilon(x)$ to $|u_*(Z)|^p(x)$ we note first that (1.9xix) implies $|u_*(Z)|^p(x) = 0$ a.e on $\{x|\, Z(x) = 0\}$. All approximate energies ${}^Z_\nu e_\varepsilon(x)$ are also zero on this set. Thus we need only verify the convergence statement on $\{x|\, Z(x) \neq 0\}$. Here we may do a $C^{1,1}$ change of coordinates from an initial local coordinate chart, turning Z into a coordinate direction. The result then follows from Lemma 1.9.5, and the uniform equivalence of the respective volume measures. □

1.10. Absolute continuity of de for $p > 1$.

Theorem 1.10. *Let* (Ω, g) *be a Riemannian domain (1.1i), and let* $1 < p < \infty$. *Let* $u \in W^{1,p}(\Omega, X)$. *Then the energy density measure* de *is absolutely continuous with respect to Lebesgue measure, i.e. there exists* $|\nabla u|_p(x) \in L^1(\Omega, \mathbb{R})$ *s.t.*

$$de = |\nabla u|_p(x) d\mu(x).$$

Proof. (We will explain the why we use a subscript rather than a superscript for p after the proof.) We reduce to the case where Ω has a local orthonormal frame, as in (1.8.1). Thus we have that equality, which we reproduce here:

$$E(f) = \int_{S^{n-1}} {}^{\omega}E(f)d\sigma(\omega). \tag{1.10i}$$

By Theorem 1.9.6, we may express the corresponding energy-density measures by

$$d({}^{\omega}e) = |u_*(\omega)|^p(x)d\mu(x),$$

so (1.10i) may be rewritten as

$$E(f) = \int_{S^{n-1}} \int_{\Omega} |u_*(\omega)|^p(x)f(x)\,d\mu(x)\,d\sigma(\omega). \tag{1.10ii}$$

It is easy to check that the non-negative functions $|u_*(\omega)|^p(x)$ are jointly measurable in (x, ω). Applying the monotone convergence theorem to an increasing sequence $\{f_i\} \subset C_c(\Omega)$ for which $f_i \to 1$ uniformly on compact subsets of Ω, we deduce

$$\int_{S^{n-1}} \int_{\Omega} |u_*(\omega)|^p(x)\,d\mu(x)\,d\sigma(\omega) \;=\; E < \infty. \tag{1.10iii}$$

Fubini's theorem then implies

$$E(f) = \int_{\Omega} f(x) \left(\int_{S^{n-1}} |u_*(\omega)|^p(x)\,d\sigma(\omega) \right) d\mu(x)$$

for the L^1 function

$$|\nabla u|_p(x) \equiv \int_{S^{n-1}} |u_*(\omega)|^p(x)\,d\sigma(\omega). \quad \square \tag{1.10iv}$$

The reason why p is placed as a subscript in the notation $|\nabla u|_p$ is to prevent confusion about the relation of different p-energies: It is not true (unless $X = \mathbb{R}$) that $|\nabla u|_p$ is equal (even up to a constant multiple) to $|\nabla u|_{p'}^{p/p'}$. (Of course if both expressions make sense, then they will be uniformly equivalent.) Because we will have a special interest in the case $p = 2$ in the second chapter, we will define

$$(1.10v)\quad |\nabla u|^2(x) = \frac{1}{\omega_n}|\nabla u|_2(x) = \frac{1}{\omega_n}\int\limits_{S^{n-1}} |u_*(\omega)|^2(x)d\sigma(\omega),$$

for $u \in W^{1,2}(\Omega, X)$. We will see in Section 2 that this definition is consistent with the usual way of defining $|du|^2$ for maps between Riemannian manifolds. (See also (1.2viii).)

1.11. The calculus of energy-density functions. We collect some useful Calculus facts about our various energy-density functions.

Theorem 1.11. *Let $u \in W^{1,p}(\Omega, X)$ for some $1 < p < \infty$. If $Z, W \in \Gamma(T\overline{\Omega})$ and if $h \in C^{0,1}(\overline{\Omega})$, then*

$$|u_*(Z+W)| \le |u_*(Z)| + |u_*(W)|$$

$$(1.11i)\quad |u_*(hZ)|^p = |h|^p|u_*(Z)|^p$$

$$|u_*(Z)|^p \le C(n)|Z|^p|\nabla u|_p.$$

If ψ is a $C^{1,1}$ diffeomorphism from $\overline{\Omega}_1$ to $\overline{\Omega}$, and if we write $v = u \circ \psi$, then $v \in W^{1,p}(\Omega_1, X)$, and the chain rule

$$(1.11ii)\quad |v_*(Z)|^p = |u_*(\psi_*(Z))|^p$$

holds. If two metrics are close, then their Sobolev energy-density functions are close. In particular, let $\Omega \subset \mathbb{R}^n$ and let δ, g be the Euclidean and a Riemannian metric on Ω. Denote the two corresponding energy-density functions of u by $|\nabla u|_p$ and $|\nabla u|_{p,g}$, and let $d\mu$ and $d\mu_g$ be the two induced volume forms. Let λ^2 and Λ^2 be the minimum and maximum eigenvalues of g (relative to δ), respectively. Then we have the estimate

$$(1.11iii)\quad \frac{\lambda^{2n}}{\Lambda^{n+p}}|\nabla u|_p d\mu \le |\nabla u|_{p,g} d\mu_g \le \frac{\Lambda^{2n}}{\lambda^{n+p}}|\nabla u|_p d\mu.$$

Proof. The statements (1.11i) are restatements of corresponding claims from Theorem 1.8.1, using the subsequent differentiability results of Theorem 1.9.6 and Theorem 1.10. Note that we have replaced the supremum norm $|Z|_\infty$ with the pointwise norm $|Z|_\infty$ in the third inequality. This is easily justified by using (1.8.i) and a partition of unity argument.

For ψ as above it is clear (say using ball averages) that $v = u \circ \psi \in W^{1,p}(\Omega_1, X)$. The chain rule (1.11ii) follows from the fact that the ε energy-density functions satisfy

$$ {}^{Z}_{\nu}e_\varepsilon(x) = {}^{\psi_* Z}_{\nu}e_\varepsilon(\psi(x)), $$

and that for appropriate ν these converge almost everywhere as $\varepsilon \to 0$ to the corresponding directional-energy functions (Theorem 1.9.6).

We prove the claim (1.11iii) as follows. Recall from 1.2 that computing an ε-energy with radius εR ball averaging corresponds to the choice

$$ d\nu(\rho) = \frac{n+p}{R^{n+p}}\rho^{n+p-1}d\rho \qquad 0 < \rho < R. $$

Call the measure above ν_R. Then the ε-approximate energy density function (with respect to a metric g) is given by

$$ {}_{\nu_R}e_{\varepsilon,g}(x) = \frac{n+p}{R^{n+p}}\frac{1}{\varepsilon^n}\int\limits_{B(x,\varepsilon R)} \frac{d^p(u(x),u(y))}{\varepsilon^p}d\mu_g(y). $$

In our case we have the inequalities

$$ \lambda^n d\mu \le d\mu_g \le \Lambda^n d\mu $$

and the ball containments

$$ B(x,\frac{\varepsilon}{\Lambda}) \subset B(x,\varepsilon)_g \subset B(x,\frac{\varepsilon}{\lambda}). $$

Letting $r = \frac{1}{\Lambda}$ and $R = \frac{1}{\lambda}$ we deduce that

$$ \frac{\lambda^n}{\Lambda^{n+p}}\,{}_{\nu_r}e_\varepsilon(x) \le {}_{\nu_1}e_{\varepsilon,g}(x) \le \frac{\Lambda^n}{\lambda^{n+p}}\,{}_{\nu_R}e_\varepsilon(x). $$

Upon integration (and applying the volume inequalities again) we see that (for any $f \ge 0$, $f \in C_c(\Omega)$)

$$ \frac{\lambda^{2n}}{\Lambda^{n+p}}\,{}_{\nu_r}E_\varepsilon(f) \le {}_{\nu_1}E_{\varepsilon,g}(f) \le \frac{\Lambda^{2n}}{\lambda^{n+p}}\,{}_{\nu_R}E_\varepsilon(f). $$

From the weak convergence properties of our ε-energy functionals, we deduce that the limit measures satisfy the same inequalities, i.e.

$$\frac{\lambda^{2n}}{\Lambda^{n+p}}|\nabla u|_p d\mu \le |\nabla u|_{p,g} d\mu_g \le \frac{\Lambda^{2n}}{\lambda^{n+p}}|\nabla u|_p d\mu.$$

This is exactly the claim (1.11iii). □

1.12. Trace theory for Lipschitz domains. In this section we develop the L^p trace theory for finite-energy maps defined on Lipschitz domains. We show that two functions have the same trace if and only if the distance between them is a real-valued function of finite energy, having zero trace on the boundary. We also prove a replacement theorem: If Ω can be partitioned into two Lipschitz subdomains, and if there are finite-energy maps for each subdomain whose traces agree on the common boundary, then together they define a global map of finite energy on all of Ω, and the total energy of this map is the sum of the energies of its components.

DEFINITION. If (Ω, g) is a Riemannian domain then the usual way of saying that $\partial\Omega$ is Lipschitz near $x \in \partial\Omega$ is to require that there be a neighborhood U of x and a local coordinate chart on U so that in these coordinates $\partial\Omega \cap U$ is the graph of a Lipschitz function above some $(n-1)$-dimensional hyperplane. It is easy to see that this definition is equivalent to the following one, which requires the existence of suitable transverse vector fields. We will say that $\partial\Omega$ is Lipschitz near $x \in \partial\Omega$ if there is a neighborhood U of x, a smooth vector field Z defined on U, and positive numbers ρ, t_0 so that for all $x \in \partial\Omega$ the flow $\overline{x}(x,t)$ in 1.1 satisfies

$$(1.12\mathrm{i}) \qquad \begin{aligned} &\overline{x}(x,t) \in \Omega && \text{if } 0 < t < t_0, \\ &\overline{x}(x,t) \notin \Omega && \text{if } -t_0 < t < 0, \\ &d(\overline{x}(x,t), \partial\Omega) > \rho|t| && \text{if } |t| < t_0. \end{aligned}$$

We will say that a compact subset Γ of $\partial\Omega$ is Lipschitz if $\partial\Omega$ is Lipschitz near each $x \in \Gamma$. It is easy to see that this is equivalent to the existence of a smooth vector field Z defined in a neighborhood of Γ which satisfies (1.12i) for some choice of positive numbers ρ, t_0. We will say that Ω itself is a Lipschitz domain if $\partial\Omega$ is Lipschitz.

Let $1 \leq p < \infty$, $u \in W^{1,p}(\Omega, X)$ if $p > 1$ ($u \in BV(\Omega, X)$ if $p = 1$). If $\Gamma \subset \partial\Omega$ is Lipschitz, and if Z is a transverse vector field satisfying (1.12i) for all $x \in \Gamma$, then there is a natural way to define a trace map $u \in L^p(\Gamma, X)$. as follows. Recall that we write $\overline{x}(x,t)$ for the flow induced by Z (1.1). By using Lemma 1.9.2 on a finite number of local coordinate charts (for which Z corresponds to the direction ω in the configuration (1.9i), we see that u has a representative so that for almost all $x \in \Gamma$ the maps $u(\overline{x}(x,t)), 0 < t, t_0$ are either Hölder continuous ($p > 1$) or of bounded variation ($p = 1$). Thus the map

$$u(x) \equiv \lim_{t \to 0^+} u(\overline{x}(x,t))$$

is defined almost everywhere (with respect to $(n-1)$-dimensional Hausdorff measure on Γ). Furthermore, we deduce from the estimate (1.9viii) and from Theorem 1.8.1 that (for a constant C depending on Ω)

(1.12ii)

$$\int_\Gamma d^p(u(x), u(\overline{x}(x,t)))d\Sigma_x^{n-1} \leq C \left(\int_{\Omega^C_{t|Z|_\infty}} de \right) (t|Z|_\infty)^{p-1}.$$

(Here we use the notation Ω_ε^C for the complement in Ω of Ω_ε, i.e. those points in Ω whose distance to $\partial\Omega$ isat most ε.) Because $u \in L^p(\Omega, X)$, for almost all t the maps $u(\overline{x}(x,t))$ are in $L^p(\partial\Omega, X)$. We conclude from (1.12ii) that the trace map is the L^p limit of the maps $u(\overline{x}(x,t))$ as $t \to 0$, so is itself an L^p map. Furthermore, if we use any representative for u it follows that the trace map is the L^p limit of almost all of the maps $u(\overline{x}(x,t)$, as $t \to 0$, and so is well-defined independently of our representative for u.

Lemma 1.12.1. *The definition of trace given above is independent of choice of transverse vector field Z. Also, if $\Gamma_1 \subset \Gamma$, then the trace of u*

on Γ_1 is the restriction to Γ_1 of the trace of u on Γ.

Proof. The second claim follows from the first one. The first claim is a consequence of Lemma 1.8.2 and the following argument. Let Z and W be two transverse vector fields satisfying (1.12i). Let $\overline{x}_1(x,t)$ and $\overline{x}_2(x,t)$ be the corresponding flows. We wish to show that the maps $u(\overline{x}_1(x,t))$ and $u(\overline{x}_2(x,t)$ (from $\partial\Omega$ to X) converge (a.e.t) to the same trace function. Denote the map which sents $\overline{x}_1(x,t)$ to $\overline{x}_2(x,t)$ by ψ. Note that ψ is a bi-Lipschitz mapping of a neighborhood of Γ to its image. Thus we may estimate

$$\text{(1.12iii)} \qquad \int_{\frac{\mu_1}{\rho}}^{\frac{\mu_2}{|Z|_\infty}} \int_{\partial\Omega} d^p(u(\overline{x}_1(x,t)), u(\overline{x}_2(x,t)))d\Sigma_x \; dt$$

$$\leq C \int_{\Omega^C_{\mu_2}} f(x) d^p(u(x), u(\psi(x)))d\mu(x),$$

where $f \in C_c(\Omega^C_{\mu_2+\delta})$ satisfies $0 \leq f \leq 1$, and $f \equiv 1$ for x satisfying

$$\mu_1 \leq d(x, \partial\Omega) \leq \mu_2.$$

Write δ for the supremum of $|\psi(x) - x|$, over the subset $U \cap \Omega^C_{\mu_2}$. Then for small μ_2 we see from continuous dependence on parameters that

$$\delta \leq C\mu_2 \text{max}_\Gamma |Z - W|.$$

Pick μ_1 to satisfy $\frac{\mu_1}{\rho} = \frac{\mu_2}{2|Z|_\infty}$. In case $|Z - W|$ is sufficiently small we may guarantee that $3\delta < \mu_1$ (for all small μ_2). In that case we may construct a suitable f for (1.11iii) which also satisfies the condition necessary to apply Lemma 1.8.2, namely $f \in C_c(\Omega_{3\delta})$. From the Lemma we see

$$\text{(1.12iv)} \qquad \int_{\Omega^C_{\mu_2}} f(x) d^p(u(x), u(\psi(x)))d\mu(x) \leq C\delta^p \int_{\Omega^C_{\mu_2+\delta}} de.$$

From the combination of (1.12iii) and (1.12iv) we see that the average values of the integrals of $d^p(u(\overline{x}_1(x,t)), u(\overline{x}_2(x,t)))$(with respect to

$\frac{\mu_2}{2|Z|_\infty} < t < \frac{\mu_2}{|Z|_\infty}$) converge to 0 as $\mu_2 \to 0$. Thus Z and W define the same trace function on Γ, if the closeness assumption above is satisfied. Since the set of strictly transverse vector fields is a positive cone (i.e. if Z and W are suitable, then so are $aZ + bW$ for positive constants a, b), it is connected. Hence it suffices to prove the lemma for sufficiently close vector fields, as we have done. □

Theorem 1.12.2. *Let (Ω, g) be a Lipschitz Riemannian domain and let $1 < p < \infty$. Any $u \in W^{1,p}(\Omega, X)$ has a well-defined trace map u (or $tr(u)$), with $tr(u) \in L^p(\partial\Omega, X)$. If the sequence $\{u_i\} \subset W^{1,p}(\Omega, X)$ has uniformly bounded energies E^{u_i}, and if $\{u_i\}$ converges in the L^p distance to a map u, then the trace functions of the u_i converge in $L^p(\partial\Omega, X)$ to the trace of u. Two functions $u, v \in W^{1,p}(\Omega, X)$ have the same trace if and only if $d(u,v) \in W^{1,p}(\Omega, \mathbb{R})$ has trace zero.*

Proof. That $u \in W^{1,p}(\Omega, X)$ has a well-defined trace $tr(u) \in L^p(\partial\Omega, X)$ follows from Lemma 1.12.1 and the remarks preceding it. Furthermore, if we fix a transverse vector field Z satisfying (1.12i), and with $|Z|_\infty < 1$, then for almost all (small) t we have the estimate (from (1.12ii)

$$\int_{\partial\Omega} d^p(tr(u)(x), u(\overline{x}(x,t))d\Sigma_x^{n-1} \le Ct^{p-1} \int_{\Omega_t^C} |\nabla u|_p d\mu. \tag{1.12v}$$

Integrating this inequality yields

(1.12vi)

$$\int_0^{t_0}\int_{\partial\Omega} d^p(tr(u)(x), u(\overline{x}(x,t)))d\Sigma_x^{n-1}dt \le C\frac{t_0^p}{p} \int_{\Omega_{t_0}^C} |\nabla u|_p d\mu.$$

We can use the L^p-triangle inequality and the estimate (1.12vi) for two Sobolev functions u, v to bound

$$\left(\int_0^{t_0}\int_{\partial\Omega} d^p(tr(u)(x), tr(v)(x))d\Sigma_x^{n-1}dt\right)^{\frac{1}{p}}.$$

After changing variables from $d\Sigma dt$ to $d\mu$ in the appropriate term, the resulting inequality is

(1.12vii)

$$\left(\int_{\partial\Omega} d^p(tr(u), tr(v))d\Sigma\right)^{\frac{1}{p}} \leq Ct_0^{1-\frac{1}{p}}\left((\int_{\Omega^C_{t_0}} |\nabla u|_p d\mu)^{\frac{1}{p}} + (\int_{\Omega^C_{t_0}} |\nabla v|_p d\mu)^{\frac{1}{p}} \right) + Ct_0^{-\frac{1}{p}}\left(\int_{\Omega^C_{t_0}} d^p(u,v)d\mu\right)^{\frac{1}{p}}.$$

This inequality shows that if the sequence $\{u_i\}$ converges to u in $L^p(\Omega, X)$, and if the $\{u_i\}$ have uniformly bounded energies, then the traces converge to the trace of u. Specifically, we know from semicontinuity that the energy of u is bounded by those of the u_i, so the first term on the right-side of (1.12vii) can be made arbitrarily small (for the function choices $u = u, v = u_i$) by choosing t_0 small. Once t_0 is chosen, the second term can be made small by choosing i large. Next we prove our characterization of equal-trace functions. Using (1.12vi) for the functions u and v, we make another choice in the L^p-triangle inequality (and change variables as above) to estimate the left-hand side below:

(1.12viii)

$$\left(\int_{\Omega^C_{\rho t_0}} d^p(u,v)d\mu\right)^{\frac{1}{p}} \leq C_p t_0 \left(\left(\int_{\Omega^C_{t_0}} |\nabla u|_p d\mu\right)^{\frac{1}{p}} + \left(\int_{\Omega^C_{t_0}} |\nabla u|_p d\mu\right)^{\frac{1}{p}} \right) + Ct_0^{\frac{1}{p}}\left(\int_{\partial\Omega} d^p(tr(u), tr(v))d\Sigma\right)^{\frac{1}{p}}.$$

If $tr(u) = tr(v)$, then (1.12viii) implies that

$$\lim_{t\to 0} t^{-p} \int_{\Omega^C_t} d^p(u(y), v(y))d\mu(y) = 0. \tag{1.12ix}$$

Write $h(y)$ for the function $d(u(y), v(y))$. Define the cut-off function $\eta_t(y)$ to be identically one inside Ω_t, to be zero in the complement of $\Omega_{\frac{t}{2}}$, and to linearly interpolate 0 and 1 on the annulus of points which are between distance $\frac{t}{2}$ and t from $\partial\Omega$. It is clear that $\eta_t h \to h$ in $L^p(\Omega, \mathbb{R})$, as $t \to 0$. Also,

(1.12x)
$$(\int_\Omega |\nabla(\eta_t h) - \nabla h|^p d\mu)^{\frac{1}{p}} \leq (\int_\Omega h^p |\nabla \eta_t|^p d\mu)^{\frac{1}{p}} + (\int_\Omega (\eta_t - 1)^p |\nabla h|^p d\mu)^{\frac{1}{p}}.$$

From (1.12ix) we see that as $t \to 0$ the first term on the right-side of (1.12x) approaches zero, and it is clear that the second one also does. Thus the functions $\eta_t h$ converge to h in the $W^{1,p}$ norm. Since each $\eta_t h$ has trace zero, we deduce (by the first part of our theorem) that h does too. (In fact, this argument shows (in case u is a real-valued function with zero trace and $v \equiv 0$), that for Lipschitz domains the set of functions with zero trace are exactly the space $W_0^{1,p}(\Omega, \mathbb{R})$, i.e. the closure under the Sobolev norm of $C_0^\infty(\Omega)$. This fact is well-known of course.)

We must now show the converse statement, that if the trace of h above is zero, then $tr(u) = tr(v)$. We may deduce from (1.12vi) with $u = h$, that since $tr(h) = 0$,

$$\int_{\Omega^C_{\rho t_0}} h^p(y) d\mu(y) \leq C \frac{t_0^p}{p} \int_{\Omega^C_{t_0}} |\nabla h|^p d\mu.$$

Using this estimate in the right-hand side of (1.12vii), and letting t_0 approach zero, we conclude that $tr(u) = tr(v)$. $\square$

Theorem 1.12.3. *Let Ω be a Lipschitz Riemannian domain which is itself a disjoint union of Lipschitz subdomains Ω_1,Ω_2 and the Lipschitz boundary $\partial\Omega_1 \cap \partial\Omega_2$. Let $1 < p < \infty$, and $u_i \in W^{1,p}(\Omega_i, X)$ for $i = 1, 2$. Suppose that the trace functions $u_1 = u_2$ on $\partial\Omega_1 \cap \partial\Omega_2$. The the map u defined by*

$$u(x) = u_i(x) \qquad \text{if } x \in \Omega_i$$

is a finite energy map, and

$$\int_{\Omega} |\nabla u|_p d\mu = \int_{\Omega_1} |\nabla u_1|_p d\mu + \int_{\Omega_2} |\nabla u_2|_p d\mu.$$

Proof. We have restricted to the case $p > 1$ in our theorem only for reasons of technical simplicity. In this case, for example, it suffices to show that u is a finite-energy map, since then its energy density will agree with u_i's energy density in Ω_i, and the absolute continuity of the energy density with respect to Lebesgue measure immediately implies the additivity of total energy claimed in the theorem. Let Z be a transverse vector field defined in a neighborhood of $\partial\Omega_1 \cap \partial\Omega_2$, satisfying (1.12i), and pointing into Ω_2. Because of our Lipschitz hypothesis, $\partial\Omega_1 \cap \partial\Omega_2$ is covered by a finite number of (relatively) open subsets Γ having the property that each Γ is the bi-Lipschitz image of a map ϕ from a radius r ball in $\mathbb{R}^{n-1}$. We will normalize (by scaling) to the case

$$\phi : B^{n-1}(0,1) \to \Gamma.$$

Then for t_0 small we use the flow induced by Z to define a bi-Lipschitz map ψ by

$$\psi(y,t) = \overline{x}(\phi(y),t) \qquad (y,t) \in B^{n-1}(0,1) \times (-t_0, t_0).$$

Define the map(s) v by $v = u \circ \psi$. Then it follows that $v = v_2$ is a finite energy map from $B \times (0, t_0)$ and $v = v_1$ is a finite energy map from $B \times (-t_0, 0)$ (because finite-energy maps are preserved under bi-Lipschitz composition). Furthermore, using the vector field Z in Ω and ∂_t in $B \times (t_0, t_0)$, we see that the two traces of v on B^{n-1} agree. If we can show that v is a finite-energy map on $B \times (-t_0, t_0)$, then it will follow that u is a finite-energy map in a neighborhood of Γ, and our theorem will follow by the remarks above.

Pick a (unit-) direction ω which is transverse to $\partial_t \equiv \partial_1$, and orient it so that $\omega^1 > 0$. Pick a representative for v so that v is Hölder continuous on almost all lines parallel to ω. This is possible by Lemma

1.9.2 and the fact that the two traces of v agree on B^{n-1}. Parameterize the ω-direction lines by $y \in B^{n-1} \times \{0\}$. We claim that for each y on which the ω-line is continuous, it has finite energy, and that this energy is the sum of the directional energies corresponding to v_1 and v_2. To see why this is so, we appeal to Lemma 1.9.2 in the one-variable case, i.e. Π is a point, and we have two finite-energy (hence Hölder continuous) maps $w_1 : (-t_0, 0) \to X$ and $w_2 : (-t_0, 0) \to X$ which have common (trace) value at $t = 0$. We let w be the resulting continuous map and wish to deduce that it has finite energy. We apply the estimate (1.9viii) and the triangle inequality to deduce that for $t < 0 < t + \varepsilon$,

(1.12xi)
$$\begin{aligned} d^p(w(t), w(t+\varepsilon)) &\leq 2^p(d^p(w(t), w(0)) + d^p(w(0), w(t+\varepsilon)) \\ &\leq 2^p\varepsilon^{p-1}\Big(\int_t^0 |\nabla w_1|_p ds + \int_0^{t+\varepsilon} |\nabla w_2|_p ds\Big). \end{aligned}$$

Of course, if t and $t + \varepsilon$ both lie on one side of 0, then the estimate corresponding to (1.12xi) also holds, in fact without the factor of 2^p. If we integrate (1.12xi) (and the corresponding estimate) between $t_1 < 0 < t_2$, and then use Fubini's Theorem we get

$$\int_{t_1}^{t_2} d^p(w(t), w(t+\varepsilon))dt \leq 2^p\varepsilon^p\Big(\int_{t_1}^{0} |\nabla w_1|_p dt + \int_0^{t_2+\varepsilon} |\nabla w_2|_p dt\Big),$$

i.e.

$$\int_{t_1}^{t_2} e_\varepsilon(t)dt \leq 2^p\Big(\int_{t_1}^{0} |\nabla w_1|_p dt + \int_0^{t_2+\varepsilon} |\nabla w_2|_p dt\Big).$$

This last inequality shows that the ε-energy contribution near $t = 0$ is arbitrarily small (depending on t_1, t_2), so we conclude that w is a finite energy map, and that its energy is the sum of the energies of w_1 and w_2. Thus we have shown that for almost all $y \in B^{n-1} \times \{0\}$ the direction-ω line has finite energy, given by the sum of its two component energies. We immediately conclude (from the converse statement in Lemma 1.9.1) that our map v has finite energy in the direction ω, equal to the sum of its two component energies. By integrating with respect to all transverse directions (i.e. almost all directions), we see that our map v satisfies

$$\int {}^{\omega}E d\sigma(\omega) < \infty.$$

Using the same reasoning as in the converse statement of 1.9.1, it is an easy consequence that $v \in W^{1,p}(B^{n-1} \times (-t_0, t_0), X)$. Thus Theorem 1.12.3 is proven. □

1.13. Precompactness. The precompactness theorem presented here is not needed for the harmonic map theory of §2, but since it is a natural part of Sobolev theory and can be proven quickly using the results we have already developed, we include it in this chapter.

Theorem 1.13. *Let (Ω, g) be a Riemannian domain, (X, d) a locally compact, complete metric space, and $1 \le p < \infty$. Let $\{u_i\} \subset W^{1,p}(\Omega, X)$ if $p > 1$ ($\{u_i\} \subset BV(\Omega, X)$ if $p = 1$) satisfy*

$$\int_{\Omega} d^p(u(x), Q) d\mu(x) + E^{u_i} \le C.$$

(Here Q is a fixed point in X, C is a fixed constant, and E^{u_i} is the total energy of the map u_i.) Then a subsequence of $\{u_i\}$ converges in $L^p(\Omega, X)$ to a finite energy map u.

Proof. The definition of locally compact which we use is that every closed ball of finite radius is compact. For $j = 0$ we define $Q_{j1} = Q$. For each $j \in \mathbb{N}$ we cover $B(Q, j)$ with a finite number of points

$$\{Q_{jr}\}_{r=1,\dots,N_j}$$

so that every $P \in B(Q, j)$ is within $\frac{1}{j}$ of one of these points. The real-valued functions

$$d(u_i(x), Q_{jr})$$

have bounded Sobolev (or BV) norm. (The L^p components of the norm are bounded depending on j, but the energy components are uniformly bounded, by Corollary 1.6.3. Thus by the standard precompactness theorems for BV and Sobolev functions, and by Cantor diagonalization, we extract a subsequence (which we also denote by $\{u_i\}$), so that for each fixed (j,r) there is an L^p function d_{jr} with

$$d(u_i(x), Q_{jr}) \to d_{jr}(x)$$

in $L^p(\Omega, \mathbb{R})$. By choice of subsequence we also assume that *a.e.x*,

$$d(u_i(x), Q_{jr}) \to d_{jr}(x) < \infty$$

for all (j,r). We claim that for such an x there is a unique point $u(x) \in X$ with

$$d(u(x), Q_{jr}) = d_{jr}(x)$$

for all (j,r). To see why this is so, fix an x as above and any j with $j > d_{01}(x)$. Then pick i_0 so that $i \geq i_0$ implies

$$|d(u_i(x), Q_{jr}) - d_{jr}(x)| < \frac{1}{j} \qquad r = 1, \dots, N_j.$$
$$d(u_i(x), Q) < j.$$

By the method in which the points $\{Q_{jr}\}$ were constructed, there is an $r = r(x,j)$ for which

$$d(u_{i_0}(x), Q_{jr(x,j)}) < \frac{1}{j}.$$

From the triangle inequality

$$d(u_i(x), Q_{jr(x,j)}) < \frac{3}{j} \qquad i \geq i_0. \tag{1.13i}$$

This estimate implies that the sequence of points $\{Q_{jr(x,j)}\}$ is Cauchy and that $u(x)$ is its limit.

The function u is clearly measurable (with separable range), and by construction

$$u_i(x) \to u(x) \qquad a.e.x.$$

By Fatou's Lemma

$$\int_\Omega d^p(u(x), Q)d\mu(x) \leq \liminf_{i\to\infty} \int_\Omega d^p(u_i(x), Q)d\mu(x) \leq C,$$

so $u \in L^p(\Omega, X)$. By lower semicontinuity (Theorem 1.6.1) u is a finite energy map, with $E^u \leq C$. By the triangle inequality

$$d^p(u_i(x), u(x)) \leq 2^p d^p(u_i(x), Q) + 2^p d^p(u(x), Q).$$

By construction the sequence on the right is convergent in $L^1(\Omega, \mathbb{R})$ to the function $2^{p+1} d^p(u(x), Q)$. By a well-known extension of the Lebesgue Dominated Convergence Theorem it follows that

$$\int_\Omega d^p(u_i(x), u(x)) d\mu \to 0,$$

so $u_i \to u$ in $L^p(\Omega, X)$. □

2. Harmonic Maps into Non-Positively Curved Metric Spaces

In §2.1 we recall what it means for a metric space to be non-positively curved (NPC). The definition is made using triangle comparisons to Euclidean space, and generalizes the Riemannian notion of non-positive sectional curvature (in the case of simply-connected manifolds). We recall some useful quadrilateral comparison inequalities which appear in a paper of Y.G. Reshetnyak [Re]. We combine these consequences of the NPC definition with the $W^{1,2}(\Omega, X)$ Sobolev-Space theory from Section 1 in order to study harmonic map questions. In §2.2 we study the Dirichlet problem, i.e. the problem of finding a $W^{1,2}(\Omega, X)$ map with given trace which is stationary (among $W^{1,2}(\Omega, X)$ maps having the same trace) for the $p = 2$ Sobolev energy. We show that there is a unique such solution, and its energy is the infimum of all admissible-function energies. In fact, the NPC hypothesis implies that the energy functional is appropriately convex with respect to natural geodesic homotopies of finite-energy maps. This convexity allows one to show that a minimizing sequence is Cauchy in $L^2(\Omega, X)$, i.e. one uses an exact generalization of Dirichlet's Principle for finding harmonic functions. As in the classical case, the convexity property also implies the uniqueness result. We find the generality, elegance, and simplicity of the proofs presented here to be an indication that we have found the proper framework for their expression. The combination of the correct definition of energy (using distances) with the distance comparison results which follow from the NPC hypothesis allows one to replace arguments which are customarily made with first derivative expressions with ones only involving (zeroth- order) distance inequalities. The existence theorem in §2.2 uses the completeness of the target but not any local compactness properties. This is important for the applications discussed in subsequent work, where for example, we consider targets $L^2(M,X)$ where M is a Riemannian manifold and X is an NPC space. (Such targets are also NPC.)

In Chapter 1 we described how a map $u \in W^{1,2}(\Omega, X)$ induces integrable directional-energy functions $|u_*(Z)|^2(x)$ for fixed vector fields $Z \in \Gamma T(\overline{\Omega})$. In §2.3 we show that it is a consequence of the NPC hypothesis that these directional-energy functions satisfy a parallelogram law, i.e.

$$|u_*(Z+W)|^2 + |u_*(Z-W)|^2 = 2|u_*(Z)|^2 + 2|u_*(W)|^2.$$

Thus there is a non-negative integrable tensor π, which generalizes the notion of the pull-back metric u^*h for maps to Riemannian targets (N,h), and so that the $p=2$ Sobolev energy-density function $|\nabla u|^2$ is given by

$$|\nabla u|^2 = g^{ij}\pi_{ij}$$

in local coordinates. The inner product π plays an important role in understanding the structure of harmonic maps to NPC spaces, and in applications to rigidity theory.

One can follow the approach of [GS] in order to prove the interior Lipschitz continuity of harmonic maps into NPC spaces, generalizing various arguments when necessary. In §2.4 we choose instead to deduce this regularity by proving a (weak) form of the classical Bochner inequality for $\Delta|\nabla u|^2$, implying that $|\nabla u|^2$ is essentially subharmonic. In the case of a harmonic function, this inequality can be obtained by a finite differencing technique applied to the energy integral. (Of course it follows formally just by computing $\Delta|\nabla u|^2$.) This technique involves retracting the solution u towards translates of itself, estimating the change in energy and noting that it must be non-negative. These ideas generalize to the NPC setting, and imply the same Bochner inequality.

In §2.5 we recall how to construct centers of mass for maps to NPC spaces. We use the quadrilateral comparisons of [Re] (discussed in (2.1)) in order to derive quantitative estimates for the distance between the centers of mass of different maps (with respect to different weightings). By integrating these estimates one can bound the energy of averages of maps in terms of averages of the energies. In §2.6 we study the equivariant harmonic map problem. We indicate how to construct initial finite-energy maps for certain configurations, and how to combine the solutions to Dirichlet problems with the averaging techniques in §2.5, in order to produce minimizing sequences with uniform (local) Lipschitz continuity control. This approach is analogous to, but technically different from, the Perron method of finding harmonic functions. We conclude the paper in §2.7, where we indicate how to solve the homotopy problem for harmonic maps to spaces with NPC universal covers; we present the natural generalization of the classical results due to Eells and Sampson [ES].

2.1. Non-positively curved metric spaces. A complete metric space (X, d) is said to be non-positively curved (NPC) if the following two conditions are satisfied:

(i) (X, d) is a length space. That is, for any two points P, Q in X, the distance $d(P, Q)$ is realized as the length of a rectifiable curve connecting P to Q. (We call such distance-realizing curves geodesics.)

(ii) For any three points P, Q, R in X and choices of geodesics $\gamma_{P,Q}$ (of length r), $\gamma_{Q,R}$ (of length p) , and $\gamma_{R,P}$ (of length q) connecting the respective points, the following comparison property is to hold: For any $0 < \lambda < 1$ write Q_λ for the point on $\gamma_{Q,R}$ which a fraction λ of the distance from Q to R. That is,

(2.1i) $$d(Q_\lambda, Q) = \lambda p, \quad d(Q_\lambda, R) = (1-\lambda)p.$$

On the (possibly degenerate) Euclidean triangle of side lengths p, q, r and opposite vertices $\overline{P}, \overline{Q}, \overline{R}$, there is a corresponding point

$$\overline{Q}_\lambda = \overline{Q} + \lambda(\overline{R} - \overline{Q}).$$

The NPC hypothesis is that the metric distance $d(P, Q_\lambda)$ (from Q_λ to its opposite vertex) is bounded above by the Euclidean distance $|\overline{P} - \overline{Q}_\lambda|$. This inequality can be written precisely as

(2.1ii)
$$d^2(P, Q_\lambda) \leq (1-\lambda)d^2(P, Q) + \lambda d^2(P, R) - \lambda(1-\lambda)d^2(Q, R).$$

Using successive subdivision one sees that knowing the comparison property just in the case $\lambda = \frac{1}{2}$ suffices to prove it for all $0 < \lambda < 1$.

It is any easy consequence of property (ii) above that geodesics in an NPC-space are unique. Indeed, if γ_1 and γ_2 are two geodesic paths from Q to R, pick P to be a point on γ_1, say at a fraction μ $(0 < \mu < 1)$ of the way from Q to R. Take $\lambda = \mu$ and construct Q_μ (on the geodesic γ_2) as above. By construction the comparison Euclidean triangle must degenerate to a line segment, so that $\overline{P} = \overline{Q}_\mu$, i.e. $d(\overline{P}, \overline{Q}_\mu) = 0$. The comparison property (ii) implies therefore that $d(P, Q_\mu) = 0$ as well. Since μ is arbitrary, we conclude that the geodesics γ_1 and γ_2 coincide. At least in the case that X is locally compact, it is an easy consequence of geodesic uniqueness that an NPC space X must be simply connected. Conversely, it is a well-known fact (proved with Jacobi Field analysis) that any complete, simply connected Riemannian manifold having non-positive sectional curvature is an example of an

NPC space. (One defines the distance between two points to be the infimum of curve lengths for paths connecting the points.) There are many examples of non-Riemannian spaces which are NPC, e.g. trees, Euclidean Buildings, Hilbert Spaces, and other infinite dimensional symmetric spaces. Also, if X is NPC and if (M, g) is a finite-volume Riemannian manifold, then the space $L^2(M, X)$ is also NPC.

A useful consequence of the NPC hypothesis is a quadrilateral comparison property, which we now describe. Theseresults appear as lemmas in a substantially more general theorem due to Reshetnyak [Re], and we include their proofs here for completeness.

Let $\{P, Q, R, S\}$ be an ordered sequence of four points in (X, d). We say that $\{P, Q, R, S\}$ is sub-embeddable into $\mathbb{R}^2$ if there exists an ordered sequence

$$\{\overline{P}, \overline{Q}, \overline{R}, \overline{S}\} \subset \mathbb{R}^2$$

so that

$$\begin{aligned} & d(P,Q) = |\overline{P} - \overline{Q}|, \;\; d(Q,R) = |\overline{Q} - \overline{R}|, \\ \text{(2.1iii)} \qquad & d(R,S) = |\overline{R} - \overline{S}|, \;\; d(S,P) = |\overline{S} - \overline{P}|, \\ & d(P,R) \leq |\overline{P} - \overline{R}|, \;\; d(Q,S) \leq |\overline{Q} - \overline{S}|. \end{aligned}$$

In the above construction we call $\{\overline{P}, \overline{Q}, \overline{R}, \overline{S}\}$ a sub-embedding for $\{P, Q, R, S\}$. (If any collection of points has a corresponding collection in $\mathbb{R}^2$ for which all pairwise distances between corresponding points are equal, we call the Euclidean collection an embedding of the metric collection.) If a metric space has the property that every ordered sequence of four points can be sub-embedded into $\mathbb{R}^2$, we say that (X, d) satisfies the sub-embedding property.

Theorem 2.1.1. [Re] *A length space is NPC if and only if it satisfies the sub-embedding property above. In fact, if (X, d) is NPC then one may always pick a sub-embedding $\{\overline{P}, \overline{Q}, \overline{R}, \overline{S}\}$ for $\{P, Q, R, S\}$ so that the Euclidean sequence forms the consecutive vertices of a convex quadrilateral.*

Proof. Assume (X, d) satisfies the sub-embedding property. Let $\{P, Q, R\} \subset X$ and $0 < \lambda < 1$. Construct Q_λ as in the discussion (ii) above. Then the sequence $\{P, Q, Q_\lambda, R\}$ has a sub-embedding $\{\overline{P}, \overline{Q}, \overline{Q}_\lambda, \overline{R}\} \subset \mathbb{R}^2$. From (2.1i) and (2.1iii) we deduce that $\overline{Q}_\lambda$ lies on the line segment connecting $\overline{Q}$ to $\overline{R}$, i.e.

$$\overline{Q}_\lambda = \overline{Q} + \lambda(\overline{R} - \overline{Q}).$$

Thus the sub-embedding hypothesis (2.1iii) that $d(P, Q_\lambda) \leq |\overline{P} - \overline{Q}_\lambda|$ is exactly the requirement for the triangle comparison property (ii) to hold. Thus (X, d) is NPC.

Conversely, let (X, d) be NPC. Let $\{P, Q, R, S\} \subset X$ be given. Construct Euclidean embeddings $\{\overline{P}, \overline{Q}, \overline{S}\}$ and $\{\overline{Q}, \overline{R}, \overline{S}\}$ for the points $\{P, Q, S\}$ and $\{Q, R, S\}$ respectively. We may construct these embeddings so that the resulting triangles share the edge $\overline{QS}$, and so that $\overline{P}$ and $\overline{R}$ lie on opposite sides of the edge.

Case I. The quadrilateral with consecutive vertices $\{\overline{P}, \overline{Q}, \overline{R}, \overline{S}\}$ is convex. In this case we claim that $d(P, R) \leq |\overline{P} - \overline{R}|$ so that we have a suitable sub-embedding. To verify this claim consider the diagonal $\overline{PR} \subset \mathbb{R}^2$. It intersects $\overline{QS}$ at a point

$$\overline{Q}_\lambda = \overline{Q} + \lambda(\overline{S} - \overline{Q}).$$

Consider the corresponding point Q_λ on the geodesic (in X) connecting Q to S. Then from the triangle inequality and the NPC hypothesis we have

$$\begin{aligned} d(P, Q) &\leq d(P, Q_\lambda) + d(Q_\lambda, R) \\ &\leq |\overline{P} - \overline{Q}_\lambda| + |\overline{Q}_\lambda - \overline{R}| \\ &= |\overline{P} - \overline{R}|. \end{aligned}$$

This proves the claim.

Case II. The quadrilateral with consecutive vertices $\{\overline{P}, \overline{Q}, \overline{R}, \overline{S}\}$ is not convex. Because the interior angles of a Euclidean quadrilateral sum to 2π, exactly one of the interior angles $\angle\overline{P},\overline{Q},\overline{R}$, $\angle\overline{R},\overline{S},\overline{P}$ is greater than π. We assume (by relettering) that $\angle\overline{P},\overline{Q},\overline{R} > \pi$. We orient our points so that $\overline{R} = (0,0)$, $\overline{P} = (0,a), a > 0$, and so that $\overline{Q}$ and $\overline{S}$ lie to the right of the y–axis, (with $\overline{S}$ farther to the right).

Case IIa. $d(P,R) \leq |\overline{P} - \overline{R}|$. In this case our points are a sub-embedding. If we reflect $\overline{Q}$ across the y–axis we increase the distance $|\overline{Q} - \overline{S}|$ and leave the other five pairwise distances unchanged. Thus we create a convex sub-embedding.

Case IIb. $d(P,R) > |\overline{P} - \overline{R}|$. In this case we do not yet have a sub-embedding. If we lift $\overline{P}$ (increase a), keeping $\overline{R} = (0,0)$ fixed, then $\overline{Q},\overline{S}$ must move uniquely if we require

$$\begin{aligned} &d(\overline{P},\overline{Q}), \quad d(\overline{Q},\overline{R}), \\ &d(\overline{P},\overline{S}), \quad d(\overline{S},\overline{R}) \end{aligned}$$

to remain fixed. Using elementary geometry one can check that the diagonal distance $|\overline{Q} - \overline{S}|$ increases as a increases. (One shows that the interior angles $\angle\overline{Q},\overline{R},\overline{S}$, $\angle\overline{S},\overline{P},\overline{Q}$ are increasing with respect to a, by studying the rates of change of the other two interior angles of the quadrilateral determined by the sequence $\{\overline{P},\overline{Q},\overline{R},\overline{S}\}$.) We continue increasing a until $d(P,R) = |\overline{P} - \overline{R}|$. This must happen by the time $\overline{Q}$ hits the y–axis, since

$$d(P,R) \leq d(P,Q) + d(Q,R) = |\overline{P} - \overline{Q}| + |\overline{Q} - \overline{R}| = |\overline{P} - \overline{R}|$$

in this case. Thus one obtains a sub-embedding for some $0 < a \leq d(P,R)$, and by reflection as in Case IIa one may assume that the points $\{\overline{P},\overline{Q},\overline{R},\overline{S}\}$ are consecutive vertices of a convex quadrilateral. □

Theorem 2.1.2. [Re] *Let (X,d) be an NPC space. Let $\{P,Q,R,S\} \subset X$ be an ordered sequence and let $\{\overline{P},\overline{Q},\overline{R},\overline{S}\} \subset \mathbb{R}^2$ be a sub-embedding. Let $0 \leq \lambda,\mu \leq 1$ be given. Define P_λ to be the point which is the fraction λ of the way from P to S (on the geodesic $\gamma_{P,S}$). Let Q_μ be the* *point which is the fraction μ of the way from Q to R (along the opposite geodesic $\gamma_{Q,R}$. Construct the corresponding Euclidean points*

$$\overline{P}_\lambda = (1-\lambda)\overline{P} + \lambda\overline{S}, \ \ \overline{Q}_\mu = (1-\mu)\overline{Q} + \mu\overline{R}.$$

Then

$$d(P_\lambda, Q_\mu) \le |\overline{P}_\lambda - \overline{Q}_\mu|.$$

Proof. This estimate follows from repeated use of the triangle comparison property. As a preliminary step we claim that

$$d(P, Q_\mu) \le |\overline{P} - \overline{Q}_\mu|.$$

In fact, we can estimate the relative lengths, using the appropriate Euclidean identity and the triangle comparison property (2.1ii):

$$\begin{aligned}|\overline{P} - \overline{Q}_\mu|^2 &= (1-\mu)|\overline{P} - \overline{Q}|^2 + \mu|\overline{P} - \overline{R}|^2 - \mu(1-\mu)|\overline{Q} - \overline{R}|^2 \\ &\ge (1-\mu)d^2(P,Q) + \mu d^2(P,R) - \mu(1-\mu)d^2(Q,R) \\ &\ge d^2(P, Q_\mu).\end{aligned}$$

Analogously,

$$d(S, Q_\mu) \le |\overline{S} - \overline{Q}_\mu|.$$

Using the comparison property a final time we estimate

$$\begin{aligned}|\overline{P}_\lambda - \overline{Q}_\mu|^2 &= (1-\lambda)|\overline{P} - \overline{Q}_\mu|^2 + \lambda|\overline{S} - \overline{Q}_\mu|^2 \quad \lambda(1-\lambda)|\overline{P} - \overline{S}|^2 \\ &\ge (1-\lambda)d^2(P, Q_\mu) + \lambda d^2(S, Q_\mu) - \lambda(1-\lambda)d^2(P,S) \\ &\ge d^2(P_\lambda, Q_\mu). \quad \square\end{aligned}$$

Corollary 2.1.3. *Abbreviate the distance function $d(T,U)$ by d_{TU}. For an ordered sequence $\{P,Q,R,S\} \subset X$, define the geodesic interpolation* *points P_λ, Q_μ as in Theorem 2.1.2. Then for any $0 \le \alpha, t \le 1$ the following estimates hold.*

(2.1iv)
$$d^2(P_t, Q_t) \le (1-t)d^2_{PQ} + td^2_{RS} - t(1-t)(\alpha(d_{SP} - d_{QR})^2 + (1-\alpha)(d_{RS} - d_{PQ})^2).$$

(2.1v)
$$d^2(Q_t, P) + d^2(Q_{1-t}, S) \le d^2_{PQ} + d^2_{RS} + t(d^2_{SP} - d^2_{QR}) + 2t^2 d^2_{QR} - t(\alpha(d_{SP} - d_{QR})^2 + (1-\alpha)(d_{RS} - d_{PQ})^2).$$

In case $t = 1$ in (2.1v) we deduce the parallelogram inequality:

$$d^2_{PR} + d^2_{QS} \le d^2_{PQ} + d^2_{QR} + d^2_{RS} + d^2_{SP} - \alpha(d_{SP} - d_{QR})^2 - (1-\alpha)(d_{RS} - d_{PQ})^2. \tag{2.1vi}$$

Proof. Given $\{P, Q, R, S\} \subset X$ we pick a subembedding $\{\overline{P}, \overline{Q}, \overline{R}, \overline{S}\} \subset \mathbb{R}^2$, and take A, B, C, D to be the oriented vectors pointing to consecutive vertices of the Euclidean quadrilateral, i.e.

$$A = \overline{Q} - \overline{P} \qquad B = \overline{R} - \overline{Q}$$
$$C = \overline{S} - \overline{R} \qquad D = \overline{P} - \overline{S}.$$

We have the Euclidean identity

(2.1vii)
$$|\overline{P}_t - \overline{Q}_t|^2 = \frac{1}{2}|t(B+D) + A|^2 + \frac{1}{2}|(1-t)(B+D) + C|^2 = t|C|^2 + (1-t)|A|^2 - t(1-t)|B+D|^2.$$

We have omitted some intermediate computations in the above identity, in which one makes repeated use of the fact that $A + B + C + D = 0$. Since $|A + C| = |B + D|$ dominates both $||C| - |A||$ and $||D| - |B||$, the sub-embedding hypothesis, Theorem 2.1.2 and (2.1vii) combine to imply (2.1iv).

Another Euclidean identity (for $A + B + C + D = 0$) is

$$|tB + A|^2 + |tB + C|^2 = |A|^2 + |C|^2 + t^2|B|^2 + t(|D|^2 - |B|^2) - t|B+D|^2. \tag{2.1viii}$$

It yields (2.1v), which then yields (2.1vi) by setting $t = 1$. □

2.2. The solution to the Dirichlet Problem. To motivate the general existence and uniqueness theorem proved below we first recall Dirichlet's variational principle in the classical case where $X = \mathbb{R}$. (The general proof will follow the samc outline.) Let (Ω, g) be a Riemannian domain. Given a map $\phi \in W^{1,2}(\Omega, \mathbb{R})$ we consider the closed convex subset

$$W_\phi^{1,2}(\Omega, \mathbb{R}) = \{u \in W^{1,2}(\Omega, \mathbb{R}) \mid u - \phi \in W_0^{1,2}(\Omega, \mathbb{R})\}.$$

($W_0^{1,2}(\Omega, \mathbb{R})$ is discussed in (1.12.2).) A harmonic function u is one that is critical for the Dirichlet integral, with respect to small $W_0^{1,2}$ perturbations. In fact, there is a unique $u \in W_\phi^{1,2}(\Omega, \mathbb{R})$ which is critical, and it is the unique energy minimizer in that class, as the following argument shows. Define

$$E_0 = \inf_{v \in W_\phi^{1,2}(\Omega, \mathbb{R})} \int_\Omega |\nabla v|^2.$$

Recall the parallelogram identity for $u, v \in W^{1,2}(\Omega, \mathbb{R})$:

(2.2i)

$$\int_\Omega |\nabla(\frac{u+v}{2})|^2 d\mu + \int_\Omega |\nabla(\frac{u-v}{2})|^2 d\mu = \frac{1}{2}\int_\Omega |\nabla u|^2 d\mu + \frac{1}{2}\int_\Omega |\nabla v|^2 d\mu.$$

Pick a minimizing sequence $\{u_i\} \subset W_\phi^{1,2}(\Omega, \mathbb{R})$, i.e. one for which the corresponding Dirichlet integrals converge to E_0. Pick $u = u_i$, $v = u_j$ in (2.2i). As $i, j \to \infty$ the right side of (2.2i) converges to E_0. Since $\frac{u+v}{2} \in W_\phi^{1,2}(\Omega, \mathbb{R})$, the first term on the left-hand side of (2.2i) is at least E_0. We deduce that for a minimizing sequence

$$\lim_{i,j \to \infty} \int_\Omega |\nabla(u_i - u_j)|^2 = 0.$$

But $u_i - u_j \in W_0^{1,2}(\Omega, \mathbb{R})$ and for $v \in W_0^{1,2}(\Omega, \mathbb{R})$ one has the Poincaré inequality

$$\int_\Omega v^2 d\mu \le C(\Omega) \int_\Omega |\nabla v|^2 d\mu. \tag{2.2ii}$$

Hence

$$\lim_{i,j\to 0} \int_\Omega (u_i - u_j)^2 d\mu = 0$$

as well. Thus $\{u_i\}$ is Cauchy in the $W^{1,2}$ norm, and it converges to $u \in W^{1,2}_\phi$ with energy E_0. If v is any other admissable function then $v_t = (1-t)u + tv$ is an admissable family and (2.2i) implies the Dirichlet integrals are a strictly convex function of t (unless $u = v$). Since this convex function has zero derivative at $t = 0$, its derivative must be non-zero at $t = 1$, so no other $v \in W^{1,2}_\phi$ can be harmonic.

We now prove the general result:

Theorem 2.2. *Let (Ω, g) be a Lipschitz Riemannian domain and let (X, d) be a NPC metric space. Let $\phi \in W^{1,2}(\Omega, X)$. Define*

$$W^{1,2}_\phi = \{u \in W^{1,2}(\Omega, X) \mid tr(u) = tr(\phi)\}.$$

Then there exists a unique $u \in W^{1,2}_\phi$ which is stationary for the $p = 2$ Sobolev energy. In fact, the energy $E^u = \int |\nabla u|^2 d\mu$ of u satisfies

$$E^u = E_0 \equiv \inf_{v \in W^{1,2}_\phi} E^v.$$

(Note that our definition of E in this chapter is consistent with (1.10v), so differs from the one in the rest of Chapter 1 by a multiplicative constant.)

Proof. Let $u, v \in W^{1,2}(\Omega, X)$. Then there is a natural analog of the map $\frac{u+v}{2}$ considered above, namely we define $w(x)$ to be the midpoint of the geodesic connecting $u(x)$ to $v(x)$. It is easy to check that $w \in L^2(\Omega, X)$.

If $x, y \in \Omega$ we consider the sequence $\{u(y), u(x), v(x), v(y)\} \subset X$. Setting $t = \frac{1}{2}$ and $\alpha = 1$ in (2.1iv) of Corollary 2.1.3 we deduce

$$\begin{aligned} 2d^2(w(x), w(y)) &\le d^2(u(x), u(y)) + d^2(v(x), v(y)) \\ &\quad - \frac{1}{2}(d(u(y), v(y)) - d(u(x), v(x)))^2. \end{aligned} \tag{2.2iii}$$

Multiplying (2.2iii) by $f(x)$ (where $f \geq 0$ and $f \in C_c(\Omega)$), then integrating and averaging on the subset $|x-y| < \varepsilon$ of $\Omega \times \Omega$ (as in (1.3)) we deduce first of all that $w \in W^{1,2}(\Omega, X)$. From Theorem 1.12.2 we conclude that w has trace ϕ, so is an admissable function. We also conclude that

$$2\int_\Omega f|\nabla w|^2 \leq \int_\Omega f|\nabla u|^2 + \int_\Omega f|\nabla v|^2 - \frac{1}{2}\int_\Omega f|\nabla d(u,v)|^2 d\mu,$$

for any $f \in C_c(\Omega), f \geq 0$. (We have used Theorem 1.6.2.) Hence we have the analog of (2.2i):

$$2E^w \leq E^u + E^v - \frac{1}{2}\int_\Omega |\nabla d(u,v)|^2 d\mu. \tag{2.2iv}$$

Now let $\{u_i\} \subset W_\phi^{1,2}$ be a minimizing sequence. From (2.2iv) we see that

$$\lim_{i,j\to\infty} \int_\Omega |\nabla d(u_i,u_j)|^2 d\mu = 0.$$

Since $d(u_i,u_j) \in W_0^{1,2}(\Omega, \mathbb{R})$ (Theorem 1.12.2), the Poincaré inequality (2.2ii) holds and we deduce that $\{u_i\}$ converges in $L^2(\Omega, X)$ to a limit function u. By Theorem 1.12.2 we have $u \in W_\phi^{1,2}$, so by semicontinuity (1.6.1) we also have $E^u = E_0$. If v is any other admissable function then we may define the one parameter family of L^2 functions u_t, by defining $u_t(x)$ to be the point which is a fraction t of the way from $u(x)$ to $v(x)$, on the connecting geodesic. From (2.1iv) with $\alpha = 1$ we have

$$\begin{aligned} d^2(u_t(x),u_t(y)) &\leq (1-t)d^2(u(x),u(y)) + td^2(v(x),v(y)) \\ &\quad - t(1-t)(d(u(y),v(y)) - d(u(x),v(x)))^2. \end{aligned} \tag{2.2v}$$

It follows as above that each $u_t \in W_\phi^{1,2}$, and that

$$E^{u_t} \leq (1-t)E^u + tE^v - t(1-t)\int_\Omega |\nabla d(u,v)|^2. \tag{2.2vi}$$

Therefore, if $u \neq v$ the function E^{u_t} is strictly convex. Since it has a minimum at $t = 0$, it is strictly increasing at $t = 1$, so no other admissable v can be harmonic. □

2.3. The pull-back inner product π. We prove the parallelogram identity discussed in the introduction, and discuss the resulting non-negative L^1 tensor π.

Lemma 2.3.1. *Let (Ω, g) be a Riemannian domain and let X be an NPC metric space. If $u \in W^{1,2}(\Omega, X)$, then for any $Z, W \in \Gamma(T\overline{\Omega})$ the parallelogram identity*

$$|u_*(Z+W)|^2 + |u_*(Z-W)|^2 = 2|u_*(Z)|^2 + 2|u_*(W)|^2$$

holds.

Proof. We integrate the pointwise parallelogram inequality (2.1vi) against a non-negative $f \in C_c(\Omega)$ as follows. For $\varepsilon > 0$ fixed, and for each x (sufficiently in) the interior of Ω, write

$$x_1(\varepsilon) = x + \varepsilon Z(x), \qquad x_2(\varepsilon) = x + \varepsilon(Z+W)(x),$$

$$x_3(\varepsilon) = x + \varepsilon W(x).$$

Then we have:

$$\int_\Omega f(x)\{\frac{d^2(u(x), u(x_2(\varepsilon)))}{\varepsilon^2} + \frac{d^2 u(x_1(\varepsilon)), u(x_3(\varepsilon)))}{\varepsilon^2}$$
$$- \frac{d^2(u(x), u(x_1(\varepsilon)))}{\varepsilon^2} - \frac{d^2(u(x_2(\varepsilon)), u(x_1(\varepsilon))}{\varepsilon^2}$$
$$- \frac{d^2(u(x_3(\varepsilon)), u(x_2(\varepsilon)))}{\varepsilon^2} - \frac{d^2(u(x_3(\varepsilon)), u(x)))}{\varepsilon^2}\} d\mu \le 0.$$

We claim that as $\varepsilon \to 0$ this expression converges to the inequality

$$\int_\Omega f\left(|u_*(Z+W)|^2 + |u_*(Z-W)|^2 - 2|u_*(Z)|^2 - 2|u_*(W)|^2\right) d\mu \le 0.$$

To see why this is true consider the second term, for example. By changing coordinates from x to $y = x_3(\varepsilon) = x + \varepsilon W(x)$ we see that it may be rewritten as

$$\int_\Omega (f(y) + o(1)) \frac{d^2(u(y), u(y + \varepsilon((Z-W)(y) + o(1)))}{\varepsilon^2} (1 + C\varepsilon) d\mu(y).$$

Here the first $o(1)$ term depends on the modulus of continuity function

$$\omega(f, \varepsilon|Z|_\infty)(y),$$

and the second one is the difference between $Z - W$ evaluated at x and y. It follows from Theorem 1.8.1 that as $\varepsilon \to 0$ this integral converges to

$$\int_\Omega f(x)|u_*(Z-W)|^2(x)d\mu(x).$$

From the argument above we conclude that

$$\text{(2.3i)} \quad |u_*(Z+W)|^2 + |u_*(Z-W)|^2 \leq 2|u_*(Z)|^2 + 2|u_*(W)|^2.$$

Applying (2.3i) to the vector fields $Z + W$ and $Z - W$ yields

$$|u_*(2Z)|^2 + |u_*(2W)|^2 \leq 2|u_*(Z+W)|^2 + 2|u_*(Z-W)|^2$$

which is exactly the reverse inequality. Thus the parallelogram identity holds. □

For $Z, W \in \Gamma T(\overline{\Omega})$ we define

$$\text{(2.3ii)} \qquad \pi(Z,W) \equiv \frac{1}{4}|u_*(Z+W)|^2 - \frac{1}{4}|u_*(Z-W)|^2.$$

Theorem 2.3.2. *The operator π defined above,*

$$\pi : \Gamma(T\overline{\Omega}) \times \Gamma(T\overline{\Omega}) \to L^1(\Omega, \mathbb{R})$$

is continuous, symmetric, bilinear, non-negative, and tensorial. Specifically,

$$\pi(Z,Z) = |u_*(Z)|^2 \geq 0$$

$$\text{(2.3iii)} \qquad \pi(Z,W) = \pi(W,Z)$$

$$\pi(Z, hV + W) = h\pi(Z,V) + \pi(Z,W) \qquad (h \in C^{0,1}(\overline{\Omega})).$$

If (Ω, g) has local coordinates

$$(x^1, x^2, \ldots, x^n)$$

and corresponding tangent basis

$$\{\partial_1, \ldots, \partial_n\},$$

we write

$$\pi_{ij} = \pi(\partial_i, \partial_j).$$

Then for $Z = Z^i\partial_i$ *and* $W = W^j\partial_j$ *we have*

(2.3iv) $$\pi(Z, W) = \pi_{ij} Z^i W^j.$$

If $\psi : \Omega_1 \to \Omega$ *is a* $C^{1,1}$ *map, then writing* $v = u \circ \psi$, *and* π_v *for the corresponding operator, we have the formula*

(2.3v) $$(\pi_v)_{ij} = \pi_{lm} \psi^l_{,i} \psi^m_{,j}.$$

Hence in local coordinates

(2.3vi) $$|\nabla u|^2 = g^{ij} \pi_{ij}$$

where $[g^{ij}]$ *is (as usual) the inverse matrix to the Riemannian metric matrix*

$$[g_{ij}] = [\langle \partial_i, \partial_j \rangle].$$

Proof. It is clear from the definition and Theorem 1.11 that the map π is continuous. Abbreviating $\pi(Z, W)$ by $\langle Z, W \rangle$, and writing $\langle Z, Z \rangle = |Z|^2$, we see from the scaling property in Theorem 1.11 that

(2.3vii) $$|hZ|^2 = |h|^2 |Z|^2$$

for any $h \in C^{0,1}(\overline{\Omega})$. In particular, $|Z|^2 = |-Z|^2$. We may write the parallelogram identity (2.3.1) in the customary way:

$$|Z + W|^2 + |Z - W|^2 = 2|Z|^2 + 2|W|^2.$$

It is well-known that the parallelogram identity is equivalent to an inner product structure, and we recall the reason why. Starting with the expression $|Z + V + W|^2$, one uses the parallelogram identity to express it in in terms of $|Z + V - W|^2$ (and squares of sums of "length" one or two), then in terms of $|Z - V - W|^2$, and finally in terms of

$|-Z-V-W|^2$, i.e. $|Z+V+W|^2$. Symmetrizing the resulting identity (via the parallelogram law) one deduces that

$$(2.3\text{viii})\qquad \begin{aligned}|Z+V+W|^2 &= 3(|Z|^2+|V|^2+|W|^2)\\ &\quad -(|Z-V|^2+|Z-W|^2+|V-W|^2).\end{aligned}$$

This reduction formula and definition (2.3ii) imply that

$$\pi(Z, V+W) = \pi(Z,V) + \pi(Z,W).$$

By successive applications of this sum-linearity and by scaling one deduces that

$$(2.3\text{ix})\qquad \pi(Z, hW) = h\pi(Z, W)$$

for any rational h. By continuity of π we deduce that (2.3ix) holds for any $h \in \mathbb{R}$. By using a partition of unity argument in Ω (as in the proof of (1.8v)) we deduce that (2.3ix) holds for $h \in C^{0,1}(\overline{\Omega})$. Thus we have verified (2.3iii).

The claim (2.3iv) follows from repeated applications of sum-linearity. To see the chain rule, we note from Theorem 1.11 that

$$\begin{aligned}\pi_v(\overline{\partial}_i, \overline{\partial}_i) = |v_*(\overline{\partial}_i)|^2 &= |u_*(\psi_*(\overline{\partial}_i))|^2\\ &= \pi(\psi^l_{,i}\partial_l, \psi^m_{,i}\partial_m)\\ &= \pi_{lm}\psi^l_{,i}\psi^m_{,i}.\end{aligned}$$

(Here we have used $\{\overline{\partial}_i\}$ for the basis vectors in Ω_1.) Thus (2.3v) holds when $i = j$ and the general result follows from the definition (2.3ii) applied to $\pi_v(\overline{\partial}_i, \overline{\partial}_j)$ and from linearity.

It remains to verify (2.3vi). We note that the function $|\nabla u|^2$ and the function $g^{ij}\pi_{ij}$ are both coordinate-invariant, the first by definition, and the second because of the tensorial-transformation law (2.3v) for π and g. Now, if g is the Euclidean metric δ_{ij}, then we have from Theorem 1.10 and the normalization (1.10v) that

$$|\nabla u|^2 = \frac{1}{\omega_n}\int_{S^{n-1}} e_\omega(x)\, d\sigma(\omega).$$

Writing $\omega = \omega^i \partial_i$ and applying (2.3iv) we see that

$$|\nabla u|^2 = \frac{1}{\omega_n} \int\limits_{S^{n-1}} \pi_{ij} \omega^i \omega^j d\sigma(\omega).$$

Since

$$\frac{1}{\omega_n} \int\limits_{S^{n-1}} \omega^i \omega^j d\sigma(\omega) = \delta^{ij}$$

we see that (2.3vi) holds in the case of a Euclidean metric. The general result follows because in small neighborhoods one can always pick almost-Euclidean local coordinates. By continuity of $|\nabla u|^2$ with respect the the metric (Theorem 1.11), we see that in such a local-coordinate chart, (2.3vi) almost holds. Thus the functions $|\nabla u|^2$ and $g^{ij}\pi_{ij}$ are arbitrarily close, so they are equal. □

2.4. Geodesic homotopies and interior Lipschitz continuity. In this section we deduce the interior Lipschitz continuity of Dirichlet Problem solutions, as a consequence of a weak version of the Bochner inequality for $\Delta|\nabla u|^2$, which states that $|\nabla u|^2$ is almost subharmonic. Using refinements of the arguments given here we will derive much more precise estimates in the sequel, not only for $\Delta|\nabla u|^2$, but for $\Delta\pi(V, V)$, where V is any Lipschitz vector field.

In order to motivate the general theorem we recall how one can use finite- difference ideas in the classical case (of harmonic functions on Euclidean domains) in order to deduce interior gradient bounds. The proof in the general case will follow the same idea but will be complicated both by the fact that the target is more general, and by the fact that the domain is non-Euclidean. Because we do not want to assume any linear structure on the target we will phrase our finite-difference argument in terms of energy inequalities between our solution and certain compactly-supported distortions of it which also make sense in the NPC setting.

Let Ω be a Euclidean domain, and let u be the solution to the Dirichlet problem, as in section 2.2. Let $\eta \in C_0^\infty(\Omega), \eta \geq 0$. Let w be a constant vector of small magnitude, and define

$$u_w(x) = u(x+w). \tag{2.4i}$$

As long as $|w| <$dist(supp$(\eta), \partial\Omega)$, the function $(1-\eta)u + \eta u_w$ is an admissable comparison function, and we have

$$\int_{\Omega} |\nabla u|^2 d\mu \leq \int_{\Omega} |\nabla((1-\eta)u + \eta u_w)|^2 d\mu. \tag{2.4ii}$$

But since u_w is also harmonic we have a symmetric inequality

$$\int_{\Omega_w} |\nabla u_w|^2 d\mu \leq \int_{\Omega_w} |\nabla((1-\eta)u_w + \eta u)|^2. \tag{2.4iii}$$

Adding these two inequalities, cancelling the zeroth order terms, and collecting terms which are first and second order in η, we get:

(2.4iv)

$$0 \leq \int -\nabla\eta \bullet \nabla(u-u_w)^2 - 2\int \eta|\nabla(u-u_w)|^2 + 2\int |\nabla\eta|^2(u-u_w)^2$$
$$+ 2\int \eta^2|\nabla(u-u_w)|^2 + \int \nabla(\eta^2) \bullet \nabla(u-u_w)^2.$$

Noting that the quadratic terms are all well-defined, we deduce (by considering variations by $t\eta$ as $t \to 0^+$)

$$0 \leq \int -\nabla\eta \bullet \nabla(u-u_w)^2 - 2\int \eta|\nabla(u-u_w)|^2. \tag{2.4v}$$

But this is just the statment that

$$\Delta(u_w - u)^2 - 2|\nabla(u_w - u)|^2 \geq 0 \tag{2.4vi}$$

weakly. (Note, that in the present case we actually deduce the equality in (2.4vi), since we are able to consider two-sided variations (t positive and negative). This will not be possible for a general NPC target.) In particular, $(u-u_w)^2$ is subharmonic, so its value at the center of a ball of radius R is bounded by its average value on the entire ball. By monotonicity of the approximate-energy functionals (or by directional derivative theory (1.9), this average value is bounded above by a dimensional constant time $|w|^2 E/R^n$ (where E is the total energy of the map u). We deduce that u is Lipschitz continuous, with constant depending on the distance to the boundary of Ω.

We remark that by considering displacements tw with $t \to 0^+$, we are able to deduce from (2.4vi) that

$$\Delta |u_*(w)|^2 \geq 0$$

weakly. By then averaging over all directions in the unit sphere (or ball), we conclude that

$$\Delta |\nabla u|^2 \geq 0$$

weakly. One could use this last inequality to deduce the Lipschitz estimate, and that is the strategy we will follow in case our domain is non-Euclidean. In that case our argument will be complicated by the fact that it is not possible to pick Killing translation fields (like w above) and so there will be error terms in our estimate.

In order apply the above arguments to maps into NPC spaces, we must first prove some lemmas showing that the analogs of the maps $(1-\eta)u + \eta u_w$ behave well in the general setting. For $u_0, u_1 \in W^{1,2}(\Omega, X)$, $0 \leq t \leq 1$ and $x \in \Omega$ define $u_t(x)$ to be the point which is the fraction t of the way along the geodesic from $u_0(x)$ to $u_1(x)$.

Lemma 2.4.1. *Let (Ω, g) be a Lipschitz Riemannian domain and let (X, d) be an NPC metric space. Let $u_0, u_1 \in W^{1,2}(\Omega, X)$ and let $\eta \in C^{0,1}(\overline{\Omega})$ be a Lipschitz function, $0 \leq \eta \leq 1$. Define*

$$u_\eta(x) \equiv ((1-\eta)u_0 + \eta u_1)(x) \equiv u_{\eta(x)}(x).$$

Then $u_\eta \in W^{1,2}(\Omega, X)$.

Proof. This lemma follows immediately from the convexity of distances under geodesic homotopy and from the triangle inequality. Indeed, we have the estimate

$$\begin{aligned} & d^2(u_{\eta(y)}(y), u_{\eta(x)}(x)) \\ & \quad \leq \; 4d^2(u_{\eta(x)}(x), u_{\eta(x)}(y)) + 4d^2(u_{\eta(x)}(y), u_{\eta(y)}(y)) \\ & \quad \leq \; 4[d^2(u_0(x), u_0(y)) \\ & \qquad + d^2(u_1(x), u_1(y)) + (\eta(y) - \eta(x))^2 d^2(u_0(y), u_1(y))], \end{aligned}$$

which implies that u_η is a finite-energy map. □

Lemma 2.4.2. *Let u_0, u_1, η be as above, with $0 \leq \eta < \frac{1}{2}$. Use subscripts on π to indicate which map is being used to compute the particular tensor. Then, considered as bilinear forms we have the following inequality between the various π's:*

$$\pi_{u_\eta} + \pi_{u_{1-\eta}} \leq \pi_{u_0} + \pi_{u_1} - \nabla\eta \otimes d^2(u_0, u_1) + Q(\eta, \nabla\eta).$$

Here $Q(\eta, \nabla\eta)$ consists of integrable terms which are quadratic in η and $\nabla\eta$.

Proof. Define

$$\eta_- = \min(\eta(x), \eta(y)) \qquad \eta^+ = \max(\eta(x), \eta(y)).$$

If $\eta_- = \eta(y)$ we consider the ordered sequence

$$\{u_{\eta_-}(y), u_{\eta_-}(x), u_{1-\eta_-}(x), u_{1-\eta_-}(y)\},$$

take

$$t = \frac{\eta(x) - \eta(y)}{1 - 2\eta(y)},$$

and apply (2.1v). In case $\eta_- = \eta(x)$ we interchange the roles of x and y and also apply (2.1v). In both cases we deduce

$$\begin{aligned} \text{(2.4vii)} \quad & d^2(u_\eta(y), u_\eta(x)) + d^2(u_{1-\eta}(y), u_{1-\eta}(x)) \\ & \leq d^2(u_{\eta_-}(y), u_{\eta_-}(x)) + d^2(u_{1-\eta_-}(y), u_{1-\eta_-}(x)) \\ & \quad - (\eta(y) - \eta(x))(d^2(u_0(y), u_1(y)) - d^2(u_0(x), u_1(x)))(1 - 2\eta(y)) \\ & \quad + 2(d^2(u_0(x), u_1(x)) + d^2(u_0(y), u_1(y))) \left(\frac{\eta(y) - \eta(x)}{1 - 2\eta(y)}\right)^2. \end{aligned}$$

It follows from the geodesic convexity statement (2.1iv) that

$$(2.4\text{viii})\quad d^2(u_{\eta_-}(y), u_{\eta_-}(x)) + d^2(u_{1-\eta_-}(y), u_{1-\eta_-}(x)) \\ \leq d^2(u_0(y), u_0(x)) + d^2(u_1(y), u_1(x)).$$

Let $Z \in \Gamma(T\overline{\Omega})$ be a Lipschitz vector field. By taking $y = \overline{x}(x, \varepsilon)$, integrating (2.4vii) against a non-negative $f \in C_c(\Omega)$, noting (2.4viii), averaging with respect to ε, and letting $\varepsilon \to 0$, we deduce

$$(2.4\text{ix})\quad |(u_\eta)_*(Z)|^2 + |(u_{1-\eta})_*(Z)|^2 \\ \leq |(u_0)_*(Z)|^2 + |(u_1)_*(Z)|^2 - \eta_*(Z)(d^2(u_0, u_1))_*(Z) + Q(\eta, \nabla\eta),$$

which is the claim of our Lemma. In this deduction we have used the fact that $d^2(u_0, u_1)$ is a real-valued Sobolev function (Corollary 1.6.3), and that if η and h are two functions in $W^{1,2}(\Omega, \mathbb{R})$, then the measure $\eta_*(Z)h_*(Z)d\mu$ is the weak limit of the expressions

$$\frac{(\eta(y) - \eta(x))(h(y) - h(x))}{\varepsilon^2} d\mu$$

(where $y = \overline{x}(x, \varepsilon)$) as $\varepsilon \to 0$. This last fact follows immediately from the identities

$$|(\eta + h)_*(Z)|^2 = |\eta_*(Z)|^2 + |h_*(Z)|^2 + 2\eta_*(Z)h_*(Z)$$

and

$$((\eta + h)(y) - (\eta + h)(x))^2 \\ = (\eta(y) - \eta(x))^2 + (h(y) - h(x))^2 + 2(\eta(y) - \eta(x))(h(y) - h(x)).$$

Thus Lemma 2.4.2 is complete. □

Remark 2.4.3. In case Ω is a Euclidean domain we can use Lemma 2.4.2 to conclude that a harmonic map is Lipschitz continuous in the interior, in direct analogy with the argument given at the beginning of this section. Indeed, we take w to be a constant vector and define $u_w(x)$ as in (2.4i). Then Lemma 2.4.1 and the minimizing property of harmonic maps imply that (2.4ii) and (2.4iii) both hold. We sum these two inequalities and use Lemma 2.4.2 to expand the right-hand side. (Take the trace of the bilinear-form inequality, i.e. apply it to the basis vectors ∂_i and sum over i.) This yields

$$\begin{aligned}\int_\Omega |\nabla u|^2 + \int_{\Omega_w} |\nabla u_w|^2 &\le \int_\Omega |\nabla u_\eta|^2 + \int_{\Omega_w} |\nabla u_{1-\eta}|^2 \\ &\le \int_\Omega |\nabla u|^2 + \int_{\Omega_w} |\nabla u_w|^2 - \int_\Omega \nabla d^2(u, u_w) \bullet \nabla\eta + \int_\Omega Q(\eta, \nabla\eta).\end{aligned}$$

Cancelling the zeroth-order terms, replacing η with $t\eta$, and letting $t \to 0$ as before, we deduce the weak subharmonicity statement which compares to (2.4vi):

$$\Delta d^2(u, u_w) \ge 0.$$

This implies the interior Lipschitz continuity. Note, what we are really showing (and using) here is the fact that whenever u_0 and u_1 are harmonic with respect to the same domain metric, then $d^2(u_0, u_1)$ is subharmonic.

Unfortunately, for a non-Euclidean domain we must estimate somewhat more carefully. In particular, it is convenient to introduce another integrable tensor. In the case of real-valued maps u_0, u_1, it corresponds to $|(1-\eta)(u_0)_*(Z) + \eta(u_1)_*(Z)|^2$.

Lemma 2.4.4. *Given $u_0, u_1 \in W^{1,2}(\Omega, X)$ and $\eta \in C(\Omega)$, with $0 \le \eta \le 1$, there exists a symmetric bilinear integrable tensor $\mathcal{P}(u_0, u_1, \eta)$ defined on the subset $\{0 < \eta < 1\} \subset \Omega$, as follows: For any $Z \in \Gamma(T\overline{\Omega})$,*

$$\frac{d^2(u_{\eta(x)}(x), u_{\eta(x)}(\overline{x}(x,\varepsilon)))}{\varepsilon^2} d\mu \rightharpoonup \mathcal{P}(u_0, u_1, \eta)(Z, Z) d\mu$$

on the subset $0 < \eta < 1$.

Proof. The reason that $\mathcal{P}$ exists is the convexity property of $d^2(P_t, Q_t)$ (2.1iv) on geodesic paths. Since d^2 is always non-negative one immediately deduces that it is Lipschitz continuous away from $t = 0, 1$. In fact, for $0 < s < t < 1$ we have the estimate

(2.4x)
$$\frac{s-t}{1-s} d^2(P_0, Q_0) \le d^2(P_t, Q_t) - d^2(P_s, Q_s) \le \frac{t-s}{1-t} d^2(P_1, Q_1).$$

Now, let $f \ge 0$ be continuous and have support in $\{\delta < \eta(x) < 1-\delta\}$, for some $\delta > 0$. Assume first that $\eta \in C_0^\infty(\Omega)$. Given any $\Delta t > 0$ we may partition a subset of the interval $(\delta, 1-\delta)$ containing the support of f, into a collection of intervals $\{(t_{i-1}, t_i]\}$ so that the t_i are regular values of η, and so that the norm of the partition (the length of the longest interval) is at most Δt.

For $Z \in \Gamma(T\overline{\Omega})$ and for ε and x given, we write y for $\overline{x}(x, \varepsilon)$. Using this convention we make the following definitions:

$$\mathcal{L}_\varepsilon(f) = \int_\Omega f(x) \frac{d^2(u_{\eta(x)}(x), u_{\eta(x)}(y))}{\varepsilon^2} d\mu$$

$$\Omega_i = \{x \mid t_{i-1} < \eta(x) \le t_i\}$$

$$\mathcal{L}_\varepsilon^i(f) = \int_{\Omega_i} f(x) \frac{d^2(u_{t_i}(x), u_{t_i}(y))}{\varepsilon^2} d\mu$$

$$\mathcal{L}^i(f) = \int_{\Omega_i} f(x) |(u_{t_i})_*(Z)|^2 d\mu$$

$$\mathcal{M}_\varepsilon(f) = \int_\Omega f(x) \frac{d^2(u_0(x), u_0(y)) + d^2(u_1(x), u_1(y))}{\varepsilon^2} d\mu$$

$$\mathcal{M}(f) = \int_\Omega f(x)(|(u_0)_*(Z)|^2 + |(u_1)_*(Z)|^2) d\mu.$$

It is clear that

$$\lim_{\varepsilon \to 0} \mathcal{M}_\varepsilon(f) = \mathcal{M}(f).$$

Because the Ω_i have smooth boundaries it is easy to show that

$$\lim_{\varepsilon \to 0} \mathcal{L}_\varepsilon^i(f) = \mathcal{L}^i(f).$$

From the continuity estimate (2.4x) we deduce

$$\sum_i \mathcal{L}^i_\varepsilon(f) - \frac{\Delta t}{\delta}\mathcal{M}_\varepsilon(f) \le \mathcal{L}_\varepsilon(f) \le \sum_i \mathcal{L}^i_\varepsilon(f) + \frac{\Delta t}{\delta}\mathcal{M}_\varepsilon(f). \tag{2.4xi}$$

Letting $\varepsilon \to 0$ in (2.4xi) yields

(2.4xii)

$$\sum_i \mathcal{L}^i(f) - \frac{\Delta t}{\delta}\mathcal{M}(f) \le \liminf_{\varepsilon\to 0} \mathcal{L}_\varepsilon(f) \le \limsup_{\varepsilon\to 0} \mathcal{L}_\varepsilon(f) \le \sum_i \mathcal{L}^i(f) + \frac{\Delta t}{\delta}\mathcal{M}(f).$$

Since Δt may be chosen arbitrarily small (2.4xii) implies the existence of a weak limit to the functionals $\mathcal{L}_\varepsilon$. The corresponding measure is absolutely continuous with respect to Lebesgue measure (from (2.1iv)), so has an L^1 density function $\mathcal{P}(u_0, u_1, \eta)(Z, Z)$. The bilinearity, symmetry and tensorial properties of $\mathcal{P}$ also follow from (2.4xii), since the functions $|(u_{t_i})_*(Z)|^2$ arise from tensors π_{t_i} possessing these properties (Lemma 2.4.1).

If η is only in $C(\Omega)$, then let f, δ be as above, and let $\overline{\eta} \in C(\Omega)$, with $|\eta - \overline{\eta}|_\infty < \delta/2$. The inequalities (2.4x) and (1.4vii) imply

(2.4xiii)

$$|\mathcal{L}^\eta_\varepsilon(f) - \mathcal{L}^{\overline{\eta}}_\varepsilon(f)| \le \frac{2|\eta - \overline{\eta}|_\infty}{\delta}\mathcal{M}_\varepsilon(f) \le \frac{2|\eta - \overline{\eta}|_\infty}{\delta}\mathcal{M}(f^C_\varepsilon).$$

With this estimate one can approximate continuous η with smooth $\overline{\eta}$ to deduce the general claim of Lemma 2.4.4. □

From the functions u_0, u_1, η as above, and the resulting tensor $\mathcal{P}$ discussed in Lemma 2.4.2, we define another auxiliary tensor $\mathcal{C} \ge 0$ by

(2.4xiv)

$$\mathcal{C}(u_0, u_1, \eta) = \pi_{u_0} + \pi_{u_1} - \mathcal{P}(u_0, u_1, \eta) - \mathcal{P}(u_0, u_1, 1 - \eta).$$

The following lemma contains the estimates which will enable us to extend our proof of Lipschitz continuity to the general setting.

Lemma 2.4.5. *Let $u_0, u_1 \in W^{1,2}(\Omega, X)$, $\eta \in C_0^1(\Omega)$ with $0 \leq \eta < \frac{1}{2}$. Using $Q(\eta, \nabla\eta)$ to represent integrable quadratic error terms as in our previous Lemma 2.4.2, we have the following estimates on $\{\eta > 0\}$:*

$$\begin{aligned}\pi_{u_\eta} + \pi_{u_{1-\eta}} \leq & \pi_{u_0} + \pi_{u_1} - \mathcal{C}(u_0, u_1, \eta) \\ & - \nabla\eta \otimes \nabla d^2(u_0, u_1) + Q(\eta, \nabla\eta)\end{aligned} \tag{2.4xv}$$

$$0 \leq \eta\pi_{u_0} + (1-\eta)\pi_{u_1} - \mathcal{P}(u_0, u_1, 1-\eta) \leq \mathcal{C}(u_0, u_1, \eta) \tag{2.4xvi}$$

$$\begin{aligned}&|\pi_{u_{1-\eta}} - \mathcal{P}(u_0, u_1, 1-\eta)| \\ &\leq C|\nabla\eta| d(u_0, u_1)(|\nabla u_0|_1 + |\nabla u_1|_1) + Q(\eta, \nabla\eta).\end{aligned} \tag{2.4xvii}$$

Proof. The inequality (2.4xv) follows from Lemma 2.4.4 (also using (2.4xiii)) and the inequality (2.4vii). Notice that (2.4xv) is an improvement on Lemma 2.4.2. The second inequality (2.4xvi) follows from properties of convex functions. To see why, let $c(t)$ be a convex function defined on the interval $[0, 1]$, and define

$$\mathcal{C}(t) = c(0) + c(1) - c(t) - c(1-t).$$

Then the inequalities

$$0 \leq tc(0) + (1-t)c(1) - c(1-t) \leq \mathcal{C}(t) \tag{2.4xviii}$$

hold, since the first one follows immediately from convexity and the second one reduces to the other convexity statement

$$c(t) \leq (1-t)c(0) + tc(1).$$

Taking $t = \eta(x)$, $y = \overline{x}(x, \varepsilon)$ $c(t) = d^p(u_t(x), u_t(y))$, integrating and averaging against suitable f, and applying (2.4xviii) gives the conclusion (2.4xvi).

We are left to show the final estimate (2.4xvii). From the triangle inequality we have

$$\begin{aligned}&d(u_{1-\eta(x)}(x), u_{1-\eta(x)}(y)) - |\eta(y) - \eta(x)| d(u_0(y), u_1(y)) \\ &\leq d(u_{1-\eta(x)}(x), u_{1-\eta(y)}(y)) \\ &\leq d(u_{1-\eta(x)}(x), u_{1-\eta(x)}(y)) + |\eta(y) - \eta(x)| d(u_0(y), u_1(y)).\end{aligned} \tag{2.4xix}$$

The inequality (2.4xvii) follows from (2.4xix) upon squaring, integrating and averaging, and letting $\varepsilon \to 0$. Letting the reader check the details, we note that the argument uses the estimates

$$d(u_t(x), u_t(y)) \leq d(u_0(x), u_0(y)) + d(u_1(x), u_1(y)),$$
$$|\eta(y) - \eta(x)| \leq \varepsilon |Z|_\infty (|\nabla \eta(x)| + o_\varepsilon(1)),$$

and the fact that the $p = 1$ approximate energy functions of a $p = 2$ Sobolev mapping, when integrated against an L^2 function h, converge to the integral of the limit $p = 1$ energy density times h. This completes the estimates of Lemma 2.4.5. □

Theorem 2.4.6. *Let (Ω, g) be a Lipschitz Riemannian domain, and let u solve the Dirichlet problem, as in Theorem 2.2. Then u is a locally Lipschitz continuous function in the interior of Ω, where the local Lipschitz constant is bounded above by*

$$C\left(\frac{E}{\min(1, dist(x, \partial\Omega)^n)}\right)^{\frac{1}{2}},$$

where C is a constant which depends only on dimension n and on the regularity of the metric g, and where E is the total energy of the map u.

Proof. Let w be a unit vector field defined on a local coordinate chart of Ω. Write $u_{sw}(x) = u(\overline{x}(x, s))$. The function u_{sw} is harmonic with respect to the pull-back metric $g_{sw} = \overline{x}^*(g)$. Let $\eta \in C_0^2(\Omega)$, with $0 \leq \eta < \frac{1}{2}$. Assume (for technical reasons) that $\partial\{\eta > 0\}$ is a set of zero Lebesgue measure. Take $u_0 = u, u_1 = u_{sw}$ and define u_η as above. Assuming that s is sufficiently small we may write the analog of the sum of (2.4ii) and (2.4iii) as

$$\int_\Omega (\pi_{u_0})_{ij} g^{ij} d\mu + \int_\Omega (\pi_{u_1})_{ij} (g^{ij} d\mu)_{sw}$$
$$\leq \int_\Omega (\pi_{u_\eta})_{ij} g^{ij} d\mu + \int_\Omega (\pi_{u_{1-\eta}})_{ij} (g^{ij} d\mu)_{sw}.$$

We may regroup the terms as follows:

$$\text{(2.4xx)}\quad \int_\Omega (\pi_{u_0}+\pi_{u_1})_{ij} g^{ij}\,d\mu \le \int_\Omega (\pi_{u_\eta}+\pi_{u_{1-\eta}})_{ij} g^{ij}\,d\mu + \int_\Omega (\pi_{u_{1-\eta}}-\pi_{u_1})_{ij}((g^{ij}\,d\mu)_{sw} - g^{ij}\,d\mu)$$

If we define the functions $\mathcal{P}$ and $\mathcal{C}$ to be zero on $\{\eta = 0\}$, then because of our assumption that $\partial\{\eta > 0\}$ has measure zero, we may deduce from Lemma 2.4.5 that

$$\text{(2.4xxi)}\quad \int_\Omega (\pi_{u_\eta}+\pi_{u_{1-\eta}})_{ij} g^{ij}\,d\mu \le \int_\Omega (\pi_{u_0}+\pi_{u_1})_{ij} g^{ij}\,d\mu - \int_\Omega \nabla\eta \bullet \nabla d^2(u, u_{sw})\,d\mu - \int_\Omega \mathcal{C}(u_0,u_1,\eta)_{ij} g^{ij}\,d\mu + Q(\eta,\nabla\eta).$$

To estimate the other terms in (2.4xx) we write

(2.4xxii)
$$\pi_{u_{1-\eta}} - \pi_u = \pi_{u_{1-\eta}} - \mathcal{P}(u_0,u_1,1-\eta) + \mathcal{P}(u_0,u_1,1-\eta) - \eta\pi_{u_0} - (1-\eta)\pi_{u_1} + \eta(\pi_{u_1}-\pi_{u_0}).$$

The last term of (2.4xxii) contributes a term in (2.4xx) which may be rewritten by changing variables in part of the expression, from x to $\overline{x}(x,s)$:

$$\text{(2.4xxiii)}\quad \int_\Omega \eta(\pi_{u_{sw}}-\pi_u)_{ij}((g^{ij}\,d\mu)_{sw} - g^{ij}\,d\mu) = \int_\Omega \eta(\pi_u)_{ij}(2g^{ij}\,d\mu - (g^{ij}\,d\mu)_{sw} - (g^{ij}\,d\mu)_{-sw}) + \int_\Omega (\pi_u)_{ij}(\eta_{-sw}-\eta)(g^{ij}\,d\mu - (g^{ij}\,d\mu)_{-sw}).$$

Combining (2.4xx)-(2.4xxiii) with the estimates of Lemma 2.4.5 we see

(2.4xxiv)

$$\begin{aligned}0 \leq &\int_\Omega d^2(u, u_{sw})\Delta\eta d\mu - \int_\Omega \mathcal{C}(u_0, u_{sw}, \eta)_{ij} g^{ij} d\mu \\ &+ Cs \int_\Omega |\nabla\eta| d(u, u_{sw})(|\nabla u|_1 + |\nabla u_{sw}|_1) d\mu \\ &+ Cs \int_\Omega |\mathcal{C}(u_0, u_{sw}, \eta)| d\mu + Cs^2 \int_\Omega (\eta + |\nabla\eta|)|\nabla u|^2 d\mu.\end{aligned}$$

Dividing the inequality (2.4xxiv) by s^2, averaging over a frame of unit vectors w, and letting $s \to 0$ we deduce

$$\int_\Omega |\nabla u|^2(\Delta\eta + C|\nabla\eta| + C\eta) d\mu \geq 0. \tag{2.4xxv}$$

(Note that we may remove the technical condition that $\partial\{\eta > 0\}$ have zero Lebesgue measure by an approximation argument, once (2.4xxv) is established for functions satisfying this requirement.) This last inequality says that $|\nabla u|^2$ is essentially subharmonic. It is well known (see e.g. [Mo]) that this kind of differential inequality implies that the essential supremum of $|\nabla u|^2$ in a ball of radius $\frac{R}{2}$ is bounded by a constant times its average value on the concentric radius R ball, as long as $R \leq 1$. (One can also show this by using radially-symmetric test functions in (2.4xxv) to deduce a differential inequality for the integral averages of $|\nabla u|^2$ over radius r spherical shells centered at points in the $\frac{R}{2}$ ball.) Hence the directional derivative energies $|u_*(Z)|^2$ are similarly bounded (for bounded vector fields Z.) Working with the local-coördinate field ∂_1 we construct a representative of u which is locally Lipschitz continuous in the ∂_1 direction, using the techniques and results of Lemma 1.9.1. Working inductively in successive directions $\partial_2, \ldots, \partial_n$, we construct a representative of u which is Lipschitz continuous in the $\frac{R}{2}$ ball, with Lipschitz constant bounded by a multiple of the bound for $|\nabla u|$. For a given $x \in \Omega$ we may choose R to be the minimum of $d(x, \partial\Omega)$ and 1. The resulting estimate for the Lipschitz constant is exactly the claim of our Theorem. □

Remark 2.4.7. Wherever the boundary of Ω is smooth and the boundary data are C^α $(0 < \alpha < 1)$, T. Serbinowski has shown that the solution u extends to be C^α up to the boundary, with the C^α norm depending on the boundary and on the total energy of the map [Se].

2.5. Center of mass constructions. We recall that averaging works well for maps to NPC spaces (X, d), and that distances between centers of mass can be estimated from above in terms of the average distances between the maps. These distance estimates follow from the quadrilateral comparison lemmas of [Re] which we discussed in §2.1. In the next section we will convert the distance estimates into Lipschitz and energy bounds for various auxiliary Sobolev maps related to the study of equivariant harmonic maps.

Note that the definition of L^2 maps given in (1.1) really only requires the domain to be a measure space, as long as one restricts to separable maps for which inverse images of open sets are measurable. In this generality we consider a domain measure space $\mathcal{M}$, and a probability measure ν defined on $\mathcal{M}$ (i.e. ν is non-negative with total mass 1).

Lemma 2.5.1. *Let $(\mathcal{M}, \nu)$ be a probability measure space, let (X, d) be an NPC space, and let $f \in L^2(\mathcal{M}, X)$. Then there exists a unique center of mass $\overline{f} = \overline{f}_\nu$ for f, defined as the point in X which minimizes the integral*

$$I_{f,\nu}(Q) = \int_{\mathcal{M}} d^2(f(m), Q)\, d\nu(m).$$

Proof. We claim that the integral above is uniformly convex in Q, so that any minimizing sequence converges to a (unique) limit. Indeed, if P_0, P_1 are two points in X, with midpoint $P_{\frac{1}{2}}$, then triangle comparison (2.1ii) yields

$$d^2(f(m), P_{\frac{1}{2}}) \leq \frac{1}{2} d^2(f(m), P_0) + \frac{1}{2} d^2(f(m), P_1) - \frac{1}{4} d^2(P_0, P_1).$$

Integrating over $\mathcal{M}$ we obtain

$$\frac{1}{4}d^2(P_0, P_1) \leq \frac{1}{2}[I(P_0) + I(P_1)] - I(P_{\frac{1}{2}}).$$

Thus any minimizing sequence $\{P_i\}$ is Cauchy, so the integral attains its minimum for a unique point in X. □

Proposition 2.5.2. *Let $\mathcal{M}$ be a measure space, and let ν, ν' be two probability measures on $\mathcal{M}$. Suppose f, h are in $L^2(\mathcal{M}, X)$ for both measure choices. Write $\overline{f}$ for $\overline{f}_\nu$ and $\overline{h}$ for $\overline{h}_{\nu'}$. Then for any $0 \leq \alpha \leq 1$ we have the estimate*

$$\begin{aligned} &d^2(\overline{f}, \overline{h}) \\ &\leq \int_{\mathcal{M}} d^2(f,h)d\nu - \alpha \int_{\mathcal{M}} [d(f,h) - d(\overline{f}, \overline{h})]^2 d\nu \\ &\quad -(1-\alpha) \int_{\mathcal{M}} [d(f,\overline{f}) - d(h,\overline{h})]^2 d\nu + 2d(\overline{f},\overline{h}) \int_{\mathcal{M}} d(h,\overline{h})|d\nu - d\nu'|. \end{aligned}$$

Proof. Consider the geodesic from $Q = \overline{f}$ to $R = \overline{h}$, and write Q_t for the point which is a fraction t of the distance between Q and R. Since

$$I_{f,\nu}(\overline{f}) \leq I_{f,\nu}(Q_t), \qquad I_{h,\nu'}(\overline{h}) \leq I_{h,\nu'}(Q_{1-t})$$

we have

$$\text{(2.5i)} \quad \begin{aligned} \int d^2(f,\overline{f}) + d^2(h,\overline{h}) \; d\nu \leq &\int d^2(f,Q_t) + d^2(h,Q_{1-t})\, d\nu \\ &+ \int [d^2(h,\overline{h}) - d^2(h,Q_{1-t})](d\nu - d\nu'). \end{aligned}$$

Estimate the integrand which appears in the first term of the right side by using Euclidean distance comparison: For each $m \in \mathcal{M}$ we construct the quadrilateral with consecutive vertices $f(m), \overline{f}, \overline{h}, h(m)$ and apply (2.1v):

(2.5ii)

$$\begin{aligned} d^2(f,Q_t) + d^2(h,Q_{1-t}) \leq &d^2(f,\overline{f}) + d^2(h,\overline{h}) + t[d^2(f,h) - d^2(\overline{f},\overline{h})] \\ &- t(\alpha[d(f,h) - d(\overline{f},\overline{h})]^2 \\ &+ (1-\alpha)[d(f,\overline{f}) - d(h,\overline{h})]^2) + 2t^2 d^2(\overline{f},\overline{h}). \end{aligned}$$

Integrating (2.5ii) with respect to ν yields a bound for the first term on the right of (2.5i). We may bound the second term by writing

$$d^2(h,\overline{h}) - d^2(h, Q_{1-t})$$

as a difference of squares, using the fact that

$$d(\overline{h}, Q_{1-t}) = td(\overline{f}, \overline{h}),$$

and applying the triangle inequality in the difference term. The resulting upper bound is

$$td(\overline{f},\overline{h}) \int [d(h,\overline{h}) + d(h, Q_{1-t})]|d\nu - d\nu'|.$$

Use these estimates in (2.5i) and note that the terms which are order zero in t cancel. Divide the resulting inequality by t. Proposition 2.5.2 is the limit inequality as $t \to 0$. □

Remark 2.5.3. Note that Proposition 2.5.2 is the natural generalization of the distance convexity statement (2.2iii) which lies at the heart of the solution to Dirichlet's problem. Indeed, (2.2iii) is the case of Proposition 2.5.2 in which the measure space consists of two points a, b, each of mass $\frac{1}{2}$ ($\nu = \nu'$) and $\alpha = 0$. The maps f,h are given by

$$f(a) = u(x), f(b) = v(x), \overline{f} = w(x)$$
$$h(a) = v(y), h(b) = v(y), \overline{h} = w(y).$$

More generally, if $u \in L^2(M \times \mathcal{M}, X)$ is any parameterized family of maps, then one can construct the average map analogously. In this case we take any probability measure $\nu = \nu'$ defined on $\mathcal{M}$, take $f(\lambda) = u(x, \lambda)$ and $h(\lambda) = u(y, \lambda)$. Applying Proposition 2.5.2 with $\alpha = 0$ we deduce (after integrating over the set $|x - y| = \varepsilon$, averaging, and letting $\varepsilon \to 0$) the generalization of (2.2vi):

$$\text{(2.5iii)}\quad E^{\overline{u}} \le \int_{\mathcal{M}} E^{u_\lambda} d\nu(\lambda) - \int_{\mathcal{M}}\int_{M} |\nabla d(u_\lambda(x), \overline{u}(x))|^2 d\mu(x) d\nu(\lambda).$$

(We have written $u_\lambda(x)$ for $u(x, \lambda)$.)

Finally, we recall the distance decreasing property of projection onto convex subsets of NPC spaces X. This fact is well-known, at least in the case of Riemannian NPC X. In metric space generality it is easiest to prove with quadrilateral comparisons.

Proposition 2.5.4. *Let K be a closed, geodesically convex subset of the NPC space X. Then there is a well-defined nearest-point projection map $\pi : X \to K$ so that*

$$d(\pi(P_0), \pi(P_1)) \leq d(P_0, P_1)$$

for all $P_0, P_1 \in X$. In particular, if $\mathcal{M}, \nu, f$ are as in Lemma 2.5.1, and if the range of f lies in K, then so does the center of mass $\overline{f}$.

Proof. The last claim, that the center of mass $\overline{f}$ lies in K whenever of the range of f does, follows from the existence of π: Because π fixes the range of f, its distance decreasing property immediately implies

$$I_{f,\nu}(\pi(Q)) \leq I_{f,\nu}(Q)$$

for all $Q \in X$. The uniqueness of $\overline{f}$ then proves this claim.

The existence of the projection map is the fact that every $Q \in X$ has a unique nearest point in K: If P_0, P_1 are in K then so is $P_{\frac{1}{2}}$, and so the reader may verify (as in Lemma 2.5.1) that triangle comparison (2.1ii) forces a (distance to Q) minimizing sequence $\{P_i\} \subset K$ to be Cauchy.

We now show the distance-decreasing property of π. Consider the quadrilateral with consecutive verticies $P_0, \pi(P_0), \pi(P_1), P_1$. Let Q_t be the point a fraction t of the way along the geodesic from $\pi(P_0)$ to $\pi(P_1)$. Applying (2.1v) (as in Proposition 2.5.2) yields

$$\begin{aligned} d^2(P_0, Q_t) + d^2(P_1, Q_{1-t}) \leq\ & d^2(P_0, \pi(P_0)) + d^2(P_1, \pi(P_1)) \\ & + t[d^2(P_0, P_1) - d^2(\pi(P_0), \pi(P_1)] \\ & + 2t^2 d^2(\pi(P_0), \pi(P_1)). \end{aligned}$$

Combine this with the competing statement

$$d^2(P_0, \pi(P_0)) + d^2(P_1, \pi(P_1)) \leq d^2(P_0, Q_t) + d^2(P_1, Q_{1-t}).$$

Note that the terms which are order zero in t cancel, divide the resulting inequality by t, and let $t \to 0$ (just as in the proof of Proposition 2.5.2) to deduce Proposition 2.5.4. □

2.6. Equivariant mapping problems. Let (M, g) be a metrically complete Riemannian manifold, possibly with smooth compact boundary ∂M. Denote the fundamental group $\pi_1(M)$ by Γ and the universal cover of M by $\tilde{M}$. Let X be a metric space, and $\rho : \Gamma \to isom(X)$ a homomorphism. Such ρ are also called representations of Γ. We will write $\rho(\gamma)x$ for $\rho(\gamma)(x)$. A particular case of this configuration is the (identity) representation of Γ to $isom(\tilde{M})$, in which Γ acts via deck transformations.

A map $u : \tilde{M} \to X$ is said to be $\Gamma-$equivariant if

$$u(\gamma x) = \rho(\gamma) u(x)$$

for all $x \in \tilde{M}$ and $\gamma \in \Gamma$.

For a $\Gamma-$equivariant map u the real-valued functions $d(u(x), u(y))$ are invariant with respect to the domain action. If the map u is locally a Sobolev map, then it follows that the directional and Sobolev energy densities considered in Chapter 1 are $\Gamma-$invariant, so we may think of them as being defined on the quotient M.

An equivariant map u is said to be harmonic if it is (locally) a $p = 2$ Sobolev map, and if it is stationary for the $p = 2$ total energy, defined for locally Sobolev, equivariant $v : \tilde{M} \to X$ by

$$E^v \equiv \int_M |\nabla v|^2 d\mu. \tag{2.6i}$$

This integral is well-defined as long as M has finite volume. In our case of NPC target X it is clear that the energy convexity statement

(2.2vi) holds:

$$E^{u_t} \leq (1-t)E^u + tE^v - t(1-t)\int_M |\nabla d(u,v)|^2. \tag{2.6ii}$$

(In fact, the more general statement (2.5iii) holds.) Thus stationary is equivalent to minimizing. (Note, the Dirichlet problem considered in §2.2 is a special case of equivariant theory, if one takes the homomorphism ρ to be trivial.)

A strategy in studying group representations which has proven useful is to construct harmonic Γ−equivariant maps, since in many cases the resulting Euler equations (or Bochner formulas) for the maps allow one to deduce information about the representation. (For example, in many cases one can prove that the map must be constant and therefore the map's equivariance implies that $\rho(\Gamma)$ has a fixed point.) One consequence of Proposition 2.6.1 below is that in case of compact M the class of finite-energy equivariant maps is non-empty, so the direct method of energy minimization has a chance of producing a harmonic equivariant map.

Assume that Γ is finitely generated, say by $\gamma_1, \dots, \gamma_p$. (If M is compact this is always the case.) For $P \in X$ define

$$\delta(P) = \max_{i=1,\dots,p} d(\rho(\gamma_i)P, P). \tag{2.6iii}$$

(It is clear that δ is a positive function on X if and only if the representation ρ of Γ has no fixed points.)

Proposition 2.6.1. *Let M, Γ, ρ be as above with $\partial M = \emptyset$ and suppose that X is NPC. Then there exists a ρ−equivariant locally Lipschitz map $u : \tilde{M} \to X$. Let $P \in X$ satisfy $\delta(P) = \delta'$. If M is compact then u can be constructed to be globally Lipschitz, and there is a constant $C = C(M)$ so that the global bound on the Lipschitz constant L of u is of the form*

$$L \leq C\delta'.$$

If M is complete (but not compact) the local Lipschitz constant $L(x)$ of $u(x)$ is bounded by

$$L(x) \leq C(x)\delta'$$

for a locally bounded function $C(x)$, which depends (only) on the domain manifold M.

Proof. Before constructing the map u note that the equality $\delta(P) = \delta'$ implies that we may estimate $d(\rho(\gamma)P, P)$ for any $\gamma \in \Gamma$, in terms of the word length of γ with respect to the generating set $\gamma_1, \ldots, \gamma_p$: For example, write ρ_i for $\rho(\gamma_i)$ and bound

$$\begin{aligned} d(\rho_i\rho_j P, P) &= d(\rho_j P, \rho_i^{-1} P) \\ &\leq d(\rho_j P, P) + d(P, \rho_i^{-1} P) \\ &= d(\rho_j P, P) + d(\rho_i P, P) \\ &\leq 2\delta'. \end{aligned}$$

It follows inductively that if γ has word length $|\gamma| \leq k$ with respect to the generating set, then

$$d(\rho(\gamma)P, P) \leq k\delta'. \tag{2.6iv}$$

We construct an initial equivariant map v which is piecewise constant on $\tilde{M}$, and we first consider the case of compact M: Pick a compact fundamental domain M_0 for M, in $\tilde{M}$, so that its boundary has zero measure. Then the action of $\gamma \in \Gamma$ on $\tilde{M}$ moves M_0 to a different fundamental domain γM_0, and these images define a partitioning of $\tilde{M}$ except on the zero-dimensional set formed by their boundaries. Define the piecewise constant function v by the equivariant extension of $v(M_0) = P$:

$$v(\gamma M_0) \equiv \rho(\gamma)P.$$

We define the (mollified) map $u(x)$ to be the average of v, taken over $B(x, 1)$. In the formalism of Proposition 2.5.2 take the measure space

$\mathcal{M}$ to be the natural numbers $\mathbb{N}$. Let $\gamma_1, \gamma_2, \ldots$ be a denumeration of Γ which extends $\gamma_1, \ldots, \gamma_p$. Define the mapping f from $\mathbb{N}$ to X by

$$f(i) = \rho(\gamma_i)P.$$

For $x \in M_0$ define the probability measure $\nu = \nu_x$ on $\mathbb{N}$ by

$$\nu_x(i) = \frac{\mu(B(x,1) \cap \gamma_i(M_0))}{\mu(B(x,1))}.$$

(Here μ is the Riemannian volume measure on $\tilde{M}$ obtained by lifting the corresponding measure from M.) Define

$$u(x) = \overline{f}_{\nu_x}.$$

The map u is Γ-equivariant because the measures ν_x are Γ-invariant, and because f is equivariant with respect to the natural action of Γ on $\mathbb{N}$.

Because M_0 is compact there is a finite bound k so that whenever the "translation" γM_0 has points within distance one of M_0, then the word length $|\gamma| \leq k$. It follows that (2.6iv) holds with this choice k, for any such γ. Hence

$$\text{(2.6v)} \qquad d(u(x), f(i)) \leq 2k\delta'$$

whenever $\gamma = \gamma_i$ is as above. This is because all such $f(i)$ are within $k\delta'$ of P so the center of mass $u(x)$ (which depends only on these values $f(i)$) is also within the $k\delta'$ ball about P (e.g. Proposition 2.5.4).

If $x, y \in \tilde{M}$ it is clear that we also have the estimate

$$\text{(2.6vi)} \qquad \sum_i |\nu_x(i) - \nu_y(i)| \leq C_1|x-y|$$

for some constant C_1 depending only on M.

We now apply Proposition 2.5.2 in the case $f = h$, $\nu = \nu_y$ and $\nu' = \nu_x$. The first term in the resulting estimate is zero, we ignore the second one since it is non-positive, and we estimate the final one using (2.6v,2.6vi) above. The result is the inequality

$$\text{(2.6vii)} \qquad d^2(u(x), u(y)) \leq 2d(u(x), u(y))2k\delta' C_1|x-y|,$$

which implies the desired uniform Lipschitz constant for u. (Since the Lipschitz constant of u is $\Gamma-$invariant it suffices to estimate it on M_0.)

For non-compact M the proof is essentially the same, except now the fundamental domain M_0 is only locally compact. We replace the global constant k in (2.6vii) with a number $k(x)$ which measures the maximum word length $|\gamma|$ over γ satisfying

$$B(x,1) \cap \gamma M_0 \neq \emptyset.$$

The constant C_1 must now be chosen to depend on x as well, so that (2.6vi) holds for all $y \in B(x,1)$. The inequality then holds for all y if C_1 is taken to be ≥ 2. (If M had sectional curvature bounded from below then C_1 could still be chosen globally.) The result of these modifications is the inequality

$$\text{(2.6viii)} \quad d^2(u(x), u(y)) \leq 2d(u(x), u(y)) 2k(x)\delta' C_1(x)|x-y|$$

for all $x, y \in \tilde{M}$. This proves Proposition 2.6.1. □

Remark 2.6.2. Assume the configuration of Proposition 2.6.1. If we select a "small" closed set $\mathcal{C}_\varepsilon$ in M, by which we mean one that lifts to a compactly contained subset of the interior of M_0, then we can find an equivariant map u which is piecewise constant on each lift of $\mathcal{C}_\varepsilon$, and which has the same Lipschitz constant control as that given in Theorem 2.6.1. The way to construct this function in terms of our previous calculations is to first scale the domain metric so that the lift of $\mathcal{C}_\varepsilon$ is distance at least 1 away from ∂M_0. In this case the function u constructed in the proposition will have the desired properties.

For general complete M it is not clear whether the maps u constructed in Proposition 2.6.1 will have finite energy; the calculation for a particular M depends on an interplay between the deterioration of the Lipschitz constant and the decay of the volume, as one approaches ∞ on M.

Even when the set of candidate maps for the equivariant problem (without boundary) can be shown to be non-empty, there is no Poincaré inequality so the question of convergence is much more delicate than it was for the Dirichlet problem. For example, it is possible for a minimizing sequence to approach ∞ in X. In the case of non-locally compact target it may even happen that a sequence which is uniformly bounded has no convergent subsequence. It is true and somewhat surprising, however, that the sequence of tensors π for any minimizing sequence does converge, to a unique limit tensor:

Proposition 2.6.3. *Let M be a metrically complete Riemannian manifold, possibly with compact Lipschitz boundary ∂M, and with fundamental group $\pi_1(M) = \Gamma$. let X be an NPC space, and let $\rho : \Gamma \to isom(X)$ be a homomorphism. If the corresponding set S of of equivariant ($p = 2$) Sobolev maps is non-empty, then for any energy minimizing sequence $\{v_i\} \subset S$ we have*

$$\lim_{i,j\to\infty} \int_M \int_{S^{n-1}} \left|\, |(v_i)_*(\omega)| - |(v_j)_*(\omega)| \,\right|^2 d\sigma(\omega) d\mu(x) = 0.$$

In particular, there is a unique integrable tensor π so that

$$\lim_{i\to\infty} \int_M \int_{S^{n-1}} |\pi(\omega,\omega) - \pi_{v_i}(\omega,\omega)| d\sigma(\omega) d\mu(x) = 0.$$

Proof. For two maps u, v and their midpoint map w, we apply quadrilateral comparison (2.1iv) with $t = \frac{1}{2}$ and (for the first time in this paper) $\alpha = 0$:

$$\begin{aligned} d^2(w(x), w(y)) &\le \frac{1}{2} d^2(u(x), u(y)) + \frac{1}{2} d^2(v(x), v(y)) \\ &\quad - \frac{1}{4}[d(u(x), u(y)) - d(v(x), v(y))]^2. \end{aligned}$$

Multiplying this inequality by a non-negative f having support in the interior of M, ball-averaging as in (1.2vii), and letting $\varepsilon \to 0$, yields

(2.6ix)

$$\limsup_{\varepsilon\to 0} \frac{n+2}{4\omega_n} \iint\limits_{|x-y|<\varepsilon} f(x) \left(\frac{d(u(x),u(y))}{\varepsilon} - \frac{d(v(x),v(y))}{\varepsilon} \right)^2 \frac{d\mu(x)d\mu(y)}{\varepsilon^n}$$
$$\leq \frac{1}{2}E^u(f) + \frac{1}{2}E^v(f) - E^w(f).$$

(We have normalized the functionals $E^u(\bullet)$ to be consistent with our total energy definition (1.10v).) Using the techniques of Chapter 1 it is possible to show that the expression on the left of (2.6ix) approaches $\frac{1}{4\omega_n}$ times

$$\int\limits_{S^{n-1}} \int\limits_{M} f(x) | \, |u_*(\omega)| - |v_*(\omega)| \, |^2 d\mu(x) d\sigma(\omega)$$

as $\varepsilon \to 0$. We leave the details to the reader, but sketch the idea: Use a partition of unity to approximate the integral in (2.6ix) with a sum of ones involving ε directional energies, so that the new limits are $\Omega \times S^{n-1}$. Then use the fact that for directional energies, the $p = 1$ approximate energy density functions of $p = 2$ Sobolev maps converge in L^2_{loc} to the $p = 1$ energy functions. The claimed limit will then follow from the Lebesgue dominated convergence theorem.

Pick an increasing sequence $\{f_k\}$ with compact support in the interior of M, so that the f_k converge to the constant 1. The right side of (2.6ix) converges to the number

$$\frac{1}{2}E^u + \frac{1}{2}E^v - E^w.$$

We may therefore deduce the first limit claim of Proposition 2.6.3 by choosing $u = v_i, v = v_j$ in (2.6ix), for the functions $\{f_k\}$. The second claim follows from the fact that

$$|\pi_u(\omega,\omega) - \pi_v(\omega,\omega)| \leq | \, |u_*(\omega| - |v_*(\omega)| \cdot | \, | \, |u_*(\omega)| + |v_*(\omega)| \, |$$

and the Cauchy-Schwartz inequality. □

In case one is attempting to understand the behavior of a minimizing sequence, it is helpful to know that the modulus of continuity is under control. The next theorem allows one to construct Lipschitz minimizing sequences. The idea is to modify a given minimizing sequence as follows: Use its values as boundary data for a family of Dirichlet problems, and then use the averaging techniques of §2.5 to piece these Dirichlet problem solutions into a Lipschitz sequence which is still minimizing. In spirit this is like the Perron method for constructing harmonic functions, but the technical ideas here are quite different since they are not based on the maximum principle.

Theorem 2.6.4. *Let M be a complete Riemannian manifold with finite volume (and without boundary), and let X be an NPC metric space. Let $\rho : \Gamma \to isom(X)$ be a representation of $\pi_1(M) = \Gamma$. If M is compact there is an energy minimizing $\rho-$equivariant sequence $\{u_i\}$ with the property that all the u_i are (uniformly) Lipschitz continuous. In fact, there exists a C depending only on M so that the Lipschitz constants of each u_i can be bounded by*

$$C\delta(P),$$

where δ is the displacement function (2.6iii) *and P is any point in X.*

If M is complete (but not compact), suppose the set of finite-energy $\rho-$ equivariant maps from $\tilde{M} \to X$ is non-empty, and that there exists one of energy $E < \infty$. Then there is an equivariant minimizing sequence $\{u_i\} : \tilde{M} \to X$, so that for any compact subset $K \subset M$ and i sufficiently large (depending on K), the u_i are Lipschitz continuous on (the lift to $\tilde{M}$ of) K, with pointwise Lipschitz constants bounded by $C(x)E^{\frac{1}{2}}$. Here $C(x)$ is a locally bounded function depending only on M.

Proof. First treat the case of compact M. Start by picking a finite covering of it with (say) balls, $\{B^j\}_{j=1,\dots,m}$. Pick the balls small enough so that for any $x \in M$ the set

$$\bigcup_{j \mid x \in B^j} B^j \tag{2.6x}$$

is simply connected. Pick a subordinate partition of unity, $\{\eta^j\}$, and compact subsets $Z^j \subset B^j$ so that the support of each η^j is contained in the interior of Z^j. Lift the functions η^j, and the sets B^j,Z^j to invariant functions $\tilde{\eta}^j$ and invariant sets $\tilde{B}^j$,$\tilde{Z}^j$ on $\tilde{M}$.

By Proposition 2.6.1 the set of admissible maps is non-empty, and the infimum of their energies E^0 is bounded by

$$E^0 \le C\delta(P)^2 \tag{2.6xi}$$

where C depends only on M. Now, let a minimizing sequence $\{v_i\}$ be given, with energies $E^{v_i} \to E^0$. In each $\tilde{B}^j$ use the (equivariant) traces of v_i as Dirichlet data, and apply Theorem 2.2 to construct equivariant harmonic maps u_i^j. Extend u_i^j outside $\tilde{B}^j$, by defining $u_i^j = v_i$ there. By Theorem 1.12.3 it follows that

$$E^{u_i^j} \le E^{v_i}. \tag{2.6xii}$$

Now define the equivariant sequence $\{u_i\}$ on $\tilde{M}$ by

$$u_i(x) = \sum_{i=1}^{m} \tilde{\eta}^j(x) u_i^j(x).$$

In the formalism of Proposition 2.5.2, this means that for the measure space $\mathcal{M} = \{1, \dots, m\}$ and for $x \in \tilde{M}$ we assign a measure ν_x with values

$$\nu_x(j) = \tilde{\eta}^j(x),$$

the map $f : \{1, \dots, m\} :\to X$ given by

$$f(j) = u_i^j(x),$$

and define $u_i(x)$ to be the center of mass $\overline{f}_{\nu_x}$.

Proposition 2.5.2 allows us to compare $u_i(x)$ to $u_i(y)$ for y near x. Defining ν_y in terms of the values $\tilde{\eta}^j(y)$ and h in terms of the values $u_i^j(y)$, we deduce

(2.6xiii)

$$d^2(u_i(x), u_i(y)) \leq \sum_{j=1}^{m} \tilde{\eta}^j(x) d^2(u_i^j(x), u_i^j(y)) + 2d(u_i(x), u_i(y)) \sum_{j=1}^{m} d(u_i^j(y), u(y)) |\tilde{\eta}^j(y) - \tilde{\eta}^j(x)|.$$

The following limit statement, which we digress to prove, quantifies the sense in which the second term in the right side of (2.6xiii) becomes negligible as $i \to \infty$:

(2.6xiv)
$$\lim_{i\to\infty} \sup_{\substack{j\in\{1,\dots,m\}\\ y\in\tilde{Z}^j}} d(u_i^j(y), u_i(y)) = 0.$$

Since $u_i(y)$ is an average of the $u_i^j(y)$ for which $y \in \tilde{Z}^j$, (2.6xiv) follows from Proposition 2.5.4 and the equation

(2.6xv)
$$\lim_{i\to\infty} \sup_{\substack{j,l\in\{1,\dots,m\}\\ y\in\tilde{Z}^j\cap\tilde{Z}^l}} d(u_i^j(y), u_i^l(y)) = 0.$$

We show (2.6xv): There is a $\delta_0 > 0$ depending on the sets Z^j so that whenever $y \in \tilde{Z}^j$, the ball $B(y, 2\delta_0) \subset \tilde{B}^j$. From the interior Lipschitz continuity (Theorem 2.4.6) and from (2.6xi) we see that for $|z-y| < \delta_0$ the Lipschitz constant of $L_i^j(z)$ of $u_i^j(z)$ is uniformly bounded (independently of y, Z^j and large i),

(2.6xvi)
$$L_i^j(z) \leq L\delta(P)^2 \equiv L'$$

(for some constant L depending on M and E^0). Thus for $y \in \tilde{Z}^j \cap \tilde{Z}^l$ and $|z - y| < \delta$ the triangle inequality yields

$$d(u_i^j(z), u_i^l(z)) \geq d(u_i^j(y), u_i^l(y)) - 2L'\delta.$$

Integrate over $B(y,\delta)$ to get

(2.6xvii)
$$\int_{B(y,\delta)} d^2(u_i^j(z), u_i^l(z))d\mu(z) \geq C\delta^n[d(u_i^j(y), u_i^l(y)) - 2L'\delta]^2.$$

The function $d(u_i^j, u_i^l)$ defined on M equals zero outside $B^j \cup B^l$, so we can bound the left side of the above inequality via the Poincaré inequality

$$\int_{B^j \cup B^l} d^2(u_i^j, u_i^l)d\mu \leq C \int_{B^j \cup B^l} |\nabla d(u_i^j, u_i^l)|^2 d\mu. \tag{2.6xviii}$$

By the energy convexity statement (2.6ii) and by (2.6xii) we have the bound

$$\int_{B^j \cup B^l} |\nabla d(u_i^j, u_i^l)|^2 d\mu \leq 4(E^{v_i} - E^0) \tag{2.6xix}$$

for the right side of (2.6xviii). The limit claim (2.6xv) follows by combining (2.6xvii), (2.6xviii), and (2.6xix) for arbitrarily small δ.

We now return to the estimate (2.6xiii) to show the uniform Lipschitz continuity of the functions u_i, for i large. By (2.6xiv) and the geometric-arithmetic mean inequality we may pick i large enough so that the second term on the right of (2.6xiii) is bounded above by

$$\frac{1}{2}d^2(u_i(x), u_i(y))$$

independently of x, y. Thus

$$d^2(u_i(x), u_i(y)) \leq 2\sum_{j=1}^{m} \tilde{\eta}^j(x) d^2(u_i^j(x), u_i^j(y)).$$

Applying the interior Lipschitz continuity estimate (2.6xvi) for the harmonic map u_i^j then gives

$$d^2(u_i(x), u_i(y)) \leq C\delta(P)^2$$

for some universal constant C depending only on M. By removing a finite number of the $\{u_i\}$ from our sequence, we deduce the Lipschitz control claimed in Theorem 2.6.4.

To show that $\{u_i\}$ is minimizing we also use (2.6xiii). Applying Young's inequality to the second term on its right side gives

$$(1-\delta)d^2(u_i(x),u_i(y)) \leq \sum_{j=1}^{m} \eta^j(x)d^2(u_i^j(x),u_i^j(y))$$
$$+\frac{1}{\delta}\left(\sum_j d^2(u_i^j(y),u_i(y))\right)\left(\sum_j |\eta^j(y)-\eta^j(x)|^2\right).$$

Averaging with respect to $|y-x|\leq\varepsilon$, integrating over M, letting $\varepsilon\to 0$, and recalling that $E^u(f)$ refers to the value of the linear functional $E^u(\bullet)$ applied to f, the above inequality implies

$$(1-\delta)E^{u_i} \leq \sum_j E^{u_i^j}(\eta^j) + \frac{C}{\delta}o_i(1). \tag{2.6xx}$$

(The $o_i(1)$ terms approaches zero as $i\to\infty$, by (2.6xiv).) Because the $\{\eta^j\}$ form a partition of unity, and because $E^u(\bullet)$ is linear,

$$(1-\delta)E^{u_i} \leq E^{v_i} + \sum_j (E^{u_i^j}(\eta^j) - E^{v_i}(\eta^j)) + \frac{C}{\delta}o_i(1).$$

Appealing to Proposition 2.6.3 we deduce thatthe sum term in the above inequality approaches zero as $i\to\infty$, so

$$\limsup_{i\to\infty}(1-\delta)E^{u_i} \leq E^0. \tag{2.6xxi}$$

Hence $\{u_i\}$ is minimizing, since δ is arbitrary. Thus the proof of Proposition 2.6.4 is complete in case M is compact.

The case of non-compact M is a modification of the above argument. For a base-point $x_0\in M$ write

$$K_m = \{x\in M : |x-x_0|\leq m\}.$$

Pick a countable covering of M by balls $\{B^j\}_{j\in\mathrm{N}}$ so that for an increasing sequence $\{j_m\}_{m\in\mathrm{N}}$ the finite collections

$$\{B^j\}_{j=1,\dots,j_m}$$

cover K_m, and so that $j>j_m$ implies $B^j\cap K_{m-1}=\emptyset$. Pick a subordinate partition of unity $\{\eta^j\}$ and sets $Z^j\subset B^j$ as in the compact case, and denote their lifts to $\tilde{M}$ as before. For fixed $m \in \mathbb{N}$ we consider the finite partition of unity

$$\{\eta^j\}_{j=1,\dots,j_m} \cup \{1 - \sum_{j=1}^{j_m} \eta^j\}.$$

Given a minimizing sequence $\{v_i\}$ as before, we again do ball replacement in $\{\tilde{B}^j\}_{j=1,\dots,j_m}$. Define the average map u_i as before: one is now averaging over a set of $j_m + 1$ points, the last of which is $v_i(x)$ itself. Because of the way they were constructed, $\{\eta^j\}_{j=1,\dots,j_{m'}}$ is a partition of unity for $K_{m'-1}$, for each $m' \leq m$. It follows that the limit statement (2.6xiv) still holds (uniformly) for $y \in \tilde{K}_{m'-1}$. Therefore one is able to deduce a uniform Lipschitz bound L for the functions u_i restricted to $\tilde{K}_{m'-1}$, when i is large. It is of the form

$$L \leq C_{m'} E^{\frac{1}{2}},$$

which is the claimed dependence. The proof that $\{u_i\}$ is minimizing proceeds as before, except now one replaces the pointwise estimate for $d(u_i^j(y), u_i(y))$ (2.6xiv) with the weaker claim

$$\lim_{i \to \infty} \int_M d^2(u_i^j(y), u_i(y)) = 0, \qquad j = 1, \dots, j_m.$$

This follows from the triangle inequality and

$$\lim_{i \to \infty} \int_M d^2(u_i^j(y), v_i(y)) = 0. \qquad j = 1, \dots, j_m.$$

Since $d(u_i^j(y), v_i(y)) = 0$ outside of a compact subset of M this last equality follows from the Poincaré inequality and energy convexity, i.e. the analogies of (2.6xvii) and (2.6xix). Thus, by a diagonalization with respect to the K_m and the v_i we are able to construct a suitable sequence $\{u_i\}$. The proof of Theorem 2.6.4 is complete. $\square$

Using minimizing sequences which have local modulus of continuity control enables one to reduce the question of global convergence to that of convergence at a point:

Proposition 2.6.5. *Let M be a complete Riemannian manifold, possibly with compact Lipschitz boundary ∂M. Let $\Gamma = \pi_1(M)$, and $\rho : \Gamma \to isom(X)$ be a homomorphism. Let $\{u_i\}$ be an equivariant minimizing sequence which has local modulus of continuity control. That is, for each $x \in \tilde{M}$ we assume there is an (equivariant) function $\omega(x,r)$ $(0 \le r < r_x)$ which is monotone increasing in r, which satisfies $\omega(x,0) = 0$ and so that*

$$\sup_i \sup_{|x-z|\le r} d(u_i(x), u_i(z)) \le \omega(x,r).$$

Then the sequence $\{u_i\}$ converges (locally uniformly and hence in L^2_{loc}) to an equivariant harmonic map u if and only if there exists an $x \in \tilde{M}$ at which the sequence of points $\{u_i(x)\}$ is convergent.

Proof. The reason this proposition is true is the energy convexity statement (2.6ii) which implies

(2.6xxii)
$$\int_M |\nabla d(u_i, u_j)|^2 d\mu \to 0.$$

Let $x \in \tilde{M}$ be a point of convergence, $\{u_i(x)\} \to P$. By the modulus of continuity estimate and the triangle inequality it follows that

$$\limsup_{i,j\to\infty} \sup_{|z-x|\le r} d(u_i(z), u_j(z)) \le 2\omega(r)$$

(where we have written $\omega(r)$ for $\omega(x,r)$. Thus for i,j sufficiently large, the function

$$d(u_i(z), u_j(z)) - 3\omega(r)$$

is negative on the set $|z-x| = r$. For such i,j we may apply the Poincaré inequality for compact exterior regions to $B(x,r)$:

(2.6xxiii)
$$\int_{B(x,R)} [(d(u_i(z), u_j(z)) - 3\omega(r))^+]^2 d\mu(z) \le C_{r,R,x} \int_{B(x,R)} |\nabla d(u_i, u_j)|^2 d\mu.$$

By (2.6xxii) the integral on the right converges to zero as $i, j \to \infty$, so we deduce from (2.6xxiii) and the modulus of continuity control that

$$\limsup_{i,j\to\infty} d(u_i(z), u_j(z)) \leq 3\omega(r)$$

for all $z \in B(x, R)$. Since R and r are arbitrary, it follows that the sequence $\{u_i\}$ converges everywhere. Because of the modulus of continuity control this convergence is uniform on lifts of compact subsets of M. In particular, the u_i converge locally in L^2 to an equivariant map u, so semicontinuity (Theorem 1.6.1) applies and the map u is harmonic. □

Remark 2.6.6. Compare the results of the last three propositions. If the class of equivariant Sobolev maps is nonempty, and if X is locally compact, then the minimizing sequences of 2.6.4 either converge to the ideal boundary (at ∞) of X, or a subsequence converges to a harmonic map. It follows from energy convexity (2.6ii)(2.5iii) that all harmonic maps are contained in a (possibly) multi-parameter family of "parallel" harmonic maps. (And from 2.6.3 they all induce the same tensor π.) A very interesting and important (unanswered) problem is to understand when harmonic maps are actually unique (up to domain or range isometries).

Even if the minimizing sequence from 2.6.4 does not have a convergent subsequence in X, it is still true that the real-valued functions $d(u_i(x), u_i(y))$ have subsequences which converge to $\Gamma-$invariant distance functions $d(x, y)$ defined on $\tilde{M} \times \tilde{M}$. In our sequels to this paper we will show that the infinitesimal metric induced by any such limit d is the unique tensor π of 2.6.3. In fact, d actually arises as an induced distance for a map to a (limit) NPC space constructed from convex subsets of X. We call these limit d's "harmonic", and their structure is the focus of much of our later work.

One way to guarantee existence of a limit is to impose Dirichlet conditions. We will make use of the following result in our sequels:

Proposition 2.6.6. *Consider the configuration of Proposition 2.6.1 and a small set C_ε in M, as defined in Remark 2.6.2. Let there be a $\rho-$ equivariant Sobolev map ψ from $\tilde{M}$ to X. Then there exists a unique locally Lipschitz $\rho-$equivariant harmonic map $u : \tilde{M} - \tilde{C}_\varepsilon :\to X$ with trace ψ on $\partial\tilde{C}_\varepsilon$. If the boundary is smooth and if ψ extends to be C^α $(0 < \alpha < 1)$ up to the boundary, then so does u.*

Proof. By energy convexity (2.6ii) and Poincaré inequalities for exterior regions (to ∂C_ε) a minimizing sequence converges in L^2_{loc}, to an equivariant harmonic map. The interior regularity follows from Theorem 2.4.6 and the boundary regularity follows from [Se] (See Remark 2.4.7.) □

2.7. Homotopy problems. As a final application of the techniques in this paper we generalize the classical Eells-Sampson harmonic map theory to the case of metric space targets. For technical simplicity we assume our domain manifold M is compact. Let N be a metric space with universal cover X which is NPC. We will say that a continuous map $u : M \to N$ is harmonic if it is locally energy minimizing. Precisely, each $x \in M$ is to have a neighborhood so that all continuous comparison maps which agree with u outside this neighborhood have no less energy. We prove:

Theorem 2.7.1. *Let M be as above, with $\partial M = \emptyset$. Let N be compact, and let $f : M \to N$ be a continuous map. Then there exists a Lipschitz harmonic map $u : M \to N$ which is homotopic to f.*

Proof. For fixed $x \in M$ the f induces the homomorphism $f_* : \pi_1(M)_x \to \pi_1(N)_{f(x)}$. We may lift f to a map $\tilde{f} : \tilde{M} \to X$, so $\tilde{f}$ is f_* equivariant. Use Proposition 2.6.1 to construct a finite energy f_*-equivariant map, and Theorem 2.6.4 to construct a uniformly Lipschitz minimizing sequence $\{\tilde{u}_i\}$. Because X is NPC one may use geodesic homotopies to deduce that all continuous f_*-equivariant maps are homotopic. (The continuity of the homotopies follows from the quadrilateral comparisons of [Re], i.e. Theorem 2.1.2.) Hence the maps $\tilde{u}_i$ are equivalently homotopic to $\tilde{f}$, and we deduce that the projections u_i are homotopic to f. Since the u_i are uniformly Lipschitz continuous, a subsequence converges uniformly to a limit map u, which is therefore also homotopic to f.

It remains to show that the map u is harmonic. Let $x \in M$ and let $\mathcal{O}$ be a simply connected neighborhood of x with Lipschitz boundary. We claim that u is minimizing with respect to continuous comparison functions which agree with u outside $\mathcal{O}$. Since $\mathcal{O}$ is simply connected the map $u|_{\mathcal{O}}$ lifts to a map $\tilde{u}$ from $\mathcal{O}$ to X, and we are reduced to showing that this map $\tilde{u}$ solves the Dirichlet problem of §2.2.

For any $\varepsilon > 0$ and i sufficiently large, we may choose the lifting $\tilde{u}$ so that

(2.7i)
$$\sup_{z\in\mathcal{O}} d(\tilde{u}_i(z), \tilde{u}(z)) < \varepsilon$$
$$\int_{\mathcal{O}} d^2(\tilde{u}_i, \tilde{u})d\mu < \varepsilon.$$

Let $\tilde{v}_i$ be the solution to the Dirichlet problem on $\mathcal{O}$, with boundary data u_i. Let $\tilde{v}$ be the solution with data $\tilde{u}$. Since $d^2(\tilde{v}_i, \tilde{v})$ is subharmonic (Remark 2.4.3), $d(\tilde{v}_i, \tilde{v})$ attains its maximum on $\partial\mathcal{O}$, so is bounded by the maximum of $d(u_i, u)$ there. In particular we see that for sufficiently large i

(2.7ii)
$$\int_{\mathcal{O}} d^2(\tilde{v}_i, \tilde{v})d\mu < \varepsilon.$$

Finally, from energy convexity (2.6ii) and the Poincaré inequality for $\mathcal{O}$, we have

(2.7iii)
$$\int_{\mathcal{O}} d^2(\tilde{u}_i, \tilde{v}_i)d\mu < \varepsilon$$

for i large. Combining the three estimates and using the L^2 triangle inequality, we deduce

$$\int_{\mathcal{O}} d^2(\tilde{u}, \tilde{v}) d\mu < 9\varepsilon.$$

Hence $\tilde{u} = \tilde{v}$ and so u is harmonic. □

If one is considering the Dirichlet homotopy problem, then one need not assume local compactness on the target space:

Theorem 2.7.2. *Let M be a compact manifold with smooth boundary, and let N be a complete metric space with NPC universal cover X. Let $f : M \to N$ be a continuous $p = 2$ Sobolev map, with the property that its trace on ∂M is C^α for some $0 < \alpha < 1$. Then there exists a unique globally minimizing harmonic map $u : M \to N$ homotopic to f and with the same boundary values. u is Lipschitz continuous in the interior of M and extends to be C^α up to the boundary.*

Proof. We may lift f to an f_* equivariant map $\tilde{f}$ as in the previous theorem. Construct the homomorphism f_* from $\pi_1(M)_x$ to $\pi_1(N)_{f(x)}$ for a base point $x \in \partial M$. By hypothesis the lifing is a finite energy map (in the sense of §2.6). It follows from energy convexity (2.6ii) and the $(M, \partial M)$ Poincaré inequality that a minimizing sequence of f_*−equivariant maps having the same boundary data as $\tilde{f}$ converges in L^2 (in the sense of §2.6), to a unique harmonic equivariant map. This map is locally Lipschitz and extends to be C^α up to the boundary (by Theorem 2.4.6 and the results of [Se]). It follows (via geodesic homotopy, as in the previous theorem) that its projection u is homotopic to f. By our choice of basepoint we see that a continuous map $v : M \to N$ which is homotopic to f relative to ∂M lifts to an f_*− equivariant map, so it follows that u is minimizing relative to all such maps. □

References

[CZ] Corlette, K. and Zimmer, R.J., *Super Rigidity for cocycles and hyperbolic geometry*, preprint.

[ES] Eells, J. and Sampson, J.H., *Harmonic mappings of Riemannian Manifolds*, Amer.J. Math., **86** (1964), 109–160.

[Fe] Federer, H., *Geometric Measure Theory*, Grundlehrer der mathematische Wissenschafte Band 153 Springer-Verlag NY 1969.

[Gi] Giusti, E., *Minimal Surfaces and Functions of Bounded Variation*, Birkhäuser,Boston 1984.

[GS] Gromov, M. and Schoen, R., *Harmonic maps into singular spaces and p– adic superrigidity for lattices in groups of rank one*, IHES Publications Mathématique **76** (1992), 165–246.

[GT] Gilbarg, D. and Trudinger, N. S., *Elliptic Partial Differential Equations of Second Order*, Second Edition, Srpinger-Verlag, New York 1983.

[Ha] Hamilton, R., *Harmonic maps of manifolds with boundary*, Lecture Notes in Math., **471** Springer, 1975.

[J] Jost, J., *Equilibrium maps between metric spaces*, Calc Var**2**(1994), 173-204.

[KS] Korevaar, N. and Schoen, R., *Sobolev spaces and harmonic maps for metric space targets*, Comm Anal Geom**1**(1993), 561-659.

[Mo] Morrey, C.B., *Multiple Integrals in the calculus of variations*, Springer-Verlag NY 1966.

[Re] Reshetnyak, Y.G., *Nonexpanding maps in a space of curvature no greater than K*, Siberian Math. Journ., **9** (1968), 918–927.

[Ru] Rudin, W., *Real and Complex Analysis*, second edition, McGraw-Hill, New York 1974.

[Sch] Schoen, R., *Analytic aspects of the harmonic map problem*, Math Sci. Res. Inst. Publ. vol. 2, Springer Berlin, 1984 321–358.

[Se] Serbinowski, T., *Boundary regularity of harmonic maps to nonpositively curved metric spaces*, Comm Anal Geom**2**(1994), 139-154.

[Wa] Wald, A., *Begründung einer koordinatenlosen Differentialgeometrie der Flächen*, Ergebnisse eines mathematischen Kooloquiums, **7** (1935), 2–46.

[Zi] Zimmer, R., *Strong rigidity for ergodic actions of semisimiple Lie groups*, Ann. of Math. **112** (1980), 511–529.

Chapter XI

Moduli Spaces of Harmonic Maps, Compact Group Actions and the Topology of Manifolds with Non-positive Curvature

Immediately after Eells and Sampson proved their famous existence theorem for harmonic maps, P. Hartman ([7]) studied the uniqueness of such maps. In this chapter, we give a more thorough study of the space of harmonic maps to a manifold N of nonpositive curvature. We then apply this to the study of compact group actions on manifolds which admit homotopically non-trivial maps into N. On the one hand, we obtain new information about (differentiable) group actions on manifolds. On the other hand, we obtain topological obstructions for manifolds to admit metrics with non-positive curvature. The results explain the main results of the second author in [13] which partially motivated the paper [12] which we follow here.

In the first part of this chapter, we describe the space of harmonic maps from M into N which are homotopic to a given map. In [7], P. Hartman showed that this space is connected and is a point if N has curvature strictly negative and the image of M is not a point or a circle. We show that the space is a compact connected totally geodesic submanifold of N and is a point if $\pi_1(N)$ has no non-trivial abelian subgroup and the image of M is not a point or a circle.Our proof is different from Hartman's and depends on the computation of the Hessian of the distance function of N considered as a function defined on $N \times N$. This last computation is of interest in its own right.

In case the harmonic maps are surjective maps from M to N and their induced maps from $\pi_1(M)$ and $\pi_1(N)$ are surjective, the space of harmonic maps is very simple and is homeomorphic to the group generated by parallel translations of N. This latter group, by a theorem proved in [8], has dimension equal to the rank of the center of $\pi_1(N)$.

In the second part, we apply the previous description of harmonic maps to the study of compact groups acting on a compact manifold M. For example if there is a map f from M into a compact manifold N with non-positive curvature so that for some class $\omega \in H^k(N, \mathbb{R})$ $f^*\omega \neq 0$, we can prove that the degree of symmetry of M in the sense of W. Y. Hsiang is $\leq (1/2)(\dim M - k)(\dim M - k + 1)$ when $k > 1$ and $\leq (1/2)(\dim M - 1)\dim M + 1$ when $k = 1$.

For finite groups G acting smoothly on M, we can get information by the following procedure. We first give M a Riemannian metric so that the compact group G acts isometrically. Then we can apply the theorem of Eells and Sampson [5] to deform f to a harmonic map which we still call f. Suppose that for some closed k-form ω $(k > 1)$ in N such that its integral over any integral cycle is an integer, we have $f^*(\omega)[M] = m \neq 0$. Then we prove that G acts locally freely on N and G^0 is a torus with dimension $\leq$ the rank of a maximal abelian subgroup of $\pi_1(N)$ or the first Betti number of M. Furthermore, the subgroup $\{g \in G \,|\, f \circ g = f\}$ is finite and has order dividing m. If we specialize to the case where $m = 1$ and where $\pi_1(N)$ has no non-trivial abelian subgroup or M has zero first Betti number, then G is finite and we prove that the natural representation of G into the outer automorphism group of $\pi_1(N)$ is injective. This shows that not only is the degree of symmetry of a compact manifold of negative curvature zero, but that the same is true for any manifold which represents a homology class (with dimension > 1) in such a manifold.

If there is a degree one map f from M onto a compact locally symmetric space which has non-positive curvature and admits no closed one or two dimensional geodesic subspace as a local direct factor, then we prove that f is homotopic to a G-equivalent map so that G acts by isometry in the image. Furthermore G acts freely on M iff G acts freely on the locally symmetric space. The same assertion is true if we replace the above locally symmetric space by a compact flat manifold but assuming further that $f_*[\pi_1(M)] = \pi_1(N)$. A rather interesting corollary is that any finite group action on a homotopy torus is equivariantly homotopic to a linear action on the standard torus. It would be nice to know a similar assertion in the topological setting.

Finally, we prove that if M is a compact spin manifold which admits a map f from M into a non-positively curved manifold N so that $(f^*\omega) \cup \Omega[M] \neq 0$ where ω is a k-dimensional cohomology class in N and Ω is the $\dim M - k$ dimensional $\hat{A}$-class of M, then any connected group acting on M is a torus acting locally freely with dimension not greater than the rank of a maximal abelian subgroup of $\pi_1(N)$ or the first Betti number of M. The last theorem was independently proved by completely different methods by W. Browder and W. C. Hsiang in the case where N is a general $K(\pi, 1)$. (However, their conclusion in our special case is weaker than ours.)

1. Computation of the Hessian of a Distance Function

Let N be a complete Riemannian manifold. Let $r : N \times N \to \mathbb{R}$ be defined by $r(u_1, u_2)$ = distance from u_1 to u_2. Let D be the diagonal of $N \times N$ and $\mathcal{O}$ an open neighborhood of D so that r is smooth on $\mathcal{O} \setminus D$, and for every pair $(u_1, u_2) \in \mathcal{O} \setminus D$ there is a unique shortest geodesic from u_1 to u_2. Let ∇ be the Riemannian connection on $N \times N$ induced by the product metric. Thus for each point $(u_1, u_2) \in \mathcal{O} \setminus D$ and X, $Y \in T_{(u_1,u_2)}N \times N$ (the tangent space of $N \times N$ at (u_1, u_2)), the Hessian of r in direction X and Y is given by

$$r_{XY} = XY(r) - (\nabla_X Y)r,$$

where Y is extended in an arbitrary manner to a smooth vectorfield in a neighborhood of (u_1, u_2). It is standard that $r_{XY} = r_{YX}$. The purpose of this section is to calculate r_{XY} in terms of the curvature of N.

For any $(u_1, u_2) \in \mathcal{O} \setminus D$, let $\gamma_{u_1 u_2}(t)$ for $0 \leq t \leq 1$ be the unique shortest geodesic from u_1 to u_2 with constant speed parametrization. Let $X = X_1 + X_2 \in T_{(u_1,u_2)}N \times N$, where X_1 and X_2 are the components tangent to the first and second factor of the product $N \times N$. Let $\sigma(s)$ denote the constant speed geodesic satisfying $\sigma(0) = (u_1, u_2)$ and $\sigma'(0) = X$. We then have $\sigma = (\sigma_1, \sigma_2)$ where $\sigma_i(0) = u_i$ and $\sigma_i'(0) = X_i$ for $i = 1, 2$. From the definition of the Hessian, we then have

$$r_{XX}(u_1, u_2) = \frac{d^2}{ds^2} r(\sigma_1(s), \sigma_2(s)) \text{ at } s = 0.$$

From the second variation of arclength formula (see [4], Page 20) we then have

$$r_{XX}(u_1, u_2) = \int_0^1 \{\|\nabla V^{\perp}\|^2 - \langle R(V,T)T, V\rangle\} \, dt$$

where V is the Jacobi field along $\gamma_{u_1u_2}$ whose value is X_i at u_i for $i = 1, 2$, $V^{\perp}$ denotes the component of V normal to the geodesic $\gamma_{u_1u_2}(t)$, and T denotes the tangent vector to $\gamma_{u_1u_2}(t)$.

We can now prove the following proposition.

Proposition 1. *Suppose the sectional curvature of N is non-positive. Let $X = X_1 + X_2 \in T_{(u_1,u_2)}N \times N$. Then $(r^2)_{XX}$ is non-negative, and is equal to zero if and only if the following is true: There is a parallel vector field V defined along $\gamma_{u_1u_2}$ such that $V(u_1) = X_1$, $V(u_2) = X_2$, and $\langle R(V,T)T, V\rangle \equiv 0$ on $\gamma_{u_1u_2}$. In particular, if the sectional curvature of N is negative, V is proportional to T.*

Proof. This follows immediately from the second variation formula above. □

2. Uniqueness of Harmonic Maps

Let M and N be two complete Riemannian manifolds. Suppose the sectional curvature of N is non-positive. we wish to study uniqueness of harmonic maps in a given free homotopy class of maps from M to N.

We first recall the definition of harmonic maps. Let $f : M \to N$ be a C^1-mapping. Let $\mathrm{d}s_N^2$ be the metric tensor of N, and consider the symmetric tensor $f^*\mathrm{d}s_N^2$ gotten by pullback. We define the energy density $e(f)$, to be the trace of $f^*\mathrm{d}s_N^2$ taken with respect to $\mathrm{d}s_M^2$, the metric tensor of M. We say that f is harmonic if for every compact domain $D \subseteq M$, and for every one-parameter family $f_t : M \to N$ for $t \in (-1,1)$ with $f_0 = f$, and $f_t \equiv f$ outside of D for all $t \in (-1,1)$, we have

$$\frac{\mathrm{d}}{\mathrm{d}t}\int_D e(f_t)\mathrm{d}V_M|_{t=0} = 0.$$

Define the energy functional $E(f)$ by

$$E(f) = \int_M e(f)\mathrm{d}V_M$$

where $\mathrm{d}V_M$ is volume element of M.

Let f and g be two harmonic maps from M to N which are freely homotopic, and such that $E(f)$ and $E(g)$ are finite. We will show that, if M has finite volume, then f and g are homotopic through harmonic maps in an explicit way (see Theorem 2 below).

Let $\tilde{M}$, $\tilde{N}$ be the universal covers of M, N respectively. Then $\pi_1(M,*)$ and $\pi_1(N,*)$ act as groups of isometries on $\tilde{M}$ and $\tilde{N}$ respectively so that $M = \tilde{M}/\pi_1(M,*)$ and $N = \tilde{N}/\pi_1(N,*)$. Let $\tilde{r} : \tilde{N}\times\tilde{N} \to \mathbb{R}$ be defined by $\tilde{r}(x,y) =$ distance from x to y. Since $\tilde{N}$ has nonpositive sectional curvatures, we know that $\tilde{r}$ is smooth on $\tilde{N}\times\tilde{N}\setminus$diagonal. Now $\pi_1(N,*)$ acts on $\tilde{N}\times\tilde{N}$ as a group of isometries by

$$\alpha(x,y) = (\alpha(x),\alpha(y)) \quad \text{for} \quad \alpha \in \pi_1(N,*).$$

Thus $\tilde{r}$ induces a functions $r : \tilde{N}\times\tilde{N}/\pi_1(N,*) \to \mathbb{R}$. Let $F : M \times [0,1] \to N$ be a homotopy of f with g so that $F(p,0) = f(p)$ and $F(p,1) = g(p)$ for all $p \in M$. We now choose a lefting $\tilde{F} : \tilde{M}\times[0,1] \to \tilde{N}$, and call $\tilde{F}(p,0) = \tilde{f}(p)$ and $\tilde{F}(p,1) = \tilde{g}(p)$ for all $p \in M$. This defines liftings $\tilde{f}$, $\tilde{g}$ of f, g. We now have that if $\gamma \in \pi_1(M,*)$, there exists $\alpha \in \pi_1(N, *)$ with

$$(1) \quad \tilde{f}(\gamma(p)) = \alpha \tilde{f}(p) \quad \text{and} \quad \tilde{g}(\gamma(p)) = \alpha \tilde{g}(p) \quad \text{for all} \quad p \in \tilde{M}.$$

Thus if we define a map $\tilde{h} : \tilde{M} \to \tilde{N} \times \tilde{N}$ by $\tilde{h}(p) = (\tilde{f}(p), \tilde{g}(p))$, we see that $\tilde{h}$ is a harmonic map, and by (1) $\tilde{h}$ induces a harmonic map

$$h : M \to N \times N/\pi_1(N, *).$$

We now define a function $\rho(f, g) : M \to \mathbb{R}$ by $\rho(f, g) = r \circ h$. The function $\rho^2(f, g)$ is smooth on M, and we will calculate its Laplacian. Before doing so we set up some terminology.

Let $f : M \to N$ be a smooth map, and let $\theta_1, \dots, \theta_m$, be an orthonormal coframe in a neighborhood of $p \in M$. Let $\omega_1, \dots, \omega_n$, be an orthonormal coframe in a neighborhood of $f(p) \in N$. Define $f_{i\alpha}$, $1 \le i \le n$, $1 \le \alpha \le m$ by

$$(2) \qquad f^*\omega_i = \sum_{\alpha=1}^{m} f_{i\alpha}\theta_\alpha \quad \text{for} \quad 1 \le i \le n.$$

We have the structure equations for N and M

$$(3) \qquad \begin{aligned} d\omega_i &= \sum_{j=1}^{n} \omega_{ij} \wedge \omega_j \quad \text{for} \quad 1 \le i \le n \\ d\theta_\alpha &= \sum_{\beta=1}^{n} \theta_{\alpha\beta} \wedge \theta_\beta \quad \text{for} \quad 1 \le \alpha \le m. \end{aligned}$$

Define $f_{i\alpha\beta}$ by

$$(4) \qquad df_{i\alpha} + \sum_{j=1}^{n} f_{j\alpha} f^*\omega_{ji} + \sum_{\beta=1}^{m} f_{j\beta}\theta_{\beta\alpha} = \sum_{\beta=1}^{m} f_{i\alpha\beta}\theta_\beta.$$

Exterior differentiating (2) and applying (3) we have

$$(5) \qquad f_{i\alpha\beta} = f_{i\beta\alpha} \quad \text{for} \quad 1 \le n, \quad 1 \le \alpha, \beta \le m.$$

If φ is a smooth function on M we define its gradient and Hessian by

$$(6)\qquad \begin{aligned} d\varphi &= \sum_{\alpha=1}^{m} \varphi_\alpha \theta_\alpha \\ d\varphi_\alpha + \sum_{\beta=1}^{m} \varphi_\beta \theta_{\beta\alpha} &= \sum_{\beta=1}^{m} \varphi_{\alpha\beta}\theta_\beta \quad \text{for} \quad 1 \le \alpha \le m \\ \varphi_{\alpha\beta} &= \varphi_{\beta\alpha}. \end{aligned}$$

Similarly, if ψ is a function on N we have

$$(7)\qquad \begin{aligned} d\psi &= \sum_{i=1}^{n} \psi_i \omega_i \\ d\psi_i + \sum_{j=1}^{n} \psi_j \omega_{ji} &= \sum_{j=1}^{n} \psi_{ij}\omega_j \quad \text{for} \quad 1 \le i \le n \\ \psi_{ij} &= \psi_{ji}. \end{aligned}$$

The condition that f be harmonic is expressed by the following system of equations

$$(8)\qquad \sum_{\alpha=1}^{m} f_{i\alpha\alpha} = 0 \quad \text{for} \quad 1 \le i \le n.$$

We now calculate the Laplacian of $\rho^2 = \rho^2(f, g)$.

$$(9)\qquad (\rho^2)_\alpha = 2\rho\rho_\alpha = 2\sum_{i=1}^{n} \rho r_i f_{i\alpha} + \rho r_{\bar{i}} g_{\bar{i}\alpha}$$

where we use $\bar{i}$ $(1 \le i \le n)$ to denote differentiation with respect to the last n coordinates of $\tilde{N} \times \tilde{N}/\pi_1(N, *)$. Now we have

$$\begin{aligned} \Delta\rho^2 &= \sum_\alpha (\rho^2)_{\alpha\alpha} = 2\sum_{\alpha,i} (\rho r_i f_{i\alpha} + \rho r_{\bar{i}} g_{\bar{i}\alpha})_\alpha \\ &= 2\sum_{\alpha,i} (r_i f_{i\alpha} + r_{\bar{i}} g_{\bar{i}\alpha})^2 + 2\sum_{\alpha,i} (\rho r_i f_{i\alpha\alpha} + \rho r_{\bar{i}} g_{\bar{i}\alpha\alpha}) \\ &\quad +2\sum_{\alpha,i,j} \rho\,(r_{ij} f_{i\alpha} f_{j\alpha} + 2r_{i\bar{j}} f_{i\alpha} g_{\bar{j}\alpha} + r_{\bar{i}\bar{j}} g_{\bar{i}\alpha} g_{\bar{j}\alpha}). \end{aligned}$$

Since f, g are harmonic, we apply (8) to obtain

$$(10)\qquad \begin{aligned} \Delta\rho^2 &= 2\sum_{\alpha,i} (r_i f_{i\alpha} + r_{\bar{i}} g_{\bar{i}\alpha})^2 \\ &\quad +2\sum_{\alpha,i,j} \rho\,(r_{ij} f_{i\alpha} f_{j\alpha} + 2r_{i\bar{j}} f_{i\alpha} g_{\bar{j}\alpha} + r_{\bar{i}\bar{j}} g_{\bar{i}\alpha} g_{\bar{j}\alpha}). \end{aligned}$$

If $e_1, \dots, e_n, \bar{e}_1, \dots, \bar{e}_n$ is an orthonormal frame on $\tilde{N} \times \tilde{N}/\pi_1(N, *)$ we define a vector X_α by

$$X_\alpha = \sum_{i=1}^{n} f_{i\alpha} e_i + \sum_{i=1}^{n} g_{i\alpha} \bar{e}_i,$$

and note that (10) may be written

$$\Delta \rho^2 = 2 \sum_{\alpha,i} (r_i f_{i\alpha} + r_{\bar{i}} g_{\bar{i}\alpha})^2 + 2 \sum_{\alpha} \rho r_{X_\alpha X_\alpha}. \tag{11}$$

Thus we may apply Proposition 1 to conclude that ρ^2 is a subharmonic function on M. We now prove a lemma.

Lemma 1. *Suppose M is a complete manifold, and φ is a function defined on M satisfying $\Delta\varphi \geq 0$ and $\int_M |\nabla\varphi| dV_M < \infty$. Then φ is harmonic.*

Proof. Fix a point $P_0 \in M$. Choose a sequence of numbers $R_i \to \infty$ so that $\int_{\partial B_{R_i}(P_0)} |\nabla\varphi| \to 0$. Such a sequence exists since $\int_M |\nabla\varphi| < \infty$. Now we have

$$\int_{B_{R_i}(P_0)} \Delta\varphi = \int_{\partial B_{R_i}(P_0)} \frac{\partial\varphi}{\partial\nu} \leq \int_{\partial B_{R_i}(P_0)} |\nabla\varphi|$$

where ν is the unit normal vector of $\partial B_{R_i}(P_0)$. Letting $i \to \infty$ we have $\int_M \Delta\varphi = 0$ and so $\Delta\varphi = 0$. □

Corollary 1. *If M has finite volume, and φ is a function on M satisfying $\Delta\varphi \geq 0$, and $\int_M |\nabla\varphi|^2 < \infty$, then φ is a harmonic function.*

Proof. This follows from Lemma 1 and the Schwartz inequality. □

Lemma 2. *Let φ be a harmonic function on a complete manifold M of finite volume. Suppose that $\int_M |\nabla\varphi|^2 < \infty$. Then φ is a constant function.*

Proof. Let ψ be any smooth function on M satisfying $\int_M \psi^2 < \infty$ and $\int_M |\nabla\psi|^2 < \infty$. Let η be a Lipschitz function so that $1 \geq \eta \geq 0$ on M,

$$\eta = \begin{cases} 1 & \text{on} \quad B_R(P_0) \\ 0 & \text{outside} \quad B_{2R}(P_0) \end{cases} \quad \text{and} \quad |\nabla\eta| \leq \frac{2}{R}.$$

We then compute

$$\begin{aligned}\int_M \nabla\varphi\cdot\nabla\psi &= \int_M \nabla\varphi\cdot\nabla((1-\eta)\psi) + \int_M \nabla\varphi\cdot\nabla(\eta\psi)\\ &= \int_M (\nabla\varphi\cdot\nabla\psi)(1-\eta) - \int_M (\nabla\varphi\cdot\nabla\eta)\psi.\end{aligned}$$

Thus

$$\begin{aligned}\left|\int_M \nabla\varphi\cdot\nabla\psi\right|^2 &\le 2\left(\int_M |\nabla\varphi||\nabla\psi|(1-\eta)\right)^2 + 2\left(\int_M |\nabla\varphi||\nabla\eta|\psi\right)^2\\ &\le 2\left(\int_{M\setminus B_R}|\nabla\varphi|^2\right)\left(\int_{M\setminus B_R}|\nabla\psi|^2\right)\\ &\quad+\frac{8}{R^2}\left(\int_M|\nabla\varphi|^2\right)\left(\int_M\psi^2\right).\end{aligned}$$

□

Letting $R\to\infty$ we conclude $\int_M \nabla\varphi\cdot\nabla\psi = 0$ for any function ψ with ψ, $|\nabla\psi|\in L^2(M)$. We now choose

$$\psi = \begin{cases} K & \text{if } \varphi > K\\ \varphi & \text{if } \varphi\in[-K,K]\\ -K & \text{if } \varphi < -K.\end{cases}$$

From this choice of ψ we conclude

$$\int_{\varphi\in[-K,K]} |\nabla\varphi|^2 = 0.$$

Letting $K\to\infty$ we conclude that φ is constant. This proves Lemma 2.

We now consider the function φ on M defined by $\varphi = (\rho^2+1)^{1/2}$. Now φ is a smooth function on M, and we see from from (9)

$$|\nabla\varphi|^2 = \rho^2\sum_\alpha \frac{\sum_{i=1}^n (r_i f_{i\alpha} + r_{\bar{i}} g_{\bar{i}\alpha})}{\rho^2+1} \le 2(e(f)+e(g)).$$

From (11) we see that $\Delta\varphi\ge 0$. If $E(f)+E(g)<\infty$, and M has finite volume, then we can apply Corollary 1 and Lemma 2 to conclude that φ is constant on M. Thus we see that $\rho(f,g)$ is a constant function on M, so (11) implies that

$$(r^2)_{X_\alpha X_\alpha} \equiv 0 \quad \text{for} \quad 1\le\alpha\le m.$$

In particular, we can lift to universal covers and conclude that $(\tilde{r}^2)_{\tilde{X}_\alpha \tilde{X}_\alpha} \equiv 0$ where

$$\tilde{X}_\alpha = \sum \tilde{f}_{i\alpha} e_i + \tilde{g}_{\bar{i}\alpha} \bar{e}_i.$$

Applying Proposition 1, we find vector fields W_α which are parallel along the unique geodesic $\tilde{\gamma}_p$ from $\tilde{f}(p)$ to $\tilde{g}(p)$ and satisfies $W_\alpha(\tilde{f}(p)) = \sum_{i=1}^{n} \tilde{f}_{i\alpha} e_i(\tilde{f}(p))$ and $W_\alpha(\tilde{g}(p)) = \sum_{i=1}^{n} \tilde{f}_{\bar{i}\alpha} \bar{e}_i(\tilde{g}(p))$. Proposition 1 also gives

$$\langle R(W_\alpha, \tilde{\gamma}_p')\tilde{\gamma}_p', W_\alpha \rangle \equiv 0 \quad \text{along} \quad \tilde{\gamma}_p. \tag{12}$$

Since $\tilde{r}$ is constant, we parametrize each $\tilde{\gamma}_p$ on $[0,1]$ proportional (independent of $p \in \tilde{M}$) to arclength. We define a one-parameter family of maps $\tilde{f}_t : \tilde{M} \to \tilde{N}$ by letting $\tilde{f}_t(p) = \tilde{\gamma}_p(t)$. Then we see that $\tilde{f}_0 = \tilde{f}$ and $\tilde{f}_1 = \tilde{g}$. It follows from the parallelism of W_α and from (12) that

$$W_\alpha(\tilde{\gamma}_p(t)) = \sum_{i=1}^{n} (\tilde{f}_t)_{i\alpha} e_i(\tilde{f}_t(p)). \tag{13}$$

Because $\pi_1(N)$ acts by isometries on $\tilde{N} \times \tilde{N}$, we see that given $\sigma \in \pi_1(M)$, there exists $\alpha \in \pi_1(N)$ so that $\tilde{f}_t \circ \sigma = \alpha \circ \tilde{f}_t$ for $0 \le t \le 1$. Thus we have induced maps $f_t : M \to N$ for $t \in [0,1]$ such that $f_0 = f$ and $f_1 = g$.

We show that each f_t is a harmonic map. Let $\gamma_p(t)$ be the geodesic from $f(p)$ to $g(p)$ in N which is the projection of $\tilde{\gamma}_p$. It follows from (13) that $\mathrm{d}f_t(v_\alpha)$ is a parallel vector field along γ_p for each $p \in M$. We therefore conclude that

$$e(f_t)(p) = \sum_{\alpha=1}^{m} |\mathrm{d}f_t(v_\alpha)|^2 \quad \text{is constant on } \gamma_p,$$

where $v_1, \dots, v_m$ is an orthonormal basis for T_pM. Therefore, for each $t \in [0,1]$ we have

$$E(f_t) = E(f). \tag{14}$$

In particular it follows that $E(f) = E(g)$; i.e. every harmonic map of finite energy homotopic to f has the same energy as f. To prove f_t is harmonic, we consider two cases. First suppose that M is compact. In this case, if f_t is not harmonic, the method of Eells and Sampson [5] shows that we can find a deformation $f_{t,s}$ of f_t so that $f_{t,0} = f_t$, and $f_{t,s}$ is a solution of the heat equation. An easy application of the maximum principle to the function $r(f_{t,s}, f)$, shows that $f_{t,s}(M)$ remains in a compact subset of N as s increases. The techniques of [6] thus show the existence of a harmonic map $f_{t\infty}$ with $E(f_{t\infty}) < E(f_t)$. Now $f_{t\infty}$ is homotopic to f, so we must have $E(f_{t\infty}) = E(f)$ in contradiction to (14).

If M is noncompact, we may solve the boundary value problem on $B_R(P_0) \subset M$ using [6] to get a harmonic map $f_{t,R} : B_R(P_0) \to N$ homotopic to f_t with $f_{t,R} \equiv f_t$ on $\partial B_R(P_0)$. since the function $r(f_{t,R}, f)$ is subharmonic in $B_R(P_0)$ and constant on $\partial B_R(P_0)$. We conclude that for a compact subset $K \subset M$, $f_{t,R}(K)$ lies in a compact subset of N (independent of R). By Chapter IX, Section 8 there is a harmonic map $f_{t,\infty}$ homotopic to f_t with $E(f_{t,\infty}) < E(f_t) = E(f)$ again giving a contradiction. Thus f_t is a harmonic map for $t \in [0,1]$. We have thus proven the following theorems.

Theorem 1. *Suppose M is complete with finite volume and N is complete with strictly negative sectional curvature. Let $f : M \to N$ be a harmonic map of finite energy. There is no other harmonic map of finite energy homotopic to f unless $f(M)$ is contained in a geodesic of N.*

Theorem 2. *Suppose M is complete with finite volume and N is complete with nonpositive sectional curvatures. If $f, g : M \to N$ are homotopic harmonic maps of finite energy, then there is a smooth one-parameter family $f_t : M \to N$ for $t \in \mathbb{R}$ of harmonic maps with $f_0 = f$ and $f_1 = g$. Moreover, for each $p \in M$, the curve $\{f_t(x) : t \in \mathbb{R}\}$ is a constant (independent of p) speed parametrization of a geodesic. Also, the map $M \times \mathbb{R} \to N$ given by $(p,t) \mapsto f_t(p)$ is harmonic with respect* *to the product metric on $M \times \mathbb{R}$. Moreover, $p \mapsto f_{t*}(\partial/\partial t)$ is a parallel section of f^*TN, the pullback of the tangent bundle of N with pulled back connection.*

Corollary 2. *Under the hypotheses of Theorem 2, it follows that M has a harmonic one-form.*

Proof. (of Corollary 2) Let $\omega_1, \ldots, \omega_n$ be an orthonormal basis for N, and let ω be the form dual to $f_*(\partial/\partial t)$. Locally, ω has the representation

$$\omega(p) = \sum_{i=1}^{n} a_i(p)\omega_i(f(p)) \quad \text{for} \quad p \in M.$$

Let $\theta_1, \ldots, \theta_m$ be an orthonormal frame on M, and define

$$da_i + \sum_{j=1}^{n} a_j f^*\omega_{ji} = \sum_{\alpha=1}^{m} a_{i\alpha}\theta_\alpha.$$

It follows from Theorem 2 that $a_{i\alpha} \equiv 0$ for $1 \leq i \leq n$, $1 \leq \alpha \leq m$. Let τ be the pullback of the form ω. Then we have $\tau = \sum_{i,\alpha} a_i f_{i\alpha}\theta_\alpha = \sum_{\alpha=1}^{m} b_\alpha\theta_\alpha$. Now,

$$b_{\alpha\beta} = \sum_i a_{i\beta} f_{i\alpha} + \sum_i a_i f_{i\alpha\beta} = \sum_i a_i f_{i\beta\alpha} = b_{\beta\alpha}$$

$$\sum_\alpha b_{\alpha\alpha} = \sum_{i,\alpha} a_{i\alpha} f_{i\alpha} + \sum_{i,\alpha} a_i f_{i\alpha\alpha} = 0$$

where we have used (5) and (8). Thus τ is a harmonic form. □

Remark. If the map $f : M \to N$ has image which contains an open subset of N, then by choosing a regular value of f in N, one sees that τ is not everywhere zero.

If M and N are compact, we have the following structure for the space of homotopic harmonic maps.

Theorem 3. *Let f be a given harmonic map from a compact manifold M to a compact real analytic manifold N having nonpositive sectional curvatures. The space of harmonic maps homotopic to f may be identified with a compact immersed totally geodesic submanifold of N. In* *fact, there is a compact manifold N_0 of nonpositive sectional curvatures and a map $F : M \to N$ so that for each $p \in M$, $F|_{\{p\}\times N_0}$ is an isometric immersion onto a totally geodesic submanifold of N, and so that*

$$\{F|_{M\times\{t\}} : t \in N_0\}$$

is the space of harmonic maps homotopic to f.

The following corollary is a direct consequence of Theorem 3.

Corollary 3. *Suppose the hypotheses of Theorem 3 hold. Suppose also that $\pi_1(N)$ has no noncyclic abelian subgroup. If $f_*(\pi_1(M))$ is not cyclic or trivial, then f is the unique harmonic map in its homotopy class.*

Corollary 3 follows because nonuniqueness implies that N_0 contains a circle whose image γ is nontrivial in $\pi_1(N)$. The assumptions imply that there is $\sigma \in f_*(\pi_1(M))$ so that γ and σ generate a subgroup of rank two which is abelian.

Proof. (of Theorem 3) Let N_0 be the space of harmonic maps homotopic to f. we first show that N_0 can be given the structure of a compact totally geodesic submanifold of N. Fix a point $p_0 \in M$, and $g \in N_0$. Let $U(g) = \{h \in N_0 : \text{dist}(g(p), h(p)) < \epsilon_0 \text{ for all } p \in M\}$ where $\epsilon > 0$ is the injectivity raduis of N. It is clear from the discussion of this section that $U(g)$ is homeomorphic in C^0 topology to $\mathcal{O}(g) = \{h(p_0) : h \in U(g)\}$. We show that $\mathcal{O}(g)$ is a totally geodesic submanifold of N. In fact, Theorem 2 implies that $\mathcal{O}(g)$ is a union of geodesic through $g(p_0)$. Let $\mathcal{L}(g) \subseteq T_{g(p_0)}N$ be the set of all multiples of tangent vectors at $g(p_0)$ to these geodesics. Thus we have

$$\mathcal{O}(g) = \{\exp_{g(p_0)} t\frac{v}{|v|} : t \in [0, \epsilon_0), \quad v \in \mathcal{L}(g) \setminus \{0\}\}.$$

We claim that $\mathcal{L}(g)$ is a linear subspace of $T_{g(p_0)}N$. To see this we first note that $\mathcal{L}(g)$ is closed because given $\{v_i\} \subseteq \mathcal{L}(g)$ with $v_u \to v \in$

$T_{g(p_0)}N$ we have harmonic maps $h_i \in U(g)$ with

$$h_i(p_0) = \exp_{g(p_0)}(\epsilon_0/2)(v_i/|v_i|).$$

Now $E(h_i) = E(f)$ is uniformly bounded, so the estimate of [5] imply that $\{h_i\}$ is C^∞ equicontinuous, and hence $h_i \to h$, a harmonic map with $h(p_0) = \exp_{g(p_0)}(\epsilon_0/2)(v/|v|)$. This shows that $\mathcal{L}(g)$ is closed. To see that $\mathcal{L}(g)$ is a linear subspace, we simply note that if v_1, $v_2 \in \mathcal{L}(g)$ with $\langle v_1, v_2\rangle \geq 0$, then $v_1 + v_2$ is the limit of tangent vectors of geodesic connecting $g(p_0)$ to a point on the short geodesic γ_t from $\exp_{g(p_0)} t(v_1/|v_1|) \exp_{g(p_0)} t(v_2/|v_2|)$. This geodesic γ_t is contained in $\mathcal{O}(g)$ since it jions points in $\mathcal{O}(g)$. Since N is real analytic, $-v \in \mathcal{L}(g)$ if $v \in \mathcal{L}(g)$, so this shows that $\mathcal{L}(g)$ is a linear subspace, and hence we have

$$\mathcal{O}(g) = \{\exp_{g(p_0)} v \,:\, v \in \mathcal{L}(g),\ |v| < \epsilon_0\}$$

is a submanifold of N. Moreover, $\mathcal{O}(g)$ is totally geodesic since it is a convex subset of N. Thus we have shown that $N_0 \to N$ given by $h \to h(p_0)$ is an immersion of N_0 onto a totally geodesic submanifold of N. We give N_0 the metric induced from this immersionand note that N_0 is compact since $E(h) = E(f)$ for all $h \in N_0$. Thus N_0 is a compact manifold of nonpositive sectional curvature, and we define a map

$$F : M \times N_0 \to N \quad \text{by} \quad F(p,h) = h(p).$$

The theorem will be finished if we can show that $F|_{\{p\}\times N_0}$ is an isometric immersion onto a totally geodesic submanifold of N, but this follows because $\operatorname{dist}(h_1(p), h_2(p)) = \operatorname{dist}(h_1(p_0), h_2(p_0))$ for all $p \in M$ and all h_1, h_2 sufficiently close. This complete the proof of Theorem 3. $\square$

Remarks. Without the assumption that N be real analytic, Theorem 3 is not strictly true as stated. For N which is only C^∞, the above proof shows that the space N_0 of homopotic harmonic maps is a connected totally geodesic submanifold of N with boundary ∂N_0. Here ∂N_0 need not be smooth, but is a Lipschitz submanifold of N.

We now consider assumptions which guarantee that a one-parameter family of harmonic maps induces a parallel vector field on N.

Theorem 4. *Suppose M, N are real analytic Riemannian manifolds. Suppose $m \geq n$ and $f : M \to N$ is a harmonic map between compact manifolds, N having nonpositive sectional curvatures. Suppose the pull-back under f of the fundamental cohomology class of N is nonzero in M. Suppose also that the induced map $f_* : \pi_1(M) \to \pi_1(N)$ is surjective. Let $g : M \to N$ be a harmonic map homotopic to f. Let $F : M \times \mathbb{R} \to N$ be the harmonic map of Theorem 2. Then there exists a parallel vector field V on N satisfying*

$$dF_{(p,0)}\left(\frac{\partial}{\partial t}\right) = V(f(p)) \quad \textit{for all} \quad p \in N.$$

Proof. Theorem 2 implies that $\mathrm{d}F_{(p,0)}\,(\partial/\partial t)$ is a parallel section of the pullback tangent bundle $f^*(TN)$. we consider a lifting $\tilde{F} : \tilde{M} \to \tilde{N}$ of F to universal covers. Let $q \in \tilde{N}$ be a regular value of $\tilde{f}$, and let $p \in \tilde{f}^{-1}(q)$. Our hypotheses imply that f is surjective, so there is a neighborhood U of p in $\tilde{M}$ which is mapped onto a neighborhood $\mathcal{O}$ of q in $\tilde{N}$. For any $\bar{q} \in \mathcal{O}$, define $V(\bar{q}) = \mathrm{d}\tilde{F}_{(\bar{p},0)}\,(\partial/\partial t)$ where $\bar{p} \in U \cap \tilde{f}^{-1}(\bar{q})$. the value of $\tilde{V}(\bar{q})$ is independent of the choice of $\bar{p}$since $U \cap \tilde{f}^{-1}(\bar{q})$ can be taken to be a connected $m-n$ dimensional manifold, and $\mathrm{d}F_{(\bar{p},0)}\,(\partial/\partial t)$ is a parallel section of $\tilde{f}^*(T_{\bar{q}}\tilde{N})$ which is a linear space. Thus $\tilde{V}$ is a parallel vector field in $\mathcal{O}$, and we extend it by parallel translation along rays from q to be a vector field on all of $\tilde{N}$. Since $\tilde{V}$ is parallel on $\mathcal{O}$, and $\tilde{N}$ is real analytic, it follows that $\tilde{V}$ is parallel on all of $\tilde{N}$. If we view $\tilde{V}$ as a section of $\tilde{f}^*(T\tilde{N})$, then we see that $\tilde{V} = \mathrm{d}F_{(\cdot,0)}\,(\partial/\partial t)$ on U, so since f is real analytic (see Morrey [10]), we have $\tilde{V} = \mathrm{d}\tilde{F}_{(\cdot,0)}\,(\partial/\partial t)$ on all of $\tilde{M}$.

We now show that $\tilde{V}$ pushes down to a vector field V on N. Let $\pi : \tilde{N} \to N$ be the projection map. Let $q \in N$, and choose a point

$\tilde{q} \in \pi^{-1}(q)$. Let γ be a closed curve in N with $\gamma(0) = \gamma(1) = q$. Define V locally near $q = \gamma(0)$ by $V = \mathrm{d}\pi(\tilde{V})$. By covering the curve γ with small neighborhoods we get a definition of V as a vector field along γ. We must show that $V(\gamma(1)) = V(\gamma(0))$. Choose a point $\tilde{p} \in \tilde{f}^{-1}(\tilde{q})$. Let $p \in M$ be the projection of $\tilde{p}$. We then know from the above that $\mathrm{d}\tilde{F}_{(\tilde{p},0)}(\partial/\partial t) = \tilde{V}(\tilde{q})$, so it follows that $\mathrm{d}F_{(p,0)}(\partial/\partial t) = V(\gamma(0))$. Since $f_* : \pi_1(M) \to \pi_1(N)$ is surjective, there is a curve α in M with $\alpha(0) = \alpha(1) = p$ and $f(\alpha) = \gamma$. Now $\mathrm{d}F_{(\alpha(s),0)}(\partial/\partial t)$ and $V(f(\alpha(s)))$ are parallel sections of $f^*(TN)$ along α which agree at $s = 0$, so we have $V(f(\alpha(1))) = V(\gamma(1)) = \mathrm{d}F_{(p,0)}(\partial/\partial t) = V(\gamma(0))$. This completes the proof of Theorem 4. $\square$

Corollary 4. *Under the hypotheses of Theorem 4 it follows that if f, g are homotopic harmonic maps then $g = \alpha \circ f$ for some isometry α in the (abelian) group of isometries generated by the parallel vector fields. In this case, the space of harmonic maps homotopic to f is parametrized by the latter group which has dimension equal to the rank of the center of $\pi_1(N)$.*

3. Harmonic Maps and Complete Manifolds

In this section, we remark that the standard existence theorem for harmonic maps between Riemannian manifolds can be generalized to non-compact manifolds in the following manner.

Theorem 5. *Let N be a complete Riemannian manifold with non-positive curvatures. Let M be any Riemannian manifold. Let f be a smooth map from M into N such that for some compact subset Ω of M, $f|_{M\setminus\Omega}$ is homotopic to a constant map and that for some k dimensional compact cycle α in M, $f_*\alpha$ is not homologous to zero in $H_k(N, N\setminus C)$ where C is some compact set in N. Then f is homotopic to some harmonic map with finite energy.*

Proof. The proof is almost the same as the one given in [11]. Only two modifications are needed. Hamilton's theorem [6] on the Dirichlet boundary-value problem can be generalized to the case where N need not be compact. This was done in Chapter IX, Section 8. In [11], we found a sequence of harmonic maps from domains of M into N. we showed the equicontinuity of this sequence. The existence of the homology class α in the hypothesis guarantees that the sequence in fact converges. □

Remark. Theorem 5 remains valid if we can find an element α in $\pi_1(M)$ so that $f_*(\alpha)$ is not trivial in $\pi_1(N, N \setminus C)$ where C is some compact set in N. From this remark, one may generalize the theorems in [11] to the case where N is not compact with suitable assumptions.

We remark here the following generalization of the flat torus theorem of [8].

Theorem. *Let M be a complete Riemannian manifold with non-positive curvature. Let A be an abelian subgroup of $\pi_1(M)$ such that for some element $\alpha \in A$ and some compact set C in M, α is not trivial in $\pi_1(M, M \setminus C)$. Then there exists a totally geodesic k dimensional flat torus on M whose fundamental group contains A as a subgroup of finite index.*

Proof. This follows easily from the above theorem and the proof of the uniqueness theorem in §2. □

4. Compact Groups Acting Smoothly on a Manifold

Let M be a complete Riemannian manifold with finite volume. Let G be the group of isometries of M. Let N be a complete Riemannian manifold with non-positive curvature. Let $A(N)$ be the group of affine transformations of N. Then by direct computation, one knows that if f is a harmonic map from M into N and g is an element in $A(N)$, then

$g \circ f$ is harmonic. we use the uniqueness theorem of §2 to study the behaviour of the action G on M.

We begin with the following

Theorem 6. *Suppose there exists a smooth map f from M into N such that f is homotopic to a constant outside a compact subset of M. Suppose for some homology class $\alpha \in H_k(M, \mathbb{Z})$, $f_*(\alpha)$ is not homologous to zero in $H_k(N, N \setminus C)$ where C is some compact set in N. Suppose that for each $g \in G$ acting on M, there is a map $\tilde{g} \in A(N)$ such that $f \circ g$ is freely homotopic to $\tilde{g} \circ f$. (We do not assume that $\tilde{g}$ is unique or non-trivial.) Then if N has negative curvature, there is a (smooth) harmonic map h homotopic to f such that either $h(M)$ is a circle or for all $g \in G$, $h \circ g = \tilde{g} \circ h$.*

Proof. By Theorem 5, we know that f is homotopic to some harmonic map, call it h. It remains to prove the last equation. For any $g \in G$, the map $h \circ g$ is clearly harmonic. As $h \circ g$ is freely homotopic to $\tilde{g} \circ h$, Theorem 1 stated in §2 shows that either $h(M)$ is a circle or $h \circ g = \tilde{g} \circ h$. □

Corollary 5. *Let M be a complete manifold with finite volume. Suppose that there exists a smooth map f from M into a complete negatively curved manifold N which is homotopic to constant outside a compact set. Suppose also that for some homology class $\alpha \in H_k(M, \mathbb{Z})$, $f_*(\alpha)$ is not homologous to zero in $H_k(N, N \setminus C)$ where C is some compact set in N. Then* $\dim G \leq (1/2)(\dim M - k)(\dim M - k + 1)$.

Proof. It is enough to restrict G to its identity component. In this case every element $g \in G$ is homotopic to the identity and $f \circ g$ is homotopic to f for all $g \in G$. Hence there is a harmonic map h from M into N so that either $h(M)$ is a circle or $h \circ g = h$ for all $g \in G$. Since the discussion for the case where $h(M)$ is a circle is the same as the later, we assume that $h(M)$ is not a circle. The existence of the homology class α clearly implies that for some point $p \in M$, the rank of h_* on $T_p(M)$ is at least k. Hence for some open set U, the rank of h_* is at least k. But, as principal orbits are dense (see [3]), there is a point $q \in U$ so that the orbit $G(q)$ is a principal orbit. The equation $h \circ g = h$ for all $g \in G$ shows that $G(q)$ is subset of the fiber $h^{-1}[h(q)]$ and has dimension not greater than $\dim M - k$. Since G acts effectively on its principal orbit, the estimate on $\dim G$ follows immediately. □

Theorem 7. *Let N be a complete real analytic manifold with non-positive curvature. Let M be a complete k-dimensional ($k > 1$) real analytic orientable manifold with a compact group G acting real analytically. Suppose for some smooth map f from M into N, $f \circ g$ is freely homotopic to f for all $g \in G$. Suppose there exists a closed k-form ω in N such that the integral of ω over any compact k-dimensional integral cycle is an integer and $f^*\omega[M]$ is a non-zero integer m. If there exists an element α in $\pi_1(M)$ so that $f_*(\alpha)$ is not trivial in $\pi_1(N, N \setminus C)$ for some compact set C in N, then there exists a real analytic map h homotopic to f with the following properties:*

- (i) *The group $\bar{G} = \{g \in G \,|\, h \circ g = h\}$ is finite and has order dividing m. Furthermore, each element of $\bar{G}$ preserves the orientation of M.*
- (ii) *For any point $p \in M$, the isotropy subgroup of G at p is finite.*
- (iii) *The image of $G(p)$ under h is a finite union of submanifolds of some embedded l dimensional totally geodesic flat manifold with l not greater than the rank of a maximal abelian subgroup of $\pi_1(N)$ or the first Betti number of M.*
- (iv) *The connected component of G is a tours with dimension $\leq l$.*
- (v) *When N has negative curvature or when $\pi_1(N)$ has no non-trivial abelian subgroup or when the first Betti number of M is zero, the group $\{g : h \circ g$ is homotopic to $h\}$ is equal to G.*

Proof. Since G acts real analytically on M, we can average a real analytic metric of M so that G acts isometrically on M with some real analytic Riemannian metric. According to Theorem 5, we can find a harmonic map h from M into N which is homotopic to f. Since both M and N are real analytic, a theorem of Morrey [10] shows that the smooth map h is in fact real analytic. By well-known theorems in real analytic geometry [9] we know that both M and N can be triangulated so that $h(M)$ is a simplicial subcomplex of N.

With respect to this h, we can define the group $\bar{G}$. We have to prove that $\bar{G}$ is finite and has order not greater than m. In fact, the assumption $f^*(\omega)[M] = h^*(\omega)[M] \neq 0$ implies that $h(M)$ is a k-dimensional simplical complex. Furthermore for point x in some open dense set U of $h(M)$, $h^{-1}(x)$ is finite and the differential of h at each point of $h^{-1}(x)$ has rank k. The equation $h \circ g = h$ shows that if $g \in \bar{G}$ and g fixes a point in $h^{-1}(x)$, g must fix an orthonormal basis in the tangent space of that point. This is possible only if g is the identity and hence $\bar{G}$ acts freely on $h^{-1}(x)$ for $x \in U$.

Let T be a tubular neighborhood (in N) of the $(k-1)$-dimensional skeleton of $h(M)$. Then the closed k-form ω restricted to T is cohomologous to zero. Hence by subtracting some exact form in N, we may assume that ω is zero in a tubular neighborhood of the $(k-1)$-dimensional skeleton of $h(M)$. For each k-dimensional simplex S of $h(M)$, we can choose a smooth k-form ω_S which has compact support in an open set U_S of S so that $U_S \subset U$, the number of components of $h^{-1}(U_S)$ is equal to the number of points in $h^{-1}(x)$ for $x \in U_S$, $h|_{h^{-1}(U_S)}$ is a covering map and $\omega_S[S] = 1$.

Clearly the closed k-form ω, restricting to $h(M)$, is cohomologous to $\sum_S a_S \omega_S$ for some real number a_S where S runs through all k-simplices of $h(M)$. Now $[h(M)]$ can be written as a sum of k dimensional cycles $C_1, \ldots, C_l$ where the support of C_i intersect the support of C_j is a complex with dimension $\leq k-1$, each k-simplex appears in exactly one C_i and no proper subcomplex of any C_i can form a cycle.

By changing s to $-s$ if necessary, we may assume that $\omega_S[C_i] = 1$ if $S \in C_i$. Then consider the closed form $\sum_{i=1}^{l}(\sum_{S\in C_i} a_S)\omega_{S_i}$ where S_i is a k-simplex randomly chosen from C_i. For each C_j, $\omega[C_j] = \sum_{i=1}^{l}(\sum_{S\in C_i} a_S\omega_S)[C_j] = \sum_{i=1}^{l}(\sum_{S\in C_i} a_S)\omega_{S_i}[C_j] = \sum_{S\in C_j} a_S$. Since C_j is an integral cycle, the assumption on ω shows that $\sum_{S\in C_j} a_S$ is an integer.

Now consider $f^*\omega[M]$. It is equal to

$$h^*\omega[M] = \sum_{i=1}\left(\sum_{S\in C_i} a_S h^*\omega_S\right)[M].$$

We claim that $h^*\omega_S[M] = h^*\omega_{\tilde{S}})[M]$ when both S and $\tilde{S}$ belongs to the same C_i. Indeed, let $T(C_i)$ be a simplical tubular neighborhood of C_i so that $H_k(T(C_i))$ is generated by $H_k(C_i)$. Since no proper subcomplex of C_i forms a cycle, $H_k(C_i)$ is cyclic. If we extend ω_S and $\omega_{\tilde{S}}$ to be both smooth k-forms in $T(C_i)$ which are zero in a neighborhood of the $(k-1)$ dimensional skeleton of C_i, then ω_S and $\omega_{\tilde{S}}$ have the same value at the generator of $H_k(T(C_i))$ and by de Rham's theorem, $\omega_S - \omega_{\tilde{S}}$ is an exact form in a neighborhood of C_i. Hence $h^*\omega_S[M] = h^*\omega_{\tilde{S}}[M]$.

Therefore we have $m = h^*\omega[M] = \sum_{i=1}^{l}(\sum_{S\in C_i} a_S)h^*\omega_{S_i}[M]$. In order to prove that the order of $\bar{G}$ divides m, it suffices to prove that it divides $h^*\omega_{S_i}[M]$. Recall that support $\omega_{S_i} \subset U_{S_i}$ where $h^{-1}(U_{S_i}) = \cup_j O_j$ so that $h|_{O_i}$ is a diffeomorphism with rank k. Hence $h^*\omega_{S_i}[M]$ can be calculated as follows. Fix a point x in N so that $\omega_{S_i} \neq 0$. Then if $\{x_1, \dots, x_q\}$ is the set of points in $h^{-1}(x)$ and $\mathrm{d}V$ is the volume form in M, $h^*\omega_{S_i}[M] = \sum_{j=1}^{q} \mathrm{sign}[h^*\omega_{S_i}(x_j)/\mathrm{d}V]$ where $\mathrm{sign}(a) = +1$ if $a > 0$ and -1 if $a < 0$.

Let $\bar{\bar{G}}$ be the subgroup of $\bar{G}$ which preserves orientation of M. Then either $\bar{\bar{G}} = \bar{G}$ or $\bar{\bar{G}}$ has index two in $\bar{G}$. We claim the latter cannot happen. Indeed, let $g \in \bar{G} \setminus \bar{\bar{G}}$ and $\{g_1, \dots, g_p, gg_1, \dots, gg_p\}$ be the complete set of elements in $\bar{G}$ with $g_i \in \bar{\bar{G}}$ for $1 \le i \le p$. Then $g_i^*\mathrm{d}V = \mathrm{d}V$ and $(gg_i)^*\mathrm{d}V = -\mathrm{d}V$ for $1 \le i \le p$. The equation $h \circ g = h$ for $g \in G$ shows that $\mathrm{sign}[h^*\omega_{S_i}(g_j(x))/\mathrm{d}V] = \mathrm{sign}[h^*\omega_{S_i}(x)/\mathrm{d}V]$ and

$\text{sign}[h^*\omega_{S_i}(gg_j(x))/\text{d}V] = -\text{sign}[h^*\omega_{S_i}(x)/\text{d}V]$ when h has rank k at x. Since $\bar{G}$ acts freely on the set $h^{-1}(x)$, it is clear from the above two equations and the formula for $h^*\omega_{S_i}[M]$ that $h^*\omega_{S_i}[M] = 0$ for all i. In particular $h^*\omega[M] = 0$ which is a contradiction. Hence $\bar{G}$ consists of orientation preserving isometries. The above proof also shows that the order of $\bar{G}$ divides $h^*\omega_{S_i}[M]$ for all i. This completes the proof of (i).

To see (ii), we note that if g_t is a one parameter family of elements in G such that $h \circ g_t(p) = h(p)$. Then as $h \circ g_t$ is freely homotopic to h, the proof of Theorem 2 shows that in fact $h \circ g_t = h$. As $\bar{G}$ is finite, g_t must be a trivial family.

To prove the other assertions, we observe that if X is a Killing field on M, then h_*X gives rise to a parallel section of h^*TN. Recall that from the proof of Theorem 3, for each point $x \in M$, there exists a totally geodesic submanifold of N passing through $h(x)$ which parametrizes the space of harmonic maps homotopic to h. We can find parallel sections $v_1, v_2, \dots, v_k$ of h^*TN so that $h_*v_1, \dots, h_*v_k$ form a basis for the totally geodesic submanifold at $h(x)$. fix a point x where h_* is injective. Since X gives rise to a one parameter family of isometries g_t which also gives rise to a one parameter family of harmonic maps $h \circ g_t$, X defines a section of h^*TN and can be written as $\sum_{i=1}^k c_i v_i$ in a neighborhood of x where c_i are smooth functions. Let τ_i be the pull back of the forms dual to v_i. Then as in Corollary 2 of Theorem 2, τ_i is harmonic and hence $\tau_i(X)$ is constant (see [13]). This shows that X is a parallel section of h^*TN in an open set of M. Unique continuation then shows that X is parallel everywhere.

If Y is any other Killing field on M, then $\nabla_{h_*X} h_*Y = 0$ and $\nabla_{h_*Y} h_*X = 0$. Hence $h_*[X,Y] = 0$. Assertion (ii) shows that $[X,Y] = 0$. Therefore G^0, the connected component of G, is abelian. As $h \circ g(p) = h(p)$ implies $h \circ g = h$ for $g \in G^0$, $h(G^0(p))$ is the image of an embedding of the torus $G^0/\bar{G} \cap G^0$. It is clearly totally geodesic and flat with dimension not greater than the rank of a maximal abelian subgroup of $\pi_1(N)$. It is also not greater than $b_1(M)$ because every Killing field defines a parallel section of h^*TN which gives rise to a harmonic 1-form on M □

Corollary 6. *Let N be a compact real analytic Riemannian manifold with non-positive curvature. Let M be a k-dimensional compact orientable real analytic manifold that represents some k-dimensional interal cohomology class ω in $H^k(N, \mathbb{Z})$, i.e. there exists a smooth map f from M into N such that $f^*\omega[M] = 1$. Suppose $\pi_1(N)$ has no non-trivial abelian subgroup or M has zero first Betti number. Then any compact group G acting real analytically on M is finite. Furthermore the natural homomorphism from G into the outer automorphism group of $\pi_1(M)$ is injective. In fact, the composite of f_* with the above map into the outer automorphim group of $\pi_1(N)$ is also injective.*

Proof. Let $\tilde{G}$ be the subgroup of elements of G so that $f \circ g$ is freely homotopic to f. Then Theorem 7 shows that $\tilde{G}$ is trivial. Since N is a $K(\pi, 1)$, the condition that $f \circ g$ is freely homotopic to f is a condition on the action of g on the fundamental group of M composed with the homomorphism $f_* : \pi_1(M) \to \pi_1(N)$. This easily implies the other half of the corollary. □

When the map from M into N is surjective, we can say a little more.

Theorem 8. *Let M and N be two compact orientable manifolds of the same dimension. Suppose N has non-positive curvature and there is a map f from M into N so that $f^*[N] - m[M] \neq 0$ and $f_*\pi_1(M) = \pi_1(N)$. Let G be a compact group that acts smoothly on M. Let $\tilde{G}$ be the subgroup of G consisting of elements $g \in G$ such that $f \circ g$ is freely homotopic to f. Let $\tilde{A}(N)$ be the subgroup of the identity component of $A(N)$ that is generated by parallel vector fields of N. Suppose for each $g \in G$, there is an (not necessary unique or non-trivial) element*

$\tilde{g} \in A(N)$ such that $f \circ g$ is freely homotopic to $\tilde{g} \circ f$. Then there is a smooth map $h : M \to N$ and a homomorphism $\gamma : G \to A(N)$ so that

(i) *$f \circ g = \gamma(g) \circ f$ for all $g \in G$.*

(ii) *The kernel of γ ($= Ker(\gamma)$) is a finite group whose order divides m.*

(iii) *$\tilde{G} = \gamma^{-1}(\tilde{A}(N))$ and $\tilde{G}/Ker(\gamma)$ is isomorphic to a subgroup of the p dimensional torus $\tilde{A}(N)$ with $p =$ rank of the center of $\pi_1(N)$.*

(iv) *The isotropy subgroup of the group $\tilde{G}$ at any point of M is a finite abelian extension of a subgroup of $Ker(\gamma)$.*

(v) *The isotropy group of $\gamma(G)$ at any point of M is finite. Furthermore if p is any prime number that divides the order of the isotropy group of $\gamma(G)$ at a point x, then p also divides m or the order of the isotropy group of G at the same point of $h^{-1}(x)$. In particular, if $m = 1$ and G acts freely on M, $\gamma(G)$ acts freely on N.*

(vi) *The group $G/\tilde{G}$ is finite and has a faithful representation into the outer automorphism group of $\pi_1(N)$ via f_*.*

(vii) *If N is a flat manifold, the rank of $\tilde{A}(N)$ is also equal to the first Betti number of N.*

Proof. (i) As in the theorem, we can assume that f is homotopic to a harmonic map h. By hypothesis, for each $g \in G$, there is an element $\tilde{g} \in A(N)$ so that $h \circ g$ is freely homotopic to $\tilde{g} \circ h$. By Theorem 4 in §2, we know that there is an element g' in $\tilde{A}(N)$ so that $h \circ g = (g' \circ \tilde{g}) \circ h$. Since $h^*[N] \neq 0$, h is surjective and $g' \circ \tilde{g}$ is uniquely defined by the above equation. Write $g' \circ \tilde{g}$ as $\gamma(g)$. Then it is clear that γ is the required group homomorphism.

(ii) The kernel of γ is the subgroup of G consisiting of elements $g \in G$ such that $h \circ g = h$. One can then apply the same argument as in Theorem 7.

(iii) If $g \in \tilde{G}$, then h and $h \circ g$ are freely homotopic. Theorem 4 shows that there is a uniquely determined $\gamma(g) \in \tilde{A}(N)$ so that $h \circ g = \gamma(g) \circ h$. Conversely if $\gamma(g) \in \tilde{A}(N)$, $\gamma(g)$ is freely homotopic to identity and hence $h \circ g$ is freely homotopic to h. The fact that $\dim \tilde{A}(N)$ is equal to the rank of the center of $\pi_1(N)$ is in Lawson-Yau [8].

(iv) This follows because $\tilde{A}(N)$ acts locally freely on N so that the isotropy group of $\tilde{A}(N)$ is a finite abelian subgroup.

(v) The first statement follows from the fact that the connected component of $\gamma(G)$ acts locally freely. To see the second statement, let F be a cyclic subgroup of $\gamma(G)$ of prime order p that fixes a point $x \in N$. Then we can find a subgroup $\tilde{F}$ of order p in $\gamma^{-1}(F)$. Let $f \in \tilde{F}$ be the generator. Then either f fixes a point in $h^{-1}(x)$ or $\tilde{F}$ acts freely in $h^{-1}(x)$. Our assertion follows easily from the argument of theorem.

(vi) This follows from the fact N is a $K(\pi, 1)$ so that existence of a homotopy between $h \circ g$ and h depends only on their action on the fundamental group.

(vii) This follows because for a compact manifold with non-negative Ricci curvature, the Bochner method (see [2]) shows that a harmonic vector field is parallel. □

In the above theorems, we assume that for each $g \in G$, there exists $\tilde{g} \in A(N)$ such that $f \circ g$ is freely homotopic to $\tilde{g} \circ f$. In this regard, it is natural to ask the following question.

Question. Let N be a compact space with non-positive curvature. Let G be a finite subgroup of the outer automorphism group of $\pi_1(N)$. Then can one find a non-positively curved metric on N so that G is induced by the action of some finite subgroup of isometries of N?

Some special cases of the above question can be answered. The following is a trivial consequence of Mostow's rigidity theorem.

Theorem 9. *Let N be a compact locally symmetric space of non-positive curvature. Suppose no closed one or two dimensional geodesic subspace of N are direct factors of N locally. Then the outer automorhpism group of N is finite and is induces by the group of isometries of N.*

In case N has zero curvature, we can also prove the following theorem.

Theorem 10. *Let N be a compact flat Riemannian manifold. Let G be any finite group in the outer automorphism group of $\pi_1(N)$. Then G is induced by a subgroup of isometries of some flat Riemannian metric on N.*

Proof. If $g \in G$, then since N is a $K(\pi, 1)$, we can realize g by a homotopy equivalence of N. By the existence of a harmonic map in a given homotopy class, we can therefore assume that g is induced by some harmonic maps $\tilde{g}$ from N into N.

Since N is flat, the only harmonic map from N to N is linear. (The curvature condition implies that the energy density is subharmonic on N. the compactness of N then imples that the energy density is in fact constant. Hence the Laplacian of the energy density is zero which implies the harmonic map is totally geodesic.) If we lift $\tilde{g}$ to a map from the euclidean space $\mathbb{R}^n$ to itself, then $\tilde{g}(x) = A_g x + B_g$ for $x \in \mathbb{R}^n$ where A_g is a constant invertible matrix and B_g is a constant vector. It is clear that the set $\{A_g \mid g \in G\}$ forms a finite subgroup of $GL(n, \mathbb{R})$. Hence we can conjugate it to a finite subgroup of $O(n, \mathbb{R}) \subseteq GL(n, \mathbb{R})$. In other words, we may choose a flat Riemannian metric on N so that the matrices A_g acts orthogonally.

With this choice of the metric, we can associate to each $g \in G$ an element $\tilde{\tilde{g}}$ in the group of isometries of N in the following way. Let $\tilde{B}_g$ be the component of B_g in the orthogonal complement of the linear space spanned by the parallel vector fields of N. Then define $\tilde{\tilde{g}} = \tilde{\tilde{A}}_g x + \tilde{B}_g$

for $x \in \mathbb{R}^n$. Since parallel vector fields act trivially on $\pi_1(N)$, $\tilde{\tilde{g}}$ induces the same action as $\tilde{g}$ on $\pi_1(N)$. furthermore by the uniqueness theorem of harmonic maps, we know that $\tilde{\tilde{g}}$ is uniquely determined by g, and the totality of $\{\tilde{\tilde{g}} \mid g \in G\}$ forms a subgroup of the group of isometries of N (with respect to the new flat metric). This finishes the proof of the theorem. □

As corollary of these last two theorems, we derive the following application for finite group actions.

Theorem 11. *Let M be a compact manifold with a finite group G acting effectively and smoothly. Suppose there is a degree one map from M onto a compact locally symmetric space N of non-positive curvature which admits no closed one or two dimensional geodesic subspaces as local direct factor. Then G has a faithful representation into the group of isometries of N. Furthermore the action of G commutes with some degree one map (homotopic to the previous one) from M into N such that G acts freely on M iff G acts freely on N.*

Remark. In this casee N cannot have a local parallel vector field so that we do not require surjectivity on fundamental group.

Theorem 12. *Let M be a compact n-dimensional manifold with a finite group G acting effectively and smoothly. Suppose there are n integral one dimensional homology classes whose dual $\alpha_1, \ldots, \alpha_n$ satisfies the equality $\alpha_1 \cup \ldots \cup \alpha_n[M] = 1$. Then there is a faithful representation of G into the group of isometries of some flat n-dimensional torus $\mathbb{T}^n$. The action of G commutes with some degree one map from M into $\mathbb{T}^n$ so that G acts freely on M iff G acts freely on $\mathbb{T}^n$. Then group $\tilde{G} = \{g \in G \mid g^*\alpha_i = \alpha_i \text{ for } i = 1, 2, \ldots, n\}$ is a normal abelian subgroup of G which acts freely on M.*

Theorem 13. *Let M be a compact manifold with a finite group G acting effectively and differentiably. Suppose we can find a degree one* *map f from M onto a compact flat Riemannian manifold N such that $f_*[\pi_1(M)] = \pi_1(N)$. If either $\pi_1(N)$ has no center or $b_1(M) = 0$, then there is a faithful representation of G into the group of isometries of N which in turn gives a faithful representation of G into the outer automorphism group of $\pi_1(N)$. The action of G commutes with a degree one map from M into N so that G acts freely on M iff G acts freely on N.*

Remark. In all these theorems, we can find similar statements if we merely assume the existence of a non-zero degree map into the corresponding "model" manifold.

So far, we have always assumed that the manifold M represents some homology class with the same dimension as M. Actually this is not quite necessary. At least when G is connected, we have the following theorem.

Theorem 14. *Let N be a complete real analytic manifold with non-positive curvature. Let ω be a k-dimensional cohomology class in N which is not cohomologous to any class in $N\backslash C$, for some compact set C in N. Let M be a compact m-dimensional real analytic orientable spin manifold with a connected compact group G acting real analytically. Suppose for some $\hat{A}$-class Ω of dimension $m-k$, $\Omega \cup f^*\omega[M] \neq 0$. Then we can find a real analytic harmonic map h homotopic to f so that*

(i) *The group $\bar{G} = \{g \in G \,|\, h \circ g = h\}$ is a finite group.*
(ii) *Any isotropy group of G acting on M is finite.*
(iii) *The image of $G(p)$ under h is a finite union of submanifolds of some embedded l dimensional totally geodesic flat manifold with l not greater than the rank of a maximal abelian subgroup of $\pi_1(N)$ and the first Betti number of M.*
(iv) *The connected component of G is a torus with dimension $\leq l$.*

(v) *Suppose $k > 1$. Then when N has negative curvature or when $\pi_1(N)$ has no non-trivial abelian subgroup or when the first Betti number of M is zero, G is trivial.*

Proof. We are going to use the terminology used in Theorem 7. As in that proof, we can find a harmonic map h homotopic to f. We can also choose the open set U defined there so that $h|_{h^{-1}(U)}$ is a fiber bundle.

To prove (i), we note that since $h^{-1}(U)$ is open, G acts effectively on $h^{-1}(U)$. By restricting U more, we may even assume that $\bar{G}$ acts effectively on $h^{-1}(x)$ for each $x \in U$. Since $h^{-1}(x)$ has trivial normal bundle, $W_2(h^{-1}(x)) = 0$ so $h^{-1}(x)$ is a spin manifold and $\Omega[h^{-1}(x)]$ is the $\hat{A}$-genus of $h^{-1}(x)$. Therefore, a theorem of Atiyah and Hirzebruch [1] shows that $\Omega[h^{-1}(x)] = 0$ if $\bar{G}$ is not finite. Since ω_S has support in U, it is clear that $\Omega \cup h^*\omega_S[M] = 0$. We can therefore arrive at the contradiction that $\Omega \cup h^*(\omega)[M] = 0$. The proofs of (ii)-(v) are the same as Theorem 7 and we omit the details of them. □

Note added in proof: Recently Prof. Eells informed us that T. Sunada has independently proved Theorem 3 for the special case of a compact locally symmetric space. His paper appeared in *Invent. Math.* **51** (1979), 297-307. There is also a related paper of Connor and Raymond in the *Proceedings of Symposia in Pure math.*, Vol. 23 (1978).

References

1. M. Atiyah and F. Hirzebruch: *Spin-manifolds and group actions*, Essays on Topology and Related Topics, Memoires dédiés à Georges de Rham, pp. 18-28. Springer-Verlag, New York-Heidelbeg-Berlin (1970).
2. S. Bochner and Y. Yano: *Curvature and Betti numbers*, Annals of Math. Studies **32** (1953).
3. G. E. Bredon: *Introduction to Compact Transformation Groups*, Academic Press, New York (1972).
4. J. Cheeger and D. Ebin: *Comparison Theorems in Riemannian Geometry*, North-Holland, Amsterdam, 1975.
5. J. Eells and J. Sampson: *Harmonic mappings of Riemannian manifolds*, Am. J. Math. **86** (1964), 109-160.

6. R. Hamilton: *Harmonic Maps of Manifolds with Boundary*, Lecture notes, Mathematics, No. 471. Springer, Berlin, Heidelberg, New York (1975).
7. P. Hartman: *On homotopic harmonic maps*, Can. J. Math. **19** (1967), 673-687.
8. H. B. Lawson and S. T. Yau: *Compact manifolds of nonpositive curvature*, J. Diff. Geom. **7** (1972), 211-228.
9. S. Lojasiewicz: *Triangulations of semianalytic sets*, Ann. Scuola Norm. Sup. di Pisa **18** (1964), 449-474.
10. C. B. Morrey: *On the analyticity of the solutions of analytic non-linear elliptic systems of partial differential equations, I and II*, Am. J. Math. **80** (1958), 198-234.
11. R. Schoen and S. T. Yau: *Harmonic maps and the topology of stable hypersurfaces and manifolds of non-negative Ricci curvature*, Comm. Math. Helv. **39** (51) (1976), 333-341.
12. R. Schoen and S. T. Yau: *Compact group actions and the topology of manifolds with nonpositive curvature*, Topology **18**(1979), 361-380.
13. S. T. Yau: *Remarks on the group of isometries of a Riemannian manifold*, Topology **16** (1977), 239-247.

Chapter XII

Harmonic Maps and the Topology of Stable Hypersurfaces and Manifolds with Non-negative Ricci Curvature

In this chapter we give a presentation of the paper [11] which uses harmonic map theory to give topological obstructions for manifolds to admit complete metrics of nonnegative curvature, or to appear as complete stable minimal hypersurfaces in manifolds of nonnegative sectional curvature. To motivate the idea, we make some observations first. It is rather easy (see §2 below) to see that every harmonic map from a compact manifold with positive Ricci curvature to a manifold with non-positive curvature is a constant map. Taking the domain manifold to be the sphere, we conclude immediately that a compact manifold with non-positive curvature is a $K(\pi, 1)$ which is a well-known consequece of a theorem of Cartan-Hadamard. On the other hand, it is also easy to see that the image of every harmonic map from a torus into a compact manifold with negative curvature is a geodesic. Coupling with the above facts, we can show that in the fundamental group of a compact manifold of negative curvature, every abelian subgroup is cyclic. This is a theorem of Preissmann.

In order to obtain new results, we turn the above arguments backward and study complete non-compact manifolds with non-negative Ricci curvature.

As far as we know, there are only a few known topological restrictions on these manifolds. Milnor [6] (subsequently, Wolf [7] and Cheeger-Gromoll [3]) showed that every finitely generated subgroup of the fundamental group of these manifolds has polynomial growth with order less than the dimension of the manifold. Then Gromoll-Meyer [5] showed that every complete non-compact manifold with positive Ricci curvature has at most one end. Using this last fact, it can be shown that for complete manifolds with positive Ricci curvature, the first cohomology group with compact support is torsion.

One of the main theorems in this chapter is to prove that given any compact domain D in a complete manifold with non-negative Ricci curvature such that ∂D is simply connected, any homomorphism from $\pi_1(D)$ into the fundamental group of a compact manifold of non-positive curvature is trivial. To see that this is not a consequence of the above topological obstructions, let M be any compact simply-connected manifold and let $\bar{M}$ be the complement of a point in $\mathbb{S}^1 \times M$. Then clearly $\bar{M}$ has only one end and the growth of $\pi_1(\bar{M})$ is small. However, according to our theorem, $\bar{M}$ does not admit a complete metric with non-negative Ricci curvature.

The second main theorem of this chapter is to apply the same idea to study the topology of a complete non-compact stable minimal hypersurface in a complete non-negatively curved manifold. We find the same topological obstruction to this class of manifolds. Even if we assume the ambient manifold is the euclidean space, our theorem seems to give the first known topological obstruction to stable hypersurfaces.

Since harmonic maps are not necessarily linear, these theorems may be considered as non-linear vanishing theorems.

1. Existence of Harmonic Maps with Finite Energy

Let M be a complete Riemannian manifold and N be a compact manifold with non-positive sectional curvature. Let $f : M \to N$ be a smooth map. Then we shall show that when $E(f) < \infty$, we can find a harmonic map $h : M \to N$ such that h is homotopic to f on each compact set of M and $E(h) < \infty$.

Let M_i be a sequence of compact manifolds (with boundary) such that $M = \cup_i M_i$. Then according to Hamilton [4] we can find harmonic maps $h_i : M_i \to N$ which are homotopic to $f|_{M_i}$. Furthermore, $E(h_i) \leq E(f|_{M_i})$. It remains therefore to prove that there is a subsequence of $\{h_i\}$ which converges uniformly on compact subsets of M.

Let p be an arbitrary point in M. Then we shall find a neighborhood U of p and a subsequence of $\{h_i\}$ which converges uniformly on U.

The first step is to produce a neighborhood of p such that the energy densities $e(h_i) = \mathrm{Tr}_M(h_i^* ds_N^2)$ of h_i are uniformly bounded on this neighborhood. Thus, let U_1 be a compact neighborhood of p and c be the infinimum of the Ricci curvuture of U_1. A straightforward computation then shows that on U_1, $\Delta e(h_i) - ce(h_i) \geq 0$.

For our purpose, $c = 0$ is the most interesting case. Hence we give a proof for this special case first.

Let (r, θ) be the geodesic polar coordinate at the point p where θ is a point in unit sphere Σ. Let $\sqrt{g} r^{n-1} dr d\theta$ be the volume element associated to this coordinate system. Then Stokes' theorem and $\Delta e(h_i) \geq 0$ shows that

$$\int_{\theta\in\Sigma} \frac{\partial e(h_i)}{\partial r}(\rho,\theta)\sqrt{g}d\theta \geq 0 \tag{1.1}$$

when ρ is less than the injectivity raduis at p.

Therefore,

$$\begin{aligned} & \frac{\partial}{\partial\rho}\left[\int_{\theta\in\Sigma} e(h_i)(\rho,\theta)\sqrt{g}d\theta\right] \\ + & \sup_{\theta\in\Sigma}\left|\frac{\partial}{\partial r}\log\sqrt{g}(\rho,\theta)\right|\int_{\theta\in\Sigma} e(h_i)(\rho,\theta)\sqrt{g}d\theta \\ \geq & \ 0 \end{aligned} \tag{1.2}$$

Integrating this inequality, we obtain

(1.3)

$$e(h_i)(p) \leq \left[\int_{\theta\in\Sigma} e(h_i)(R,\theta)\sqrt{g}d\theta\right]\exp\left[R\sup_{\theta\in\Sigma, r\leq R}\left|\frac{\partial}{\partial r}\log\sqrt{g}(\rho,\theta)\right|\right]$$

where R is less than the injectivity raduis at p.

By the mean value theorem, we see that for each R within the injectivity radius at p, there is a number $\bar{R}$ such that

$$\int_{\theta\in\Sigma} e(h_i)(\bar{R},\theta)\sqrt{g}d\theta \leq R^{-n}\int_{R\leq r\leq 2R} e(h_i) \tag{1.4}$$

Therefore, when R is within the injectivity radius at p and M_i covers tha ball of radius R,

$$(1.5)\quad e(h_i)(p) \le R^{-n} \exp\left[R \sup_{\theta\in\Sigma, r\le R} \left| \frac{\partial}{\partial r} \log \sqrt{g}(\rho,\theta) \right| \right] E(h_i|_{M_i})$$

In case $c < 0$, one can use arguments of [1]. In fact, let R be small number so that for all point q with distance not greater than nR away from p, the injectivity radius at q is greater than R where $n = \dim M$. Let ϕ be a function defined defined on the real line so that $\phi(x) = 0$ for $|x| \le R^{-n+2}$ and $\phi(x) = x$ for $|x| \ge (R/2)^{-n+2}$. Then for $n \ge 3$, define the function $F(q_1, q_2) = \phi(d(q_1, q_2)^{-n+2})$ on $M \times M$ where $d(q_1, q_2)$ is the distance between q_1 and q_2. (For $n = 2$, we replace $d(q_1, q_2)^{-n+2}$ by $\log d(q_1, q_2)$.)

As in [1] p. 142, there is a constant A such that

$$(1.6)\qquad e(h_i)(p) \le A \int_{d(p,q)\le R} (F(p,q) + 1)\, e(h_i)(q)$$

Iterating this, we obtain

$$(1.7)\qquad e(h_i)(p) \le A^k \int_{d(p,q)\le kR} (F_k(p,q) + 1)\, e(h_i)(q)$$

where F_k are defined inductively by $F_1 = F + 1$ and

$$(1.8)\qquad F_k(q_1, q_2) = \int_M F_{k-1}(q_1, x)\,(F(x, q_2) + 1)$$

When $k > n/2$, $F_k(p, q)$ is bounded for $d(p, q) \le nR$. It follows from this and (1.7) that $e(h_i)(p)$ can be bounded by $E(h_i|_M)$.

Having bounded the energy density, it is then standard (cf. [1]) to estimate the higher derivatives of h_i and hence prove that a subsequence of h_i converges uniformly in a neighborhood of p. By using the diagonal process, we have then proved the claim at the beginning of this section.

2. The Fundamental Group of Complete Manifolds with Non-negative Ricci Curvature

Let M be a complete manifold with non-negative Ricci curvature and N be a compact manifold with non-positive sectional curvature.

Let f be a harmonic map from M to N with finite energy. Then we claim that f is a constant map.

The Bochner formula says that

$$\Delta e(f) = |\beta(f)|^2 + Q(f^*) \tag{2.1}$$

where $Q(f^*) \geq 0$ under our curvature assumption.

By definition of $\beta(f)$, one can check easily

$$2e(f)|\beta(f)|^2 \geq |\nabla e(f)|^2 \tag{2.2}$$

The inequalities (2.1) and (2.2) together show that $\sqrt{e(f)}$ is subharmonic on M.

In [8], it is shown that every non-negative L^2-integrable subharmonic function on a complete Riemannian manifold must be a constant. Applying this to $e(f)$, we conclude that $e(f)$ is a constant.

On the other hand, we have also shown in [8] that when M is a complete non-compact manifold with non-negative Ricci curvature, then the volume of M is infinite (also known to E. Calabi [9]). This forces the constant $e(f)$ to be zero and f to be a constant map. Therefore, we have proved the following

Theorem 1. *Let M be a complete manifold with non-negative Ricci curvature and N be a compact manifold with non-positive curvature. Let f be any smooth map from M to N with finite energy. Then f is homotopic to a constant on each compact set.*

As an application of this theorem, we note the following

Corollary. *Let M be a complete manifold with non-negative Ricci curvature. Let D be a compact domain in M with smooth simply connected boundary. Then there is no non-trivial homomorphism from $\pi_1(D)$ into the fundamental group of a compact manifold with non-positive curvature.*

Proof. Suppose $h : \pi_1(D) \to \pi_1(N)$ is a homomorphism where N is a compact manifold with non-positive curvature. Then as N is a $K(\pi, 1)$, there is a smooth map $f : D \to N$ such that $f_* = h$. This map is homotopic to a constant map on ∂D because ∂D is simply connected. Therefore, we can extend f to be a smooth map $\tilde{f} : M \to N$ such that outside a compact set, $\tilde{f}$ is a constant map. Clearly, $\tilde{f}$ has finite energy and we can apply the theorem to see that f is homotopic to a constant map and that h is trivial. □

3. Fundamental Group of Stable Immersions

Throughout this section let M^n be a complete non-compact manifold immersed in a manifold $\bar{M}^{n+1}$ having non-negative sectional curvatures. Also, suppose the immersion is stable (i.e. minimal and non-negative second variation of area with respect to compactly supported deformations). Let $e_1, \dots, e_n, e_{n+1}$ be a local orthonormal frame on $\bar{M}$ such that $e_1, \dots, e_n$ form an orthonormal frame for M in a neighborhood of a point $x_0 \in M$. Let $\omega_1, \dots, \omega_{n+1}$ be the dual coframe. Let ω_{ij} be the connection 1-forms of $\bar{M}$ and $\omega_{n+1\,i} = \sum_{j=1}^n h_{ij}\omega_j$ when restricted to M defines the symmetric second fundamental form of M. Since the immersion is minimal, we have $\sum_{i=1}^n h_{ii} = 0$, and the stability inequality is

$$\int_M \left(\sum_{i=1}^n K_{n+1,i,n+1,i} + \sum_{i,j=1}^n h_{ij}^2 \right) \varphi^2 \mathrm{d}V_M \le \int_M |\nabla\varphi|^2 \mathrm{d}V_M$$

where φ is any Lipschitz function with compact support on M, K_{ijkl} is the curvature tensor of $\bar{M}$, and $\mathrm{d}V_M$ is the volume element of M (see [10]).

Lemma 1. *M has infinite volume.*

Proof. Let $B_R(x_0)$ be the geodesic ball of radius R in M centered at x_0. Choose φ to be a Lipschitz function with the properties

$$\varphi = \begin{cases} 1 & \text{inside} \quad B_R(x_0) \\ 0 & \text{outside} \quad B_{3R}(x_0) \end{cases} \qquad \text{and} \quad |\nabla\varphi| \le 1/R.$$

The stability inequality implies

$$\int_{B_R(x_0)} \sum_{i,j=1}^{n} h_{ij}^2 \mathrm{d}V_M \leq \frac{\mathrm{Vol}(M)}{R^2}.$$

If $\mathrm{Vol}(M) < \infty$, by letting $R \to \infty$ we find that M is totally geodesic. Thus, the sectional curvatures of M are non-negative and by a theorem of [8] we must have $\mathrm{Vol}(M) = \infty$, a contradiction. □

Let N^k be a compact manifold with non-positive sectional curvature. Let $f : M \to N$ be a harmonic map. Let $x_0 \in M$ and $\bar{e}_1, \ldots, \bar{e}_k$ be an orthonormal frame in a neighborhood of $f(x_0) \in N$. Let $\theta_1, \ldots, \theta_k$ be the dual coframe and $\theta_{\alpha\beta}$, $1 \leq \alpha, \beta \leq k$, the connection forms. Define $f_{\alpha i}$, $1 \leq \alpha \leq k$, $1 \leq i \leq n$, by $f^*\theta_\alpha = \sum_{i=1}^n f_{\alpha i}\omega_i$. Then $e(f) = \sum_{i=1}^n \sum_{\alpha=1}^k f_{\alpha i}^2$. Define $f_{\alpha ij}$ by $\mathrm{d}f_{\alpha i} + \sum_{\beta=1}^k f_{\beta i} f^*\theta_{\beta\alpha} + \sum_{j=1}^n f_{\alpha i}\omega ji = \sum_{j=1}^n f_{\alpha ij}\omega_j$. Then f is harmonic means $\sum_{i=1}^n f_{\alpha ii} = 0$ for $1 \leq \alpha \leq k$.

Lemma 2. *If $E(f) < \infty$, then f is constant.*

Proof. The stability inequality implies

$$\int_M \left(\sum_{i,j=1}^n h_{ij}^2\right) \varphi^2 \leq \int_M |\nabla\varphi|^2.$$

Replacing φ by $\sqrt{e(f)}\varphi$ we obtain

$$\begin{aligned}
&\int_M \left(\sum_{i,j=1}^n h_{ij}^2\right) e(f)\varphi^2 \\
&\leq \int_M e(f)|\nabla\varphi|^2 + 2\int_M \sqrt{e(f)}\varphi\nabla\sqrt{e(f)} \cdot \nabla\varphi \\
&\quad + \int_M \varphi^2|\nabla\sqrt{e(f)}|^2 \\
&= \int_M e(f)|\nabla\varphi|^2 - \frac{1}{2}\int_M \varphi^2\Delta e(f) + \int_M \varphi^2|\nabla\sqrt{e(f)}|^2.
\end{aligned}$$

By the Gauss curvature equation and formula (2.1) using the fact that the sectional curvatures of $\bar{M}$ are non-negative and those of N are non-positive we obtain

$$\begin{aligned}\Delta e(f) &\geq 2\sum_{\alpha,i,j} f_{\alpha ij}^2 - 2\sum_{alpha,i}\left(\sum_j h_{ij}f_{\alpha j}\right)^2\\ &\geq 2\sum_{\alpha,i,j} f_{\alpha ij}^2 - 2\left(\sum_{i,j} h_{ij}^2\right)e(f).\end{aligned}$$

Putting this into the above inequality and rearranging, we obtain

$$\int_M \varphi^2\left(\sum f_{\alpha ij}^2 - |\nabla\sqrt{e}|^2\right) \leq \int_M e|\nabla\varphi|^2.$$

Now

$$|\nabla\sqrt{e}|^2 = \frac{\sum_j(\sum_{i,\alpha} f_{\alpha i}f_{\alpha ij})^2}{e}$$

and hence

$$\begin{aligned}\sum f_{\alpha ij}^2 - |\nabla\sqrt{e}|^2 &= \frac{1}{2e}\sum_{j,i,k,\alpha,\beta}(f_{\alpha i}f_{\beta kj} - f_{\beta k}f_{\alpha ij})^2\\ &\geq \frac{1}{2e}\sum_{j,i,\alpha}(f_{\alpha i}f_{\alpha jj} - f_{\alpha j}f_{\alpha ij})^2\end{aligned}$$

where we have thrown out terms where $k \neq j$ or $\alpha \neq \beta$. Using the Schwartz inequality we obtain

$$\begin{aligned}\sum f_{\alpha ij}^2 - |\nabla\sqrt{e}|^2 &\geq \frac{1}{2nke}\sum_i\left(\sum_{j,\alpha} f_{\alpha i}f_{\alpha jj} - \sum_{j,\alpha} f_{\alpha j}f_{\alpha ij}\right)^2\\ &= \frac{1}{2nke}\sum_i\left(\sum_{j,\alpha} f_{\alpha j}f_{\alpha ji}\right)^2 = \frac{1}{2nk}|\nabla\sqrt{e}|^2\end{aligned}$$

where we have used $\sum_j f_{\alpha jj} = 0$ and $f_{\alpha ij} = f_{\alpha ji}$. Therefore, we have $\int \varphi^2|\nabla\sqrt{e}|^2 \leq 2nk\int e|\nabla\varphi|^2$. Choosing φ as in lemma 1 we obtain

$$\int_{B_R(x_0)} |\nabla\sqrt{e}|^2 \leq \frac{2nk}{R^2}E(f).$$

Letting $R \to \infty$ we see that $e =$ constant on M. In light of Lemma 1 and the fact that $E(f) < \infty$, we see that $e(f) \equiv 0$. Thus f is constant. □

Combining the existence theorem of §1 with Lemma 2 we have the following result.

Theorem 2. *Let M be a complete non-compact stably immersed hypersurface in a manifold of non-negative curvature and N be a compact manifold with non-positive curvature. Let $f : M \to N$ be a smooth map with $E(f) < \infty$. Then f is homotopic to a constant on each compact set.*

As in §2 we have the following corollary.

Corollary. *Let M be as in Theorem 2. Let D be a compact domain in M with smooth simply connected boundary. Then there is no non-trivial homomorphism from $\pi_1(D)$ into the fundamental group of a compact manifold with non-positive curvature.*

References

1. Eells, J. and Sampson, J. H., *Harmonic mappings of Riemannian manifolds*, Amer. J. Math. **86**, (1964) 109-160.
2. Cheeger, J. and Gromoll, D., *On the structure of complete manifolds of non-negative curvature*, Ann. Math. **96**, (1972) 413-443.
3. Cheeger, J. and Gromoll, D., *The splitting theorem for manifolds of non-negative Ricci curvature*, J. Diff. Geom. **6**, (1971) 119-128.
4. R. Hamilton: *Harmonic Maps of Manifolds with Boundary*, Lecture notes, Mathematics, No. 471. Springer, Berlin, Heidelberg, New York (1975).
5. Gromoll, D. and Meyer, W., *On complete open manifolds of positive curvature*, Ann. Math. **90**, (1969) 75-90.
6. Milnor, J., *A note on curvature and fundamental group*, J. Diff. Geom. **2**, (1968) 1-7.
7. Wolf, J., *Growth of finitely genrated solvable groups and curvature of Riemannian manifolds*, J. Diff. Geom. **2**, (1968) 421-446.
8. Yau, S. T., *Some function-theoretic properties of complete Riemannian manifold and their applications to geometry*, to appear in Indiana J. Math.
9. Calabi, E., *On manifolds with non-negative Ricci curvature II*, Notices A.M.S. **22**, 1975, A205.
10. Chern S. S., *Minimal submanifolds in a Riemannian manifold*, Mimeographed Lecture Notes, Univ. of Kansas, 1968.
11. Schoen, R. and Yau, S. T., *Harmonic maps and topology of stable hypersurfaces and manifolds of nonnegative Ricci curvature*, Comm Math Helv **39**(1976), 333-341.

Chapter XIII

Harmonic Maps and Superrigidity by Jürgen Jost and Shing-Tung Yau

We prove a geometric generalization of Margulis' superrigidity theorem in the cocompact case by showing that harmonic maps from compact locally symmetric spaces of noncompact type and rank ≥ 2 into Riemannian manifolds with nonpositive curvature operator are totally geodesic. Technical tools are a Matsushima type formula for harmonic maps and a detailed study of the curvature tensor of symmetric spaces.

INTRODUCTION.

The main result of this paper is the following.

Theorem A. *Let* $\tilde{M} = G/K$ *be an irreducible symmetric space of noncompact type, other than* $SO_o(p,\,1)/SO(p) \times SO(1)$, $SU\,(p,1)\,/S(U\,(p)$ $\times U\,(1))$, *(in the terminology of* [He; p. 518]).

Let Γ be a discrete cocompact subgroup of G(also called a cocompact lattice). Let $\tilde{N}$ be a complete simply connected Riemannian manifold of nonpositive curvature operator with isometry group $I\left(\tilde{N}\right)$. Let $\rho : \Gamma \to I\left(\tilde{N}\right)$ be a homomorphism for which $\rho\,(\Gamma)$ either does not have a fixed point on the sphere at ∞ of $\tilde{N}$ or if it does, it centralizes a totally geodesic flat subspace. Then there exists a totally geodesic ρ-equivariant map, $f : \tilde{M} \to \tilde{N}$.

If $\tilde{N}$ is also a symmetric space G'/K' (of noncompact or Euclidean type), then the condition on $\rho\,(\Gamma)$ means that it is a reductive subgroup of G'.

The curvature condition on $\tilde{N}$ may be weakened, see the remark after Lemma 2 below.

The corresponding result for $SO_o\,(p,1)\,/SO\,(p) \times SO\,(1)$ and $SU\,(p,1)\,/\;S\,(U\,(p) \times U\,(1))$ is false, because it would imply that compact quotients have vanishing first Betti number and there are examples of compact quotients of these spaces for which this is not the case. In the case of $SU(p,1)/S(U(p)\times U(1))$, one gets, however the existence of a pluriharmonic ρ-equivariant map, essentially a special case of a result of Siu[S].

For $Sp(p,1)/Sp(p)\times Sp(1)$ and the hyperbolic Cayley plane, the result is Corlette's theorem [C2]. For Hermitian symmetric spaces, the result is due to Mok [MSY].

A consequence of the theorem is

Corollary B. *Let $\tilde{M} = G/K$ and Γ be as above. Let H be a semisimple noncompact Lie group with trivial center, $\rho : \Gamma \to H$ a homomorphism with Zariski dense image. Then ρ extends to a homomorphism from G onto H.*

This result is due to Margulis [Mg1] for rank$(G/k) \geq 2$ and to Corlette for $Sp(p,1)/Sp(p)\times Sp(1)$ and the hyperbolic Cayley plane.

Using the constructions of the paper [GS] by Gromov and Schoen, our results extend to the nonarchimedean case to show

Theorem C. *let $\tilde{M} = G/K$ and Γ be as above.*

Let $\rho : \Gamma \to Sl(n, Q_p)$ be a homomorphism, for some $n \in N$ and some prime p. Then $\rho(\Gamma)$ is contained in a compact subgroup of $Sl(n, Q_p)$.

The result is again due to Margulis [Mg1] for rank$(G/k) \geq 2$, and to Gromov-Schoen [GS] for quaternionic hyperbolic space and the hyperbolic Cayley plane.

The conclusion of this theorem holds more generally for a homomorphism ρ from Γ into the isometry group of an F-connected complex in the sense of [GS].

Let us now explain in more detail how our results fit into the rigidity theory for locally symmetric varieties of noncompact type, or equivalently for lattices in simple noncompact algebraic groups. The basic question of the theory is whether such a lattice can be deformed, or more generally, whether it can operate in more than one way on a symmetric space.

For $G = Sl(2,\mathbf{R}) \cong SO_o(2,1)$, there exist continuous families of compact quotients, namely Riemann surfaces of a given genus $p \geq 2$. Thus, no rigidity results holds in this case. This, however, is a singular phenomenon.

The first rigidity result was obtained by Calabi-Vesentini [CV] who showed that compact quotients of any irreducible Hermitian symmetric space of noncompact type other than $Sl(2,\mathbf{R})/SO(2)$ are infinitesimally, hence locally, rigid. They showed that the relevant cohomology group rising from the theory of Kodaira and Spencer vanishes in all these cases.

Mostow [Ms] showed strong rigidity of compact quotients of irreducible symmetric spaces of noncompact type. This means that any two such lattices Γ, Γ' which are isomorphic as abstract groups are lattices in the same G and isomorphic as subgroups of G. Geometrically this means that the quotients $\Gamma\backslash G/K$ and $\Gamma'\backslash G/K$ are isometric. Here, they carry the Riemannian metric induced from the symmetric metric on G/K. (For the purpose of the present discussion, we assume that the lattices operate without fixed points on G/K, *i.e.* no element in Γ other than the central element leaves any point fixed. Then the quotient space are (locally symmetric) Riemannian manifolds. This condition on the lattices is also called torsion-free, and it is known that any lattice has a torsion-free sublattice of finite index. Therefore, this assumption gives no serious restriction. Margulis [Mg1] then showed superrigidity if $\text{rank}(G/K) \geq 2$. This means that any homomorphism $\rho : \Gamma \to H(\Gamma$ as above) extends to a homomorphism $G \to H$, if H, like G, is a simple noncompact algebraic group and if $\rho(\Gamma)$ is Zariski dense, or that $\rho(\Gamma)$ is contained in a compact subgroup of H, if H is an algebraic subgroup of some $Sl(n, Q_p)$. (If H is defined over $\mathbf{C}$, then actually either of the two cases may occur but this does not interest us here.) Margulis results and their proofs are nicely presented in

Zimmer's book [Z]. We also refer to the recent monograph of Margulis [Mg2] for more general results.

Superrigidity for quotients of quaterionic hyperbolic space and the hyperbolic Cayley plane was shown by Corlette [C2] in the Archimedean case and by Gromov-Schoen [GS] in the p-adic case, as already pointed out.

Margulis also showed that superrigidity implies arithmeticity of a lattice.

Our results then can be considered as a geometric generalization of superrigidity because we do not assume that the image is symmetric but only that it satisfies a curvature condition (which is automatically fulfilled for symmetric spaces of noncompact or Euclidean type). Therefore, such a lattice may also operate in essentially only one manner on a nonpositively curved Riemannian manifold. In particular, any homomorphism into the isometry group of such a manifold is severely restricted. (In this context, one should also mention the work of Spatzier-Zimmer [SZ] on actions of lattices preserving certain geometric structures.)

Besides this generalization of superrigidity, an important point of the present paper is that it gives a proof of the celebrated superrigidity results of Margulis which is conceptually quite different from the original approaches of Mostow and Margulis.

Our proof involves harmonic maps, and we now discuss the connection between the theory of harmonic maps and rigidity question for lattices.

We start by recalling Bochner's method for showing the vanishing of harmonic 1-forms on Riemannian manifolds satisfying suitable curvature conditions. Thus, let M be a compact Riemannian manifold. We shall employ geodesic normal coordinates $x^1, ...x^m$ so that no Christoffel symbols will occur. Let

$$\Delta = \sum_{\alpha=1}^{m} \frac{\partial^2}{(\partial x^\alpha)^2}$$

be the Laplace operator.

We consider a harmonic 1-form on M as a harmonic map $f : M \to S^1$ into the unit circle. f then satisfies the Bochner formula

$$\frac{1}{2}\Delta \sum_{\alpha} f_\alpha f_\alpha = \sum_{\alpha,\beta} f_{\alpha\beta} f_{\alpha\beta} + \sum_{\alpha,\beta} R_{\alpha\beta} f_\alpha f_\beta$$

where the subscripts for f denote partial derivatives and $(R_{\alpha\beta})_{\alpha,\beta=1,\ldots m}$ is the Ricci tensor of M. (The notation is explained in more detail below.)

If we assume M is compact, the integral over the left hand side vanishes, and so then does the integral over the right hand side. The first term on the right hand side is nonnegative and if we assume that M has nonnegative Ricci curvature then so is the second one. Both of them then have to vanish identically. In particular,

$$f_{\alpha\beta} \equiv 0$$

and the harmonic 1-form df is parallel. If the Ricci tensor is even positive, then

$$f_\alpha \equiv 0$$

and the harmonic 1-form vanishes, and so then does $b_1(M)$.

This type of reasoning does not look too helpful for manifolds with negative Ricci curvature. Matsushima $[M]$, however, discovered that on symmetric spaces of noncompact type, the term involving the $f_{\alpha\beta}$ can balance the Ricci curvature term in such a manner that for rank ≥ 2, one still gets $b_1(M) = 0$, for $M = \Gamma\backslash G/K$ as above. (For the nonhermitian cases, this was not carried out by Matsustiman himself, but by Kaneyuki-Nagano [KN]).

For simplicity of discussion, we now assume that $N = \tilde{N}/\rho(\Gamma)$ is also a compact Riemannian manifold, because this case already reveals the idea of our approach. ρ then defines a homotopy class of maps $M \to N$. One then needs to show that this homotopy class contains a totally geodesic map. For this aim, it is natural to first select a distinguished representative by a variational condition and then show that this representative is totally geodesic under the assumptions on M and N. The map that we select is the harmonic one. In the case discussed presently, its existence was shown by Eells-Sampson [ES] and Al'ber [A1], [A2]. It is also essentially unique, again due to Al'ber and also to Hartman [Ha]. It is characterized by the fact that it minimizes the L^2-norm of its energy in the given homotopy class.

If $f : M \to N$ is harmonic, one then has the Bochner type formula

$$\frac{1}{2}\Delta \sum_{\alpha} \langle f_\alpha, f_\alpha \rangle = \sum_{\alpha,\beta} \langle f_{\alpha\beta}, f_{\alpha\beta} \rangle + \sum_{\alpha,\beta} R_{\alpha\beta} \langle f_\alpha, f_\beta \rangle - \sum_{\alpha,\beta} \langle R(f_\alpha, f_\beta) f_\beta, f_\alpha \rangle$$

(In the notation explained below), where $R(..)$ is the curvature tensor of N. Since we assume that N has nonpositive curvature, the third term on the right hand side has the same sign as the first one, but the Ricci curvature term again seems to present serious obstacles for deriving vanishing results. The main technical achievement of the present paper is that it can again be balanced against the other terms so as to show vanishing of $f_{\alpha\beta}$ in the cases considered here.

We present the relevant generalization of Matsushima's formula to harmonic mappings in Lemma 1 below (§1). It should have more applications as the domain need only be Einstein but not necessarily locally symmetric. Balancing the Ricci curvature term against the other ones becomes much harder here than in Matsushima's case, because of the image curvature term. We need to estimate the two curvature terms in opposite directions, and the estimates need to be sharp for the identity map $M \to M$. For this purpose, we need subtle estimates for the curvature tensor of our locally symmetric domain, and this requires to consider various cases separately, the hardest one being $SU(p,q)/S(U(p) \times U(q))$. These estimates are carried out in §3. The theorems are proved in §2.

Let us also quote the previous applications of harmonic maps to the rigidity questions considered here.

Siu [S] derived a Bochner type identity for harmonic maps between Kähler manifolds. If the image has nonpositive curvature, it implies that the product of the Hessian of the map with the Kähler form of the domain vanishes, or in other words that the map is pluriharmonic. A detailed study of the curvature tensors of Hermitian symmetric spaces of noncompact type then allowed him to conclude that a harmonic homotopy equivalence between compact quotients of such spaces is holomorphic or antiholomorphic. It then also is a diffeomorphism. An application of Royden's version of Yau's Schwarz lemma ([R], [Y]) then shows that the map is an isometry, proving Mostow's theorem in the Hermitian case. It is interesting to note that the curvature terms to be investigated here come from the image and not from the domain. Sampson [Sa] found a different formula that applies to harmonic maps from Kählerian to Riemannian manifolds. We also quote Mok's work [Mk1], [MSY] on Archimedean superrigidity in the Hermitian case.

Corlette [C2] showed that the product of the Hessian of a harmonic map with any parallel form on the domain vanishes if the image has nonpositive curvature. For quotients of quaternionic hyperbolic space and the hyperbolic Cayley plane this allowed him to conclude that the Hessian itself vanishes, *i.e.* that the map is totally geodesic. This gives his results discussed above.

For p-adic superrigidity, the relevant harmonic map theory was developed in the paper [GS] by Gromov and Schoen. This theory allowed them to carry the preceding results over from the Archimedean to the p-adic case.

Gromov [G] suggested to use foliations of symmetric spaces by Hermitian symmetric ones in order to carry the methods of Siu and Sampson over from the Kähler case. Recent work of Siu-Yeung [MSY] achieves this for the symmetric space $SO(p,q)/SO(p) \times SO(q)$, and there are results for some other symmetric spaces as well.

In this paper, we restrict attention to locally irreducible locally symmetric spaces. Superrigidity also holds for ones which are only globally irreducible, and our method is also applicable here in principle.

The harmonic map approach to rigidity is still not complete:

First of all, so far it has been unable to derive Mostow's rigidity theorem for quotients of real hyperbolic space. Secondly, the results for spaces that are of finite volume but not compact (*i.e.* for nonuniform lattices) are still not complete.

The problem here is to show the existence of a ρ-equivariant map of finite energy. Once this is achieved there also exists such a harmonic map, and the method works. For the existence of a finite energy map one so far has only partial results. The authors [JY1], [JY2] have identified two situations where one does get a finite energy map, and these two cases combined actually cover most cases for superrigidity of nonuniform lattices. The first one is an image of rank 1, more generally with controlled negative curvature. The second one is where the volume form of the domain is decaying sufficiently fast towards the cusps to balance unbounded energy density there. It may actually be true that this can always be achieved. Besides the cases verified in [JY2] for locally Hermitian symmetric with rank ≥ 2, this was verified for quaternionic hyperbolic space and the hyperbolic Cayley plane by Corlette [C2] where the situation is easier.

The project leading to the present paper was started in the spring of 1990 when the first author was visiting Harvard University. The essential part was carried out in August 1990 during the AMS Summer Institute on Differential Geometry at UCLA. On this occasion, we discussed our approach with E. Calabi, and we thank him for explaining his geometric interpretation of Matsushima's computation to us (see [D]).

Recently (in February 1992) the second author learned that Y.T.Siu had also obtained superrigidity results by a generalization of Matsushima's method. We thank Prof. Siu for suggesting that we also write up our results. This paper was written while the authors were visiting Tsing Hua University in Hsinchu, Taiwan. We thank the members of the Math. Dept. and in particular the chairman Prof. Heng-Lun Lai, for making our visit here very pleasant. The first author acknowledges support form DFG, the second one from NSF. We would like to thank Sherry for spending a lot of time typing this paper.

Notational Conventions:

All integrations on M will be with respect to the volume form.

We shall use local coordinates $x^\alpha, \alpha = 1, ...m$, on M. The curvature tensor with respect to these coordinates is $R_{\alpha\beta\gamma\delta}$ and the Ricci tensor is

$$R_{\alpha\gamma} = \sum_\beta R_{\alpha\beta\gamma\beta}$$

In all summations, Greek minuscles will range from 1 to m. For a map f, we abbreviate

$$f_\alpha := \frac{\partial}{\partial x^\alpha} f$$

We shall use geodesic normal coordinates throughout, allowing us to freely lower indices in terms of notation. In such coordinates, the Laplace operator is

$$\Delta = \sum_\alpha \frac{\partial^2}{(\partial x^\alpha)^2}$$

On N, we shall use invariant notation. The pointwise product in the tangent bundle TN is denoted by $\langle \cdot, \cdot \rangle$. The curvature tensor of N is denoted by $R(\cdot, \cdot)$. If $f : M \to N$ is a map, the bundle $f^{-1}TN$ inherits the metric $\langle \cdot, \cdot \rangle$ from TN. The metric in $T*M \otimes f^{-1}TN$ will likewise be denoted by $\langle \cdot, \cdot \rangle$. We let $(\cdot, \cdot)$ denote the L^2-product in $T^*M \otimes f^{-1}TN$. ∇ denotes the Levi-Civita connection in $T^*M \otimes f^{-1}TN$. We abbreviate

$$\nabla_\gamma := \nabla_{\frac{\partial}{\partial x^\gamma}}.$$

1. An analog of Matsushima's formula for harmonic maps.

We first recall the usual Bochner type identity for harmonic maps

(1)
$$\frac{1}{2}\Delta\sum_{\alpha}\langle f_\alpha, f_\alpha\rangle = \sum_{\alpha\beta}\{\langle f_{\alpha\beta}, f_{\alpha\beta}\rangle + R_{\alpha\beta}\langle f_\alpha, f_\beta\rangle - \langle R(f_\alpha, f_\beta) f_\beta, f_\alpha\rangle\}$$

We compute

(2)
$$\sum(\nabla_\gamma\nabla_\delta - \nabla_\delta\nabla_\gamma)(f_\alpha dx^\alpha) = \sum_{\alpha\beta}(-R_{\beta\alpha\gamma\delta} f_\beta dx^\alpha) + \sum_{\alpha} R(f_\gamma, f_\delta) f_\alpha dx^\alpha$$

From (2),

(3)
$$\begin{aligned}
&\sum_{\alpha,\beta,\gamma,\delta}\langle(\nabla_\gamma\nabla_\delta - \nabla_\delta\nabla_\gamma) f_\alpha dx^\alpha, (\nabla_\gamma\nabla_\delta - \nabla_\delta\nabla_\gamma) f_\beta dx^\beta\rangle \\
&\quad = \sum_{\alpha,\beta,\gamma,\delta,\eta} R_{\beta\alpha\gamma\delta} R_{\eta\alpha\gamma\delta}\langle f_\beta, f_\eta\rangle + \sum_{\alpha,\gamma,\delta}\langle R(f_\gamma, f_\delta) f_\alpha, R(f_\gamma, f_\delta) f_\alpha\rangle \\
&\quad\quad - \sum_{\alpha,\beta,\gamma,\delta} 2R_{\alpha\beta\gamma\delta}\langle R(f_\gamma, f_\delta) f_\beta, f_\alpha\rangle \\
&\sum_{\alpha,\beta,\gamma,\delta}\langle(\nabla_\gamma\nabla_\delta - \nabla_\delta\nabla_\gamma) f_\alpha dx^\alpha, (\nabla_\gamma\nabla_\delta - \nabla_\delta\nabla_\gamma) f_\beta dx^\beta\rangle \\
&\quad = \sum_{\alpha,\beta,\gamma,\delta,\eta}\langle -R_{\beta\alpha\gamma\delta} f_\beta dx^\alpha, (\nabla_\gamma\nabla_\delta - \nabla_\delta\nabla_\gamma) f_\eta dx^\eta\rangle \\
&\quad\quad + \sum_{\alpha,\beta,\gamma,\delta}\langle R(f_\gamma, f_\delta) f_\alpha dx^\alpha, (\nabla_\gamma\nabla_\delta - \nabla_\delta\nabla_\gamma) f_\beta dx^\beta\rangle \\
&\quad = 2\sum_{\alpha,\beta,\gamma,\delta,\eta}\langle -R_{\beta\alpha\gamma\delta} f^\beta dx^\alpha, \nabla_\gamma\nabla_\delta f_\eta dx^\eta\rangle \\
&\quad\quad + \sum_{\alpha\gamma\delta}\langle R(f_\gamma, f_\delta) f_\alpha, R(f_\gamma, f_\delta) f_\alpha\rangle \\
&\quad\quad - \sum_{\alpha,\beta,\gamma,\delta} R_{\alpha\beta\gamma\delta}\langle R(f_\gamma, f_\delta) f_\beta, f_\alpha\rangle.
\end{aligned}$$

Integrating the first term by parts, we get

$$= 2\int_M \sum_{\alpha\beta\gamma\delta} \langle \frac{\partial}{\partial\gamma}\left(R_{\beta\alpha\gamma\delta} f_\beta\right), f_{\alpha\delta}\rangle$$
$$+ \int_M \sum_{\alpha\gamma\delta} \langle R\left(f_\gamma, f_\delta\right) f_\alpha, R\left(f_\gamma, f_\delta\right) f_\alpha\rangle$$
$$- \int_M \sum_{\alpha\beta\gamma\delta} R_{\alpha\beta\gamma\delta} \langle R\left(f_\gamma, f_\delta\right) f_\beta, f_\alpha\rangle .$$

Now

$$\sum_\gamma \left\langle \frac{\partial}{\partial\gamma}\left(R_{\beta\alpha\gamma\delta} f_\beta\right), f_{\alpha\delta}\right\rangle = \sum_\gamma R_{\beta\alpha\gamma\delta}\langle f_{\beta\gamma}, f_{\alpha\delta}\rangle + \sum_\gamma R_{\beta\alpha\gamma\delta,\gamma}\langle f_\beta, f_{\alpha\delta}\rangle$$

The second term vanishes for any Einstein metric, in particular a locally symmetric one, since

$$\sum_\gamma R_{\beta\alpha\gamma\delta,\gamma} = \sum_\gamma \left(R_{\delta\gamma\alpha\gamma,\beta} - R_{\delta\gamma\beta\gamma,\alpha}\right)$$

by the Bianchi identity, and this vanishes if the Ricci tensor is parallel.

We obtain for an Einstein metric on M

(4)
$$\sum_{\alpha\beta\gamma\delta} \left(\left(\nabla_\gamma\nabla_\delta - \nabla_\delta\nabla_\gamma\right) f_\alpha dx^\alpha, \left(\nabla_\gamma\nabla_\delta - \nabla_\delta\nabla_\gamma\right) f_\beta dx^\beta\right)$$
$$= -2\int_M \sum_{\alpha\beta\gamma\delta} R_{\alpha\beta\gamma\delta}\langle f_{\alpha\delta}, f_{\beta\gamma}\rangle + \int_M \sum_{\alpha\gamma\delta}\langle R\left(f_\gamma, f_\delta\right) f_\alpha, R\left(f_\gamma, f_\delta\right) f_\alpha\rangle$$
$$- \int_M \sum_{\alpha\beta\gamma\delta} R_{\alpha\beta\gamma\delta}\langle R\left(f_\gamma, f_\delta\right) f_\beta, f_\alpha\rangle$$

From (1), (3), (4) we get the analogue of Matsushima's formula for maps between Riemannian manifolds.

Lemma 1. *Let $f : M \to N$ be a harmonic map between Riemannian* *manifolds where M is compact and Einstein. Then for any $\lambda \in \mathbf{R}$*

(5)

$$\lambda \int_M \sum_{\alpha\beta} \langle f_{\alpha\beta}, f_{\alpha\beta} \rangle + 2 \int_M \sum_{\alpha\beta\gamma\delta} R_{\alpha\beta\gamma\delta} \langle f_{\alpha\delta}, f_{\beta\gamma} \rangle$$
$$= -\lambda \int_M \sum_{\alpha\beta} R_{\alpha\beta} \langle f_\alpha, f_\beta \rangle - \int_M \sum_{\alpha\beta\gamma\delta\eta} R_{\alpha\beta\gamma\delta} R_{\eta\beta\gamma\delta} \langle f_\alpha, f_\eta \rangle$$
$$+ \lambda \int_M \sum_{\alpha\beta} \langle R(f_\alpha, f_\beta) f_\beta, f_\alpha \rangle + \int_M \sum_{\alpha\beta\gamma\delta} R_{\alpha\beta\gamma\delta} \langle R(f_\gamma, f_\delta) f_\beta, f_\alpha \rangle$$

2. Rigidity Theorems For Locally Symmetric Spaces Of Noncompact Type.

Theorem 1. *Let $\tilde{M} = G/K$ be an irreducible symmetric space of noncompact type and of rank ≥ 2. Let Γ be a discrete cocompact subgroup of G. Let $\tilde{N}$ be a complete simply connected Riemannian manifold of nonpositive curvature operator, with isometry group $I\left(\tilde{N}\right)$. Let $\rho : \Gamma \to I\left(\tilde{N}\right)$ be a homomorphism for which $\rho(\Gamma)$ either does not have a fix point on the sphere at infinity of $\tilde{N}$ or if it does, it centralizes a totally geodesic flat subspace. Then there exists a totally geodesic ρ-equivariant map*

$$f : \tilde{M} \to \tilde{N}$$

Remark. The condition on ρ may be generalized to reductivity in the sense of Labourie [*L*]. In order to prove Theorem 1 we start with a harmonic ρ-equivariant map

$$\rho : \tilde{M} \to \tilde{N}$$

(In case Γ is torsion free and $\rho(\Gamma)$ is also torsion free, cocompact and discrete, the existence of f was proved by Eells-Sampson [ES] and Al'ber [A1], [A2]. In the general case, it was proved by Jost-Yau[JY3] and Labourie [L], with earlier results by Diederich-Ohsawa [DO], Donaldson [Dn], and Corlette [C1]. If $\tilde{N}$ is also a symmetric space G'/K', the condition on ρ means that $\rho(\Gamma)$ is a reductive subgroup of G'. Labourie [L] defined a notion of reductive subgroup of $I\left(\tilde{N}\right)$ for $\tilde{N}$ as in Theorem 1 and the result continues to hold if $\rho(\Gamma)$ is reductive in his sense (see [L]).) Theorem 1 then follows from

Theorem 2. *let $\tilde{M}$, $\tilde{N}$ be as in Theorem 1, $\rho : \Gamma \to I\left(\tilde{N}\right)$ a homomorphism. Then any ρ-equivariant harmonic map*

$$f : M \to N$$

is totally geodesic.

The proof of Theorem. 2 will be based on Lemma 1.

Lemma 1 generalizes to the situation of Theorem. 2 by taking M as a fundamental region for Γ. The integration by parts in the proof of Lemma 1 then is justified by the ρ-equivariance of f. The aim then is to find a value for λ for which the left hand side of (5) is nonnegative, perhaps even positive unless all $f_{\alpha\beta} = 0$, whereas the right hand side is nonpositive. Thus, we have to find λ with

$$2\sum_{\alpha\beta\gamma\delta} R_{\alpha\beta\gamma\delta}\langle f_{\alpha\delta}, f_{\beta\gamma}\rangle \geq -\lambda\langle f_{\alpha\beta}, f_{\alpha\beta}\rangle \tag{6}$$

$$\sum_{\alpha\beta\gamma\delta\eta} R_{\alpha\beta\gamma\delta}R_{\eta\beta\gamma\delta}\langle f_\alpha, f_\eta\rangle \geq -\lambda\sum_{\alpha\eta} R_{\alpha\eta}\langle f_\alpha, f_\eta\rangle \tag{7}$$

$$\sum_{\alpha\beta\gamma\delta} R_{\alpha\beta\gamma\delta}\langle R(f_\gamma, f_\delta) f_\beta, f_\alpha\rangle \leq -\lambda\langle R(f_\alpha, f_\beta) f_\beta, f_\alpha\rangle \tag{8}$$

Now as in [M; §9]

$$\sum_{\alpha\beta\gamma\delta} R_{\alpha\beta\gamma\delta}\langle f_{\alpha\delta}, f_{\beta\gamma}\rangle \geq \lambda_1 \sum_{\alpha\beta}\langle f_{\alpha\beta}, f_{\alpha\beta}\rangle \tag{9}$$

where λ_1 was computed by Calabi-Vesentini [CV] and Borel [B] in the hermitian cases (see[M; §9] for a table), and by Kaneyuki-Nagano [KN] for the other cases.

We let $c^\lambda_{\alpha\beta}$ be the structure constants of the symmetric space $\tilde{M}$ covering M, *i.e.*

(10) $$[X_\alpha, X_\beta] = \sum_\lambda c^\lambda_{\alpha\beta} X_\lambda$$

where $M = G/K$, with associated Lie algebra decomposition

$$\mathcal{G} = k \oplus p$$

and bases $X_1, ...X_m$ of p, $X_{m+1}, ...X_{m+k}$ of k satisfying

(11) $$\begin{aligned} B\,(X_\alpha, X_\beta) &= \delta_{\alpha\beta}, \quad \alpha, \beta = 1, ...m \\ B\,(X_\lambda, X_\mu) &= -\delta_{\lambda\mu}, \lambda, \mu = m+1, ...m+k \end{aligned}$$

here B is the killing form of $\mathcal{G}$. In the sequel, such a base will be called orthonormal for the killing form.

(12) $$\sum_{\beta,\lambda} c^\lambda_{\alpha\beta} c^\lambda_{\gamma\beta} = \frac{1}{2}\delta_{\alpha\gamma}$$

and

(13) $$R_{\alpha\beta\gamma\delta} = -\sum_\lambda c^\lambda_{\alpha\beta} c^\lambda_{\gamma\delta}$$

Since $[p,\ p] \subset k$, in the previous formulae, λ only ranges from $m + 1, ...m + k$, if $\alpha, \beta \in \{1, ...m\}$. In particular, we have for the Ricci tensor

(14) $$R_{\alpha\beta} = -\frac{1}{2}\delta_{\alpha\beta}$$

Let

$$k = z \oplus k_1 \oplus ... \oplus k_l$$

where z is the center of k, and $k_1, ...k_l$ are the simple ideals. We choose our basis (X_λ) of k to respect this decomposition, *i.e.* each X_λ is contained in one of the summands. Then

(15) $$\sum_{\alpha,\beta} c^\lambda_{\alpha\beta} c^\mu_{\alpha\beta} = a_i \delta_{\lambda\mu}$$

for $X_\lambda \in k_i$, $i = 1, ... l$ with $0 < a_i < 1$

(16) $$\sum_{\alpha,\beta} c^\lambda_{\alpha\beta} c^\mu_{\alpha\beta} = \delta_{\lambda\mu}$$

for $X_\lambda \in z$. For unity of notation, we sometimes put $k_o := z$; then $a_o = 1$, in case $z \neq \{0\}$; if z is trivial, we put $a_o = 0$. We also put for $i = 0, ... l$

(17) $$\sum_{\beta, X_\lambda \in k_i} c^\lambda_{\alpha\beta} c^\lambda_{\eta\beta} = b_i \, \delta_{\alpha\eta}$$

by (12),

(18) $$\sum_{i=0}^{l} b_i = 1/2$$

We compute

(19) $$\sum_{\alpha\beta\gamma\delta\eta} R_{\alpha\beta\gamma\delta} R_{\eta\beta\gamma\delta} \langle f_\alpha, f_\eta \rangle = \sum_{\alpha\beta\gamma\delta\eta} c^\lambda_{\alpha\beta} c^\lambda_{\gamma\delta} c^\mu_{\eta\beta} c^\mu_{\gamma\delta} \langle f_\alpha, f_\eta \rangle$$
$$= \sum_{\alpha,i} a_i b_i \langle f_\alpha, f_\alpha \rangle = -2 \sum_i a_i b_i \sum_{\alpha\eta} R_{\alpha\eta} \langle f_\alpha, f_\eta \rangle$$

where we use (14). We put

(20) $$\mu = 2 \sum_{i=0}^{l} a_i b_i$$

we then have equality in (7) for $\lambda = \mu$. In the next section, we shall show that (8) holds for $\lambda = \mu$. From the computations of Matsushima [M] and Kaneyuki-Nagano [KN], one knows

(21) $$\mu > -2\lambda_1 \text{ if rank}\,(G/K) \geq 2$$

and

(22) $$\mu = -2\lambda_1 \text{ if rank}\,(G/K) = 1.$$

If rank$(G/K) \geq 2$, we may choose any λ with $-\lambda_1 < \lambda < \mu$ in (5) and conclude from (6), (7), (8)

(23) $$f_{\alpha\beta} \equiv 0 \text{ for all } \alpha, \beta$$

and also

$$\sum_{\alpha\beta} R_{\alpha\beta}\langle f_\alpha, f_\beta\rangle = \sum_{\alpha\beta}\langle R(f_\alpha, f_\beta) f_\beta, f_\alpha\rangle \tag{24}$$

Since we are employing geodesic normal coordinates, (23) is equivalent to f being totally geodesic.

This proves Theorem 2.

An application is the following result of Margulis [Mg1]

Corollary 1. *Let Γ be as in Theorem 1; let H be a semisimple non-compact Lie group with discrete center, $\rho : \Gamma \to H$ a homomorphism with Zariski dense image. Then ρ extends to a homomorphism $G \to H$.*

The derivation of Corollary 1 from Theorem 2 is standard.

If rank $(G/K) = 1$, one may still conclude that equality has to hold in (9). Unless G/K is real hyperbolic space, this still allows to draw conclusions; these will also be given in the next section for the individual cases.

In particular, one may conclude Corlette's theorem:

Theorem 3. *The results of Theorem 1 and 2 and Cor.1 also hold for $G/K = Sp(p,1)/Sp(p) \times Sp(1)$ and the hyperbolic Cayley plane.*

Our method is also applicable to representations in p-adic groups, and we obtain the following result of Margulis [Mg1]:

Theorem 4. *Let $\tilde{M} = G/K$ and Γ be as in Theorem 1. Let $\rho : \Gamma \to Sl(n, Q_p)$ be a homomorphism, for some $n \in \mathbf{N}$ and some prime p. Then $\rho(\Gamma)$ has compact closure in $Sl(n, Q_p)$.*

P.f.: Let X be the Euclidean benlding of Bruhat-Tits associated to $Sl(n, Q_p)$. As explained in [GS], $Sl(n, \mathbf{Q}_p)$ acts isometrically on X, and there exists a ρ-equivariant map

$$g : \tilde{M} \to X$$

Since $\rho(\Gamma)$ either does not have fixed point at ∞ or centralizes some totally geodesic flat, there then exists a ρ-equivariant harmonic map

$$f : \tilde{M} \to X$$

as proved in [GS], and this harmonic map has sufficient regularity properties to justify the application of the Bochner type formula of Lemma 1, see [GS; §6]. Therefore, f has to be totally geodesic by the argument of the proof of Theorem 2. As observed in [GS; pf. of Theorem 7.4], any such totally geodesic map is constant. Thus $f\left(\tilde{M}\right) = p$, for some $p \in X$. This implies that $\rho(\Gamma)$ is contained in the isotropy group of ρ, a compact group.

As in Theorem 3, our method again applies also to $Sp(p,1)/Sp(p) \times Sp(1)$ and the hyperbolic Cayley plane to prove the theorem of Gromov-Schoen [GS]:

Theorem 5. *The result of Theorem 4 also holds for*

$$G/K = Sp(p,1)/Sp(p) \times Sp(1)$$

and the hyperbolic Cayley plane.

Theorem 4 treats harmonic maps into the Euclidean building X associated to $Sl(n, \mathbf{Q}_p)$. The reasoning works more generally for harmonic maps into F-connected complexes (see [GS] for the definition), showing that ρ-equivariant harmonic maps into such a space are constant if G/K and Γ are as in Theorem 1.

3. Various Cases.

In this section, we want to show (8) with $\lambda = \mu$, where μ is defined in (20). We shall need to consider various cases. We shall also present further consequences of (5) for the various types of symmetric spaces of rank 1.

We start with a general observation, We put

$$P_{\alpha\beta\gamma\delta} = -\langle R(f_\alpha, f_\beta) f_\delta, f_\gamma \rangle$$

where R is the curvature tensor of N. The properties of R imply

Lemma 2. $P_{\alpha\beta\gamma\delta}$ *satisfies*

$$P_{\alpha\beta\gamma\delta} = -P_{\beta\alpha\gamma\delta}$$

$$P_{\alpha\beta\gamma\delta} = P_{\gamma\delta\alpha\beta}$$

$$P_{\alpha\beta\gamma\delta} + P_{\alpha\gamma\delta\beta} + P_{\alpha\delta\beta\gamma} = 0$$

If N has nonpositive curvature operator, then P is positive semidefinite in the sense that

$$2P_{\alpha\beta\gamma\delta} \leq P_{\alpha\beta\alpha\beta} + P_{\gamma\delta\gamma\delta} \text{ for all } \alpha, \beta, \gamma, \delta$$

Remark. The assertions of Lemma 2 for P are the only properties of the image curvature tensor that we shall use in the sequel. Therefore, instead of requiring that N has nonpositive curvature operator, it suffices that

$$-\langle R(f_\alpha, f_\beta) f_\delta, f_\gamma \rangle$$

is positive semidefinite in the sense described in Lemma 2.

3.1. We consider the case where k is simple.

Then

$$\sum_{\alpha,\beta} c^\lambda_{\alpha\beta} c^\mu_{\alpha\beta} = a\delta_{\lambda\mu}$$

for all λ, μ with $0 < a < 1$. (15) here becomes

$$\begin{aligned} \sum_{\alpha\beta\gamma\delta\eta} R_{\alpha\beta\gamma\delta} R_{\eta\beta\gamma\delta} \langle f_\alpha, f_\eta \rangle &= \sum_{\alpha\beta\gamma\delta} c^\lambda_{\alpha\beta} c^\lambda_{\gamma\delta} c^\mu_{\eta\beta} c^\mu_{\gamma\delta} \langle f_\alpha, f_\eta \rangle \\ &= -a \sum_{\alpha,\eta} R_{\alpha\eta} \langle f_\alpha, f_\eta \rangle \end{aligned}$$

Now

$$-\sum_{\alpha\beta\gamma\delta} R_{\alpha\beta\gamma\delta} P_{\alpha\beta\gamma\delta} = \sum_{\alpha\beta\gamma\delta\lambda} c^\lambda_{\alpha\beta} c^\lambda_{\gamma\delta} P_{\alpha\beta\gamma\delta} \leq a \sum_{\alpha,\beta} P_{\alpha\beta\alpha\beta}$$

from the Schwarz inequality, since the $c^\lambda_{\alpha\beta}$, $\lambda = m+1, ...m+k$, are pairwise orthogonal and of the same squared norm a.

This is the desired estimate.

3.2. We consider the case where the derived ideal of k is simple, but k is not simple itself. Then

$$k = z \oplus k_1$$

where $k_1 = [k, k]$ is simple, and the center z is 1-dimensional. We choose our basis X_λ so that

$$X_{m+1} \in z$$
$$X_{m+2}, \ldots X_{m+k} \in k_1$$

and so that the matrix $\left(c_{\alpha\beta}^{m+1}\right)_{\alpha,\beta=1,\ldots m}$ is of the form

$$(2.1) \qquad \frac{1}{\sqrt{m}} \begin{bmatrix} 0 & 1 & & & & 0 \\ -1 & 0 & & & & \\ & & \cdot & & & \\ & & & \cdot & & \\ & & & & 0 & 1 \\ 0 & & & & -1 & 0 \end{bmatrix}$$

which is possible because $ad_p X_{m+1}$ is skew-symmetric (see [M; §5]). We compute

$$\begin{aligned} &\sum_{\alpha,..\eta} R_{\alpha\beta\gamma\delta} R_{\eta\beta\gamma\delta} \langle f_\alpha, f_\eta \rangle \\ &\quad = \sum_{\alpha,..\eta} c_{\alpha\beta}^{\lambda} c_{\gamma\delta}^{\lambda} c_{\eta\beta}^{\mu} c_{\gamma\delta}^{\mu} \langle f_\alpha, f_\eta \rangle \\ &\quad = \sum_{\alpha\beta\eta} \left(c_{\alpha\beta}^{m+1} c_{\eta\beta}^{m+1} + a_1 \sum_{i=2}^{k} c_{\alpha\beta}^{m+i} c_{\eta\beta}^{m+i} \right) \langle f_\alpha, f_\eta \rangle \\ &\quad = \sum_{\alpha\beta\eta} (1 - a_1)\, c_{\alpha\beta}^{m+1} c_{\eta\beta}^{m+1} \langle f_\alpha, f_\eta \rangle - \sum_{\alpha\eta} a_1 R_{\alpha\eta} \langle f_\alpha, f_\eta \rangle \\ &\quad = -\left(\frac{2}{m} (1 - a_1) + a_1 \right) \sum_{\alpha,\eta} R_{\alpha\eta} \langle f_\alpha, f_\eta \rangle \\ &\quad = \left(\frac{1}{m} (1 - a_1) + a_1/2 \right) \sum_{\alpha} \langle f_\alpha, f_\alpha \rangle \end{aligned}$$

We thus have

$$\mu = \frac{2}{m} (1 - a_1) + a_1$$

Next

$$\begin{aligned}
-\sum_{\alpha,..\delta} R_{\alpha\beta\gamma\delta}P_{\alpha\beta\gamma\delta} &= \sum_{\alpha,..\delta} c^{\lambda}_{\alpha\beta}c^{\lambda}_{\gamma\delta}P_{\alpha\beta\gamma\delta} \\
&= \sum_{\alpha,..\delta} (1-a_1)\, c^{m+1}_{\alpha\beta}c^{m+1}_{\gamma\delta}P_{\alpha\beta\gamma\delta} \\
&\qquad + \sum_{\alpha,..\delta}\left(a_1\; c^{m+1}_{\alpha\beta}c^{m+1}_{\gamma\delta} + \sum_{i=2}^{k} c^{m+i}_{\alpha\beta}c^{m+i}_{\gamma\delta}\right) P_{\alpha\beta\gamma\delta} \\
&\leq \sum_{\alpha,..\delta} (1-a_1)\, c^{m+1}_{\alpha\beta}c^{m+1}_{\gamma\delta}P_{\alpha\beta\gamma\delta} + \sum_{\alpha,\beta} a_1\; P_{\alpha\beta\alpha\beta}
\end{aligned}$$

again since the $a_1^{1/2}c^{m+1}_{\alpha\beta}$, $c^{m+2}_{\alpha\beta}$, $...c^{m+k}_{\alpha\beta}$ are orthogonal and of the same squared norm a_1. We adopt the following convention:

For $\alpha = 1, ...m$, we put

$$\alpha^* = \alpha - (-1)^{\alpha}$$

Then

$$\begin{aligned}
\sum_{\alpha,..\delta} c^{m+1}_{\alpha\beta}c^{m+1}_{\gamma\delta}P_{\alpha\beta\gamma\delta} &= \frac{1}{m}\sum_{\alpha,\gamma} P_{\alpha\alpha^*\gamma\gamma^*} && \text{by (2.1)} \\
&= \frac{1}{m}\sum_{\alpha,\gamma} (P_{\alpha\gamma\gamma^*\alpha^*} + P_{\alpha\gamma^*\alpha^*\gamma}) && \text{by the Bianchi identity} \\
&\leq \frac{1}{2m}\sum_{\alpha,\gamma} (P_{\alpha\gamma\alpha\gamma} + P_{\alpha^*\gamma^*\alpha^*\gamma^*} + P_{\alpha\gamma^*\alpha\gamma^*} + P_{\alpha^*\gamma\alpha^*\gamma}) \\
&= \frac{2}{m}\sum_{\alpha,\beta} P_{\alpha\beta\alpha\beta}
\end{aligned}$$

Altogether

$$-\sum_{\alpha,..\delta} R_{\alpha\beta\gamma\delta}P_{\alpha\beta\gamma\delta} \leq \left(\frac{2}{m}(1-a_1) + a_1\right)\sum_{\alpha,\beta} P_{\alpha\beta\alpha\beta} = \mu \sum_{\alpha,\beta} P_{\alpha\beta\alpha\beta}$$

This is the desired estimate.

3.3. We consider the case of $SO_o(p,q)\,/\,(SO(p)\times SO(q))$. The Lie algebra $so(p,q)$ is the space of matrices of the form

$$\begin{bmatrix} x' & y \\ y^t & x'' \end{bmatrix}$$

where x', x'' are $(p \times p)$ and $(q \times q)$ resp, skew-symmetric matrices, and Y is an arbitrary $(p \times q)$ matrix.

The Killing form of $so\,(p,q)$ is given by

$$B\,(M_1, M_2) = (p+q-2)\; tr\,(M_1, M_2)$$

we let E^{xy} be the matrix with 1 in the intersection of row x and column y and 0 otherwise. We put

$$\begin{aligned}
X_{ij} &= \frac{1}{\sqrt{2(p+q-2)}}\left(E^{ij} + E^{ji}\right) \text{ for } i = 1, ...p, j = p+1, ...p+q \\
X_{a'b'} &= \frac{1}{\sqrt{2(p+q-2)}}\left(E^{a'b'} - E^{b'a'}\right) \text{ for } a', b' = 1, ...p, a' < b' \\
X_{a''b''} &= \frac{1}{\sqrt{2(p+q-2)}}\left(E^{a''b''} - E^{b''a''}\right) \text{ for } a'', b'' = p+1, ...p+q, a'' < b''
\end{aligned}$$

These matrices form an "orthonormal" basis for B in the sense that

$$\begin{aligned}
B\,(X_{ij}, X_{kl}) &= \delta_{ik}\delta_{jl} \\
B\,(X_{a'b'}X_{c'd'}) &= -\delta_{a'c'}\delta_{b'd'} \\
B\,(X_{a''b''}, X_{c''d''}) &= -\delta_{a''c''}\delta_{b''d''}
\end{aligned}$$

and all other products vanish.

With respect to this basis, the structure constants of $so\,(p,q)$ are

$$\begin{aligned}
c_{ij,kl}^{a'b'} &= (p+q-2)\; tr([X_{ij}, X_{kl}], X_{a'b'}) \\
&= \frac{1}{\sqrt{2(p+q-2)}}\delta_{jl}\left(\delta_{a'k}\delta_{b'i} - \delta_{a'i}\delta_{b'k}\right) \\
c_{ij,kl}^{a''b''} &= \frac{1}{\sqrt{2(p+q-2)}}\delta_{ik}\left(\delta_{a''l}\delta_{b''j} - \delta_{a''j}\delta_{b''l}\right)
\end{aligned}$$

In the sequel, summation over i, k, m, r, v will be from 1 to p, over j, l, n, s, w from $p+1$ to q, over a$'$, b' from 1 to p, over a$''$, b'' from $p+1$

to q, subject to the restrictions $a' < b'$, $a'' < b''$. One computes

$$\sum_{ijkl} c^{a'b'}_{ij,kl} c^{d'e'}_{ij,kl} = \frac{q}{p+q-2}\delta_{a'd'}\delta_{b'e'}$$

$$\sum_{ijkl} c^{a''b''}_{ij,kl} c^{d''e''}_{ij,kl} = \frac{p}{p+q-2}\delta_{a''d''}\delta_{b''e''}$$

$$\sum_{ijkl} c^{a'b'}_{ij,kl} c^{d''e''}_{ij,kl} = 0$$

$$\sum_{k,l,a'<b'} c^{a\prime b\prime}_{ij,kl} c^{a\prime b\prime}_{mn,kl} = \frac{p-1}{2(p+q-2)}\delta_{jn}\delta_{im}$$

$$\sum_{k,l,a'',b''} c^{a\prime\prime b\prime\prime}_{ij,kl} c^{a\prime\prime b\prime\prime}_{mn,kl} = \frac{q-1}{2(p+q-2)}\delta_{jn}\delta_{im}$$

We put

$$a_1 = \frac{q}{p+q-2}, \qquad a_2 = \frac{p}{p+q-2}$$

$$b_1 = \frac{p-1}{2(p+q-2)}, \qquad b_2 = \frac{q-1}{2(p+q-2)}$$

and note

$$b_1 + b_2 = \frac{1}{2}$$

The curvature tensor of $SO_o(p,q)/(SO(p)\times SO(q))$ is given by

$$R_{ij,kl,mn,rs} = -\left(\sum_{a'<b'} c^{a'b'}_{ij,kl} c^{a'b'}_{mn,rs} + \sum_{a''<b''} c^{a''b''}_{ij,kl} c^{a''b''}_{mn,rs}\right)$$

In particular, the Ricci tensor is

$$R_{ij,kl} = -\frac{1}{2}\delta_{ik}\delta_{jl}$$

We put

$$f_{ij} = X_{ij}(f)$$

We compute

$$\begin{aligned}
&\sum_{klmnrs} R_{ij,kl,mn,rs} R_{vw,kl,mn,rs} \langle f_{ij}, f_{vw}\rangle \\
&= \frac{q(p-1)+p(q-1)}{2(p+q-2)^2}\sum_{i,j}\langle f_{ij}, f_{ij}\rangle \\
&= -\frac{q(p-1)+p(q-1)}{2(p+q-2)^2}\sum_{ijvw} R_{ij,vw} f_{ij} f_{vw}
\end{aligned}$$

Thus

$$\mu := \frac{q(p-1)+p(q-1)}{2(p+q-2)^2} = 2a_1b_1 + 2a_2b_2$$

$$\begin{aligned}
&\sum_{i,..s,a'<b'} c^{a'b'}_{ij,kl} c^{a'b'}_{mn,rs} P_{ij,kl,mn,rs} \\
&= \frac{1}{2(p+q-2)} \sum \delta_{jl}\delta_{ns} \left(\delta_{a'k}\delta_{b'i} - \delta_{a'i}\delta_{b'k}\right)\left(\delta_{a'r}\delta_{b'm} - \delta_{a'm}\delta_{b'r}\right) \\
&= \frac{1}{2(p+q-2)} \sum \delta_{jl}\delta_{ns} \left(\delta_{kr}\delta_{im}\left(1-\delta_{ik}\delta_{mr}\right) - \delta_{km}\delta_{ir}\left(1-\delta_{ik}\delta_{mr}\right)\right) \\
&= \frac{1}{2(p+q-2)} \sum \left(P_{ij,kj,in,kn} - P_{ij,kj,kn,in}\right) \\
&= \frac{1}{p+q-2} \sum_{ijkl} P_{ij,kj,il,kl}
\end{aligned}$$

and

$$\sum_{i,..s,a''<b''} c^{a''b''}_{ij,kl} c^{a''b''}_{mn,rs} P_{ij,kl,mn,rs} = \frac{1}{p+q-2} \sum_{ijkl} P_{ij,il,kj,kl}$$

By the Bianchi identity

$$P_{ij,kj,il,kl} = P_{ij,il,kj,kl} + P_{ij,kl,il,kj}$$

Now

$$\sum_{ijkl} P_{ij,kl,il,kj} = \sum_{ijk} P_{ij,kj,ij,kj} - \sum_{ijl} P_{ij,il,ij,il} + \hat{P}_{ij,kl,il,kj}$$

with

$$\hat{P}_{ij,kl,il,kj} = \sum_{i\neq k, j\neq \ell} P_{ij,kl,il,kj}$$

From the preceding formulae, we conclude

$$\begin{aligned}
&-\sum_{i,..s} R_{ij,kl,mn,rs} P_{ij,kl,mn,rs} \\
&\quad = \sum \left(c^{a'b'}_{ij,kl} c^{a'b'}_{mn,rs} + c^{a''b''}_{ij,kl} c^{a''b''}_{mn,rs} \right) P_{ij,kl,mn,rs} \\
&\quad = \frac{1}{p+q-2} \left(\sum P_{ij,kj,il,kl} + \sum P_{ij,il,kj,kl} \right) \\
&\quad = \frac{1}{p+q-2} \left(\lambda \sum_{ijkl} P_{ij,kj,il,kl} + (2-\lambda) \sum_{ijkl} P_{ij,il,kj,kl} + \right. \\
&\qquad \left. (1-\lambda) \sum_{ijk} P_{ij,kj,ij,kj} - (1-\lambda) \sum_{ije} P_{ij,il,ij,il} + (1-\lambda)\, \hat{P}_{ij,kl,il,kj} \right)
\end{aligned}$$

for any $\lambda \in \mathbf{R}$.

Since $P_{ij,kl,mn,kl}$ is positive semidefinite,

$$\begin{aligned}
\sum_{ijkl} P_{ij,kj,il,kl} &\leq q \sum_{ijk} P_{ij,kj,ij,kj} \\
\sum_{ijkl} P_{ij,il,kj,kl} &\leq p \sum_{ijl} P_{ij,il,ij,il} \\
\hat{P}_{ij,kl,il,kj} &\leq \hat{P}_{ij,kl,ij,kl}
\end{aligned}$$

For $0 \leq \lambda \leq 2$, we then obtain

$$\begin{aligned}
&-\sum R_{ij,kl,mn,rs} P_{ij,kl,mn,rs} \\
&\quad \leq \frac{1}{p+q-2} \left((\lambda q + 1 - \lambda) \sum_{ijk} P_{ij,kj,ij,kj} + \right. \\
&\qquad \left. ((2-\lambda)\, p + \lambda - 1) \sum_{ijl} P_{ij,il,ij,il} + |1-\lambda| \hat{P}_{ij,kl,il,kl} \right)
\end{aligned}$$

We put

$$\lambda = 1 + \frac{p-q}{p+q-2}$$

Then $0 \leq \lambda \leq 2$, and moreover

$$\lambda q + (1 - \lambda) = \frac{2pq - p - q}{(p + q - 2)}$$
$$(2 - \lambda)\, p + \lambda - 1 = \frac{2pq - p - q}{(p + q - 2)}$$
$$|1 - \lambda| \leq \frac{2pq - p - q}{(p + q - 2)}$$

We then obtain

$$\begin{aligned} &- \sum R_{ij,kl,mn,rs} P_{ij,kl,mn,rs} \\ &\leq \frac{2pq - p - q}{(p + q - 2)^2} \left(\sum_{ijk} P_{ij,kj,ij,kj} + \sum_{ijl} P_{ij,il,ij,il} + \hat{P}_{ij,kl,ij,kl} \right) \\ &= \mu \sum_{ijkl} P_{ij,kl,ij,kl} \end{aligned}$$

This is the desired estimate.

In the case of $SO_o\,(p, q) \,/ SO\,(p) \times SO\,(q)$, equality in (9) only implies

$$\sum_a f_{\alpha\alpha} = 0$$

This, however, follows anyway, as f is harmonic. Further conclusions cannot be drawn in this case and are also not expected.

3.4. We consider

$$SU\,(p, q)\,/ S\,(U\,(p) \times U\,(q))$$

The Lie algebra $su\,(p, q)$ is the space of matrices of the form

$$\begin{bmatrix} A' & B \\ \bar{B}^t & A'' \end{bmatrix}$$

where $A\prime, A''$ are $(p \times p)$ and $(q \times q)$ resp.skew hermitian matrices, with $tr\,(A' + A'') = 0$ and B is an arbitrary complex matrix.

The Killing form of $su\,(p, q)$ is given by

$$B\,(M_1, M_2) = 2\,(p + q)\, tr\,(M_1\ M_2)$$

We define the matrices E^{xy} as for $so\,(p, q)$.

We let

$$Z := \begin{pmatrix} qi & & & & & \\ & \ddots & & & & \\ & & qi & & & \\ & & & -pi & & \\ & & & & \ddots & \\ & & & & & -pi \end{pmatrix}$$

Z generates the center of $s\left(u\left(p\right) \times u\left(q\right)\right)$

$$X_o := \frac{1}{(p+q)\sqrt{2pq}} Z$$
$$X_{a'b'} := \frac{1}{2\sqrt{p+q}} \left(E^{a'b'} - E^{b'a'}\right)$$
$$Y_{a'b'} := \frac{i}{2\sqrt{p+q}} \left(E^{a'b'} + E^{b'a'}\right)$$

for $a', b' = 1, ...p$, $a' \neq b'$ and similarly $X_{a''b''} Y_{a''b''}$ for $a'', b'' = p + 1, ...p+q$, $a'' \neq b''$

$$X_{a\prime} := \frac{i}{\sqrt{2\left(p+q\right)}} \sqrt{\frac{a'}{a'+1}} \left(\frac{1}{a\prime} \sum_{c=1}^{a\prime} E^{cc} - E^{a\prime+1, a\prime+1}\right)$$

for $a\prime = 1, ...p-1$ and similarly $X_{a''}$ for $a'' = p+1, ...p+q-1$. The matrices $X_{a'}, X_{a'b}, Y_{a'b'}$ form an orthonormal basis of $su\left(p\right)$.

$$X_{ij} = \frac{1}{2\sqrt{p+q}} \left(E^{ij} + E^{ji}\right),$$
$$Y_{ij} = \frac{i}{2\sqrt{p+q}} \left(E^{ij} - E^{ji}\right)$$

for $i = 1, ...p$, $j = p+1, ...p+q$.

All the preceding matrices together form an orthonormal basis of $su\left(p, q\right)$. We now compute the structure constants with respect to this basis. Indices with a $\tilde{}$ will refer to the corresponding Y-matrix, those without to X-matrices. Thus *e.g.*

$$c_{ij,\tilde{kl}}^{a\prime b\prime} = 2\left(p+q\right) tr\left(\left[X_{ij}, Y_{kl}\right] X_{a\prime b\prime}\right)$$

We get

$$c^{a'b'}_{\widetilde{ij,kl}} = \frac{1}{2\sqrt{p+q}}\delta_{jl}\left(\delta_{ib'}\delta_{ka'} - \delta_{ia'}\delta_{kb'}\right) = c^{\widetilde{a'b'}}_{ij,kl}$$

$$c^{a''b''}_{ij,kl} = \frac{1}{2\sqrt{p+q}}\delta_{ik}\left(\delta_{jb''}\delta_{la''} - \delta_{ja''}\delta_{lb''}\right) = c^{\widetilde{a''b''}}_{\widetilde{ij,kl}}$$

$$c^{\widetilde{a'}\widetilde{b'}}_{\widetilde{ij,kl}} = \frac{-1}{2\sqrt{p+q}}\delta_{jl}\left(\delta_{kc'}\delta_{ib'} + \delta_{ia'}\delta_{kb'}\right) = -c^{\widetilde{a'b'}}_{\widetilde{ij,kl}}$$

$$c^{\widetilde{a''}\widetilde{b''}}_{\widetilde{ij,kl}} = \frac{-1}{2\sqrt{p+q}}\delta_{ik}\left(\delta_{jb''}\delta_{la''} + \delta_{ja''}\delta_{lb''}\right) = -c^{\widetilde{a''b''}}_{\widetilde{ij,kl}}$$

$$c^{a'}_{\widetilde{ij},kl} = \frac{-1}{\sqrt{2(p+q)}}\sqrt{\frac{a'}{a'+1}}\delta_{jl}\left(\frac{1}{a'_c}\sum_{c\le a'}\delta_{ic}\delta_{kc} - \delta_{i,a'+1}\delta_{k,a'+1}\right) = -c^{a'}_{ij,\widetilde{kl}}$$

$$c^{a''}_{\widetilde{ij},kl} = \frac{-1}{\sqrt{2(p+q)}}\sqrt{\frac{a''}{a''+1}}\delta_{ik}\left(\frac{1}{a''_c}\sum_{c=p+1}^{a''}\delta_{jc}\delta_{lc} - \delta_{j,a''+1}\delta_{l,a''+1}\right)$$

$$= -c^{a''}_{ij,\widetilde{kl}}$$

$$c^{0}_{\widetilde{ij},kl} = \frac{-1}{\sqrt{2pq}}\delta_{ik}\delta_{jl} = -c^{0}_{ij,\widetilde{kl}}$$

The remaining ones vanish.

In the sequel, summation over i, k will be from $1,\dots p$, over j, l from $p+1,\dots p+q$, over a', b' from 1, $\dots p-1$ with $b' > a'$, over a$''$, b'' from $p+1$ to $p+q-1$, with $b'' > a''$. We get for $a' \neq b'$

$$\sum_{ijkl}\left(c^{a'b'}_{ij,kl}c^{a'b'}_{ij,kl} + c^{\widetilde{a'b'}}_{\widetilde{ij,kl}}c^{\widetilde{a'b'}}_{\widetilde{ij,kl}}\right) = \frac{q}{p+q} =: a_1$$

$$\sum_{ijkl}\left(c^{a'}_{ij,\widetilde{kl}}c^{a'}_{ij,\widetilde{kl}} + c^{a'}_{\widetilde{ij},kl}c^{a'}_{\widetilde{ij},kl}\right) = \frac{q}{p+q} = a_1$$

for $a'' \neq b''$

$$\sum_{ijkl}\left(c^{a''b''}_{ij,kl}c^{a''b''}_{ij,kl} + c^{\widetilde{a''b''}}_{\widetilde{ij,kl}}c^{\widetilde{a''b''}}_{\widetilde{ij,kl}}\right) = \frac{p}{p+q} =: a_2$$

$$\sum_{ijkl}\left(c^{a''}_{ij,\widetilde{kl}}c^{a''}_{ij,\widetilde{kl}} + c^{a''}_{\widetilde{ij},kl}c^{a''}_{\widetilde{ij},kl}\right) = \frac{p}{p+q} = a_2$$

$$\sum_{ijkl}\left(c^{0}_{ij,\widetilde{kl}}c^{0}_{ij,\widetilde{k}} + c^{0}_{\widetilde{ij},kl}c^{0}_{\widetilde{ij},kl}\right) = 1 =: a_o$$

(the last equation also follows from general grounds as X_o generates the center of $s(u\,(p) \times u\,(q))$. Moreover,

$$\sum_{k,l,a'<b'} \left(c^{a'b'}_{ij,kl} c^{a'b'}_{mn,kl} + c^{\widetilde{a'}\widetilde{b'}}_{ij,\widetilde{kl}} c^{\widetilde{a'}\widetilde{b'}}_{mn,\widetilde{kl}} + c^{a'}_{ij,\widetilde{kl}} c^{a'}_{mn,\widetilde{kl}} \right)$$

$$= \frac{\delta_{jn}\delta_{im}}{2\,(p+q)} \frac{(p-1)\,(p+1)}{p} =: b_1$$

$$\sum_{k,l,a''<b''} \left(c^{a''b''}_{ij,kl} c^{a''b''}_{mn,kl} + c^{\widetilde{a''}\widetilde{b''}}_{ij,\widetilde{kl}} c^{\widetilde{a''}\widetilde{b''}}_{mn,\widetilde{kl}} + c^{a''}_{ij,\widetilde{kl}} c^{a''}_{mn,\widetilde{kl}} \right)$$

$$= \frac{\delta_{jn}\delta_{im}}{2\,(p+q)} \frac{(q-1)\,(q+1)}{q} =: b_2$$

$$\sum_{k,l} c^{0}_{ij,\widetilde{kl}} c^{0}_{mn,\widetilde{kl}} = \frac{\delta_{im}\delta_{jn}}{2pq} =: b_o$$

Then

$$2b_o + 2b_1 + 2b_2 = 1$$

which can also be derived from general grounds (cf. (18)).

The curvature tensor of $SU\,(p,q)\,/S\,(U\,(p) \times U\,(q))$ has components given by

$$R_{ij,kl,mn,rs} = -\left({}_a \sum_{a'<b'} c^{a'b'}_{ij,kl} c^{a'b'}_{mn,rs} +_a \sum_{a''<b''} c^{a''b''}_{ij,kl} c^{a''b''}_{mn,rs} \right)$$

$$R_{ij,\widetilde{kl},mn,\widetilde{rs}} = -\left(\sum_{a'} c^{a'}_{ij,\widetilde{kl}} c^{a'}_{mn,\widetilde{rs}} + \sum_{a''} c^{a''}_{ij,\widetilde{kl}} c^{a''}_{mn,\widetilde{rs}} + c^{0}_{ij,\widetilde{kl}} c^{0}_{mn,\widetilde{rs}} \right) \qquad \text{etc.}$$

We have by (19)

$$\sum_{\alpha,..\eta} R_{\alpha\beta\gamma\delta} R_{\eta\beta\gamma\delta} \langle f_\alpha, f_\eta \rangle = -\,(2a_o b_o + 2a_1 b_1 + 2a_2 b_2) \sum_{\alpha,\eta} R_{\alpha\eta} \langle f_\alpha, f_\eta \rangle$$

and thus

$$\mu := 2a_o b_o + 2a_1 b_1 + 2a_2 b_2 = \frac{2\,(1+pq)}{(p+q)^2}$$

Using symmetries, we get

(4.1)

$$
\begin{aligned}
-R\,P = & \sum_{ijkl} 4\, c^0_{ij,\widetilde{kl}} c^0_{mn,\widetilde{rs}} P_{ij,\widetilde{kl},mn,\widetilde{rs}} \\
& + \left\{ \sum_{ijkl,b'>a'} \left(c^{a'b'}_{ij,kl} c^{a'b'}_{mn,rs} P_{ij,kl,mn,rs} + c^{\widetilde{a'}\widetilde{b'}}_{\widetilde{ij},\widetilde{kl}} c^{\widetilde{a'}\widetilde{b'}}_{\widetilde{mn},\widetilde{rs}} P_{\widetilde{ij},\widetilde{kl},\widetilde{mn},\widetilde{rs}} \right.\right. \\
& + 2 c^{a'b'}_{ij,kl} c^{\widetilde{a'}\widetilde{b'}}_{\widetilde{mn},\widetilde{rs}} P_{ij,kl,\widetilde{mn},\widetilde{rs}} + \left. 4 c^{\widetilde{a'}\widetilde{b'}}_{ij,\widetilde{kl}} c^{\tilde{a}\tilde{b}}_{mn,\widetilde{rs}} P_{ij,\widetilde{kl},mn,\widetilde{rs}} \right) \\
& + \left. \sum_{ijkl,a'} 4 c^{a'}_{ij,\widetilde{kl}} c^{a'}_{mn,\widetilde{rs}} P_{ij,\widetilde{kl},mn,\widetilde{rs}} \right\} + \left\{ \sum_{ijkl,b''>a''} + \ldots \right\} \\
=: & \; a_o\|P_o\| + a_1\|P_1\| + a_2\|P_2\|
\end{aligned}
$$

Here, $\|P_o\|$ is the length of the projection of P onto the space spanned by the $c^0_{ij,\widetilde{kl}}$, whereas $\|P_1\|$ is the length of the projection of the space spanned by $c^{a'b'}_{ij,kl}$, ...$c^{a'}_{ij,\widetilde{kl}}$, and similarly $\|P_2\|$ with a$'$, b' replaced by a$''$, b''. Then

(4.3)

$$
\sum_{ijkl} \left(P_{ij,kl,ij,kl} + P_{\widetilde{ij},\widetilde{kl},\widetilde{ij},\widetilde{kl}} + 2P_{ij,\widetilde{kl},ij,\widetilde{kl}} \right) = \|P\| = \|P_o\| + \|P_1\| + \|P_2\|
$$

We want to derive the inequality

$$
-R \cdot P \leq \mu \|P\|, \tag{4.3}
$$

or equivalently,

$$
\begin{aligned}
\text{(4.4)} \quad & a_o\|P_o\| + a_1\|P_1\| + a_2\|P_2\| \\
& \leq (2b_o a_o + 2b_1 a_1 + 2b_2 a_2)\left(\|P_o\| + \|P_1\| + \|P_2\|\right)
\end{aligned}
$$

We recall

$$
\begin{aligned}
a_o &= 1 \\
a_1 &= \frac{q}{p+q} \\
a_2 &= \frac{p}{p+q} \\
b_o &= \frac{1}{2pq} \\
b_1 &= \frac{1}{2\,(p+q)} \frac{(p-1)\,(p+1)}{p} \\
b_2 &= \frac{1}{2\,(p+q)} \frac{(q-1)\,(q+1)}{q}
\end{aligned}
$$

We may assume without loss of generality

$$q \geq p$$

We compute

(4.5)

$$
a_o \|P_o\| = \sum_{ijkl} 4 c^0_{ij,\widetilde{kl}} c^0_{mn,\widetilde{rs}} P_{ij,\widetilde{kl},mn,\widetilde{rs}} = \frac{2}{pq} \sum_{ijkl} P_{ij,\widetilde{ij},kl,\widetilde{kl}}
$$

$$
\sum_{ijkl,b'>a'} \left(c^{a'b'}_{ij,kl} c^{a'b'}_{mn,rs} P_{ij,kl,mn,rs} + \ldots + 4 c^{\tilde{a}'\tilde{b}}_{ij,\widetilde{kl}} c^{\tilde{a}'\tilde{b}}_{mn,\widetilde{rs}} P_{ij,\widetilde{kl},mn,\widetilde{rs}} \right)
$$

$$
\begin{aligned}
&= \frac{1}{2\,(p+q)} \Bigg\{ \sum_{ijkl} \Big(P_{ij,kj,il,kl} + P_{\widetilde{ij},\widetilde{kj},\widetilde{il},\widetilde{kl}} + 2 P_{ij,kj,\widetilde{il},\widetilde{kl}} \\
&\qquad + 2 P_{ij,\widetilde{kj},il,\widetilde{kl}} + 2 P_{ij,\widetilde{kj},kl,\widetilde{il}} \Big) - 4 \sum_{ijk} P_{ij,\widetilde{ij},kj,\widetilde{kj}} \Bigg\}
\end{aligned}
$$

$$
\sum_{ijkl} 4 c^{a'}_{ij,\widetilde{kl}} c^{a'}_{mn,\widetilde{rs}} P_{ij,\widetilde{kl},mn,\widetilde{rs}}
$$

$$
= \frac{1}{2\,(p+q)} \left\{ 4 \sum_{ijk} P_{ij,\widetilde{ij},kj,\widetilde{kj}} - \frac{4}{p} \sum_{ijkl} P_{ij,\widetilde{ij},kl,\widetilde{kl}} \right\}
$$

hence

$$
\begin{aligned}
a_1\|P_1\| &= \frac{1}{2\,(p+q)}\Bigg(\sum_{ijkl}\Big(P_{ij,kj,il,kl} + P_{\widetilde{ij},\widetilde{kj},\widetilde{il},\widetilde{kl}} \\
&\quad + 2P_{ij,kj,\widetilde{il},\widetilde{kl}} + 2P_{ij,\widetilde{kj},il,\widetilde{kl}} + 2P_{ij,\widetilde{kj},kl,\widetilde{il}}\Big) \\
&\quad - \frac{2}{p\,(p+q)}\sum_{ijkl} P_{ij,\widetilde{ij},kl,\widetilde{kl}}
\end{aligned}
\tag{4.6}
$$

and similarly

(4.7)

$$
\begin{aligned}
a_2\|P_2\| &= \frac{1}{2\,(p+q)}\sum_{ijkl}\Big(P_{ij,il,kj,kl} + P_{\widetilde{ij},\widetilde{il},\widetilde{kj},\widetilde{kl}} + 2P_{ij,il,\widetilde{kj},\widetilde{kl}} \\
&\quad + 2P_{ij,\widetilde{il},kj,\widetilde{kl}} + 2P_{ij,\widetilde{il},kl,\widetilde{kj}}\Big) - \frac{2}{q\,(p+q)}\sum_{ijkl} P_{ij,\widetilde{ij},kl,\widetilde{kl}}
\end{aligned}
$$

In particular

(4.8)

$$
\begin{aligned}
-R\cdot P &= a_o\|P_o\| + a_1\|P_1\| + a_2\|P_2\| \\
&= \frac{1}{2\,(p+q)}\sum_{ijkl}\Big\{P_{ij,kj,il,kl} + P_{ij,il,kj,kl} + P_{\widetilde{ij},\widetilde{kj},\widetilde{il},\widetilde{kl}} + P_{\widetilde{ij},\widetilde{il},\widetilde{kj},\widetilde{kl}} \\
&\quad + 2P_{ij,kj,\widetilde{il},\widetilde{kl}} + 2P_{ij,il,\widetilde{kj},\widetilde{kl}} + 2P_{ij,\widetilde{kj},il,\widetilde{kl}} + 2P_{ij,\widetilde{il},kj,\widetilde{kl}} \\
&\quad + 2P_{ij,\widetilde{kj},kl,\widetilde{il}} \quad + 2P_{ij,\widetilde{il},kl,\widetilde{kj}}\Big\}
\end{aligned}
$$

This formula may also be computed with the help of the root space decompositions associated to the Cartan subalgebras of $su\,(p)$ and $su\,(q)$ generated by the $X_{a'}$ and $X_{a''}$ respectively.

We put $\nu = \frac{1}{2(p+q)}$ recall $\mu = \frac{2(1+pq)}{(p+q)^2}$ put

$$
\begin{aligned}
L_1\,(p) &= \sum_{ijkl}\Big\{P_{ij,kj,il,kl} + P_{ij,il,kj,kl} + P_{\widetilde{ij},\widetilde{kj},\widetilde{il},\widetilde{kl}} + P_{\widetilde{ij},\widetilde{il},\widetilde{kj},\widetilde{kl}} + 2P_{ij,kj,\widetilde{il},\widetilde{kl}} \\
&\quad + 2P_{ij,il,\widetilde{kj},\widetilde{kl}} + 2P_{ij,\widetilde{kj},il,\widetilde{kl}} + 2P_{ij,\widetilde{il},kj,\widetilde{kl}} + 2P_{ij,\widetilde{kj},kl,\widetilde{il}} + 2P_{ij,\widetilde{il},kl,\widetilde{kj}}\Big\} \\
L_2\,(p) &= \sum_{ijkl}\Big\{P_{ij,kl,ij,kl} + P_{\widetilde{ij},\widetilde{kl},\widetilde{ij},\widetilde{kl}} + 2P_{ij,\widetilde{kl},ij,\widetilde{kl}}\Big\}
\end{aligned}
$$

Thus, L_2 is the sum of all the sectional curvature terms. We have to show that

$$L_1(P) \leq \mu L_2(P) \tag{4.9}$$

for any positive (semi) definite curvature tensor P, or equivalently

$$L(P) = \mu L_2(P) - \nu L_1(P) \geq 0$$

for any such P. L assumes its minimum on the intersection of the convex cone C_+ of positive semidefinite curvature tensors and the set of curvature tensors of norm k. We want to show that this minimum value is zero and achieved at the tensor P^o corresponding to the identity map of M(*put* $k = norm$ of P^o for compatibility). P^o lies on the boundary of C_+, since for example most components with all indices i, j, k, l different vanish.

Since L is linear in P and C_+ is convex, it suffices to show that any variation of P^o not leaving C_+ will keep L nonnegative. In order to verify this, we shall investigate the effect of variations of the individual terms in L_1 and L_2. We first exhibit some other inequalities implying (4.9) which are likewise sharp for the identity map.

By Bianchi's identity and positive definitenes

(4.10)
$$\begin{aligned}
a_o\|P_o\| &= \frac{2}{pq}\sum_{ijkl} P_{ij,\widetilde{ij},kl,\widetilde{kl}} \\
&= \frac{2}{pq}\Bigg(\sum_{ijkl}\left(P_{ij,kl,\widetilde{ij},\widetilde{kl}} + P_{ij,\widetilde{kl},kl,\widetilde{ij}}\right) \\
&\leq \frac{1}{pq}\sum_{ijkl}\left(P_{ij,kl,ij,kl} + P_{\widetilde{ij},\widetilde{kl},\widetilde{ij},\widetilde{kl}} + 2P_{ij,\widetilde{kl},ij,\widetilde{kl}}\right) \\
&= \frac{1}{pq}\|P\| \\
&= 2b_o\|P\|
\end{aligned}$$

By Bianchi's identity

$$P_{ij,kj,il,kl} = P_{ij,il,kj,kl} + P_{ij,kl,il,kj}$$

$$P_{\widetilde{ij},\widetilde{kj},\widetilde{il},\widetilde{kl}} = P_{\widetilde{ij},\widetilde{il},\widetilde{kj},\widetilde{kl}} + P_{\widetilde{ij},\widetilde{kl},\widetilde{il},\widetilde{kj}}$$

$$P_{ij,kj,\widetilde{il},\widetilde{kl}} = P_{ij,\widetilde{il},kj,\widetilde{kl}} + P_{ij,\widetilde{kl},\widetilde{il},kj}$$

$$P_{ij,\widetilde{kj},il,\widetilde{kl}} = P_{ij,il,\widetilde{kj},\widetilde{kl}} + P_{ij,\widetilde{kl},il,\widetilde{kj}}$$

$$P_{ij,\widetilde{kj},kl,\widetilde{il}} = P_{ij,\widetilde{il},kl\widetilde{kj}} + P_{ij,kl,\widetilde{kj},\widetilde{il}}$$

Therefore, for $\lambda \in \mathbf{R}$,

$$
\begin{aligned}
& L_1(P) \\
&= \sum_{ijkl} \Big\{ \lambda \Big(P_{ij,kj,il,kl} + P_{\widetilde{ij},\widetilde{kj},\widetilde{il},\widetilde{kl}} + 2P_{ij,kj,\widetilde{il},\widetilde{kl}} + 2P_{ij,\widetilde{kj},il,\widetilde{kl}} + 2P_{ij,\widetilde{kj},kl,\widetilde{il}} \Big) \\
&\quad + (2-\lambda) \Big(P_{ij,il,kj,kl} + P_{\widetilde{ij},\widetilde{il},\widetilde{kj},\widetilde{kl}} \\
&\qquad + 2P_{ij,il,\widetilde{kj},\widetilde{kl}} + 2P_{ij,\widetilde{il},kj,\widetilde{kl}} + 2P_{ij,\widetilde{il},kl,\widetilde{kj}} \Big) \Big\} \\
&\quad + (1-\lambda) \Bigg\{ \sum_{ijk} P_{ij,kj,ij,kj} - \sum_{ijl} P_{ij,il,ij,il} + \sum_{i\neq k, j\neq l} P_{ij,kl,il,kj} \\
&\quad + 2\sum_{ijk} P_{ij,\widetilde{kj},\widetilde{ij},kj} - 2\sum_{ijk} P_{ij,\widetilde{il},ij,\widetilde{il}} + 2\sum_{i\neq k, j\neq l} P_{ij,\widetilde{kl},\widetilde{il},kj} - 2\sum_{i,j} P_{ij,\widetilde{il},\widetilde{il},ij} \\
&\quad + \sum_{ijk} P_{\widetilde{ij},\widetilde{kj},\widetilde{ij},\widetilde{kj}} - \sum_{ijl} P_{\widetilde{ij},\widetilde{il},\widetilde{ij},\widetilde{il}} + \sum_{i\neq k, j\neq l} P_{\widetilde{ij},\widetilde{kl},\widetilde{il},\widetilde{kj}} + 2\sum_{ijl} P_{ij,\widetilde{il},il,\widetilde{ij}} \\
&\quad + 2\sum_{ijk} P_{ij,\widetilde{kj},ij,\widetilde{kj}} - 2\sum_{i,j} P_{ij,\widetilde{ij},ij,\widetilde{ij}} + 2\sum_{i\neq k, j\neq l} P_{ij,\widetilde{kl},il,\widetilde{kj}} + 2\sum_{ijl} P_{ij,il,\widetilde{ij},\widetilde{il}} \\
&\quad + 2\sum_{ijk} P_{ij,kj,\widetilde{kj},\widetilde{ij}} + 2\sum_{i\neq k, j\neq l} P_{ij,kl,\widetilde{kj},\widetilde{il}} \Bigg\} \\
&=: \lambda\, L_1'(P) + (2-\lambda)\, L_1''(P) + (1-\lambda)\, L_1^r(P)
\end{aligned}
$$

Using Bianchi's identity again,

$$
\begin{aligned}
L_1^r(P) &= \sum_{ijk} P_{ij,kj,ij,kj} - \sum_{ijl} P_{ij,il,ij,il} + \sum_{ijk} P_{\widetilde{ij},\widetilde{kj},\widetilde{ij},\widetilde{kj}} - \sum_{ijl} P_{\widetilde{ij},\widetilde{il},\widetilde{ij},\widetilde{il}} \\
&\quad + \sum_{ijk} 2P_{ij,\widetilde{kj},ij,\widetilde{kj}} - \sum_{ijl} 2P_{ij,\widetilde{il},ij,\widetilde{il}} + \sum_{i\neq k, j\neq l} \Big(P_{ij,kl,il,kj} + P_{\widetilde{ij},\widetilde{kl},\widetilde{il},\widetilde{kj}} \\
&\quad + 2P_{ij,\widetilde{kl},\widetilde{il},kj} + P_{ij,\widetilde{kl},il,\widetilde{kj}} + 2P_{ij,kl,\widetilde{kj},\widetilde{il}} \Big)
\end{aligned}
$$

Then we conclude

(4.11)

$$a_o\|P_o\| + a_1\|P_1\| + a_2\|P_2\| = a_o\|P_o\| + \lambda a_1\|P_1\| + (2-\lambda)\, a_2\|P_2\|$$
$$+ \left(-\frac{2}{pq} + \frac{2\lambda p + 2(2-\lambda)\, q}{pq\,(p+q)}\right) \sum_{ijkl} P_{ij,\widetilde{ij},kl,\widetilde{kl}} + \frac{(1-\lambda)}{2\,(p+q)} L_1^r(P)$$

We put

$$\lambda = \frac{2p}{p+q}$$

Then

$$\lambda a_1 = (2-\lambda)\, a_1 = \frac{2pq}{(p+q)^2}$$

by (4.10)

Altogether

$$R\cdot P \le \frac{2pq}{(p+q)^2}\left(\|P_o\| + \|P_1\| + \|P_2\|\right)$$
$$+ \frac{2}{(p+q)^2}\|P\| + \frac{4\,(p-q)^2}{pq\,(p+q)^2} \sum_{ijkl} P_{ij,\widetilde{ij},kl,\widetilde{kl}} + \frac{q-p}{2\,(p+q)^2} L_1^r(P)$$

Therefore, since $\mu = \frac{2pq}{(p+q)^2} + \frac{2}{(p+q)^2}$, the estimate

$$RP \le \mu\|P\|$$

is implied by

(4.13)

$$\frac{4\,(p-q)^2}{pq\,(p+q)^2} \sum_{ijkl} P_{ij,\widetilde{ij}kl,\widetilde{kl}} + \frac{q-p}{2\,(p+q)^2} \Bigg\{ \Bigg(\sum_{ijk} P_{ij,kj,ij,kj} - \sum_{ijl} P_{ij,il,ij,il}$$
$$+ \sum_{ijk} P_{\widetilde{ij},\widetilde{kj},\widetilde{ij},\widetilde{kj}} - \sum_{ijl} P_{\widetilde{ij},\widetilde{il,ij,il}} + \sum_{ijk} 2P_{ij,\widetilde{kj},ij,\widetilde{kj}} - \sum_{ijl} 2P_{ij,\widetilde{il},ij,\widetilde{il}} \Bigg)$$
$$+ \sum_{i\neq k,\, j\neq l} \Big(P_{ij,kl,il,kj} + P_{\widetilde{ij},\widetilde{kl,il,kj}} + 2P_{ij,\widetilde{kl,il},kj}$$
$$+ 2P_{ij,\widetilde{kl},il,\widetilde{kj}} + 2P_{ij,kl,\widetilde{kj},\widetilde{il}} \Big) \le 0$$

Again, this estimate is sharp for the identity map.

Since

$$
\begin{aligned}
&2(2-\lambda)P_{ij,\widetilde{ij},il,\widetilde{il}} + 2\lambda P_{ij,\widetilde{ij},kj,\widetilde{kj}} + (1-\lambda)P_{ij,kj,ij,kj} - (1-\lambda)P_{ij,il,ij,il} \\
&+ (1-\lambda)P_{\widetilde{ij},\widetilde{kj},\widetilde{ij},\widetilde{kj}} - (1-\lambda)P_{\widetilde{ij},\widetilde{il},\widetilde{ij},\widetilde{il}} \\
&+ 2(1-\lambda)P_{ij,\widetilde{kj},ij,\widetilde{kj}} - 2(1-\lambda)P_{ij,\widetilde{il},ij,\widetilde{il}} \\
&\le P_{ij,kj,ij,kj} + P_{\widetilde{ij},\widetilde{kj},\widetilde{ij},\widetilde{kj}} + 2P_{ij,\widetilde{kj},ij,\widetilde{kj}} + P_{ij,il,ij,il} + P_{\widetilde{ij},\widetilde{il},\widetilde{ij},\widetilde{il}} + 2P_{ij,\widetilde{il},ij,\widetilde{il}}
\end{aligned}
$$

(4.13) in turn is implied by

(4.14)

$$
\begin{aligned}
&\frac{4(p-q)^2}{pq(p+q)^2}\sum_{ijkl} P_{ij,\widetilde{ij},kl,\widetilde{kl}} \\
&\quad + \frac{q-p}{2(p+q)^2}\left(\sum_{ijk} P_{ij,kj,ij,kj} + P_{\widetilde{ij},\widetilde{kj},\widetilde{ij},\widetilde{kj}} + 2P_{ij,\widetilde{kj},ij,\widetilde{kj}}\right. \\
&\quad \left. + \sum_{ijl} P_{ij,il,ij,il} + P_{\widetilde{ij},\widetilde{il},\widetilde{ij},\widetilde{il}} + 2P_{ij,\widetilde{il},ij,\widetilde{il}}\right) \\
&\quad + \frac{q-p}{2(p+q)^2}\sum_{i\neq k, j\neq l}\Big(P_{ij,kl,il,kj} + P_{\widetilde{ij},\widetilde{kl},\widetilde{il},\widetilde{kj}} \\
&\quad +2P_{ij,\widetilde{kl},\widetilde{il},kj} + 2P_{ij,\widetilde{kl},il,\widetilde{kj}} + 2P_{ij,kl,\widetilde{kj},\widetilde{il}}\Big) - \frac{2p}{(p+q)^2}\sum_{ijk} P_{ij,\widetilde{ij},kj,\widetilde{kj}} \\
&\quad - \frac{2q}{(p+q)^2}\sum_{ijk} P_{ij,\widetilde{il},ij,\widetilde{il}} \le 0
\end{aligned}
$$

which again is sharp for the identity map.

The estimate we want to show is recalled as

(4.15)

$$
\begin{aligned}
&\frac{1}{2(p+q)}\sum_{ijkl}\Big\{P_{ij,kj,il,kl} + P_{ij,il,kj,kl} + P_{\widetilde{ij},\widetilde{kj},\widetilde{il},\widetilde{kl}} + P_{\widetilde{ij},\widetilde{il},\widetilde{kj},\widetilde{kl}} + 2P_{ij,kj,\widetilde{il},\widetilde{kl}} \\
&+2P_{ij,il,\widetilde{kj},\widetilde{kl}} + 2P_{ij,\widetilde{kj},il,\widetilde{kl}} + 2P_{ij,\widetilde{il},kj,\widetilde{kl}} + 2P_{ij,\widetilde{kj},kl,\widetilde{il}} + 2P_{ij,\widetilde{il},kl,\widetilde{kj}}\Big\} \\
&\qquad \le \frac{2(1+pq)}{(p+q)^2}\sum_{ijkl}\left(P_{ij,kl,ij,kl} + P_{\widetilde{ij},\widetilde{kl},\widetilde{ij},\widetilde{kl}} + 2P_{ij,\widetilde{kl},ij,\widetilde{kl}}\right)
\end{aligned}
$$

and it follows from either (4.13) or (4.14).

All there estimates are sharp at the identity, and we are going to show that any variation of the tensor P^o associated with the identity not leaving the cone C_+ preserves at least one of these three inequalities, hence in particular (4.15). As explained above, this implies (4.15), the desired estimate.

We have to study several types of variations:

i) Increase of $|P_{ij,kj,il,kl}|$ for $i \neq k$, $j \neq l$: By positive definiteness

$$|P_{ij,kj,il,kl}| \leq \frac{1}{2} P_{ij,kj,ij,kj} + \frac{1}{2} P_{il,kl,il,kl}$$

Since this inequality is sharp at the identity, in order not to leave the cone C_+, an increase of $|P_{ij,kj,il,kl}|$ has to be matched by the increase of sectional curvature terms, making (4.15) strict since $\mu > \nu$. Alternatively, we observe that such terms do not occur in (4.13) or (4.14), hence an increase of them leaves these inequalities unaffected. This latter observations also holds for terms like $P_{ij,\widetilde{kj},il,\widetilde{kl}}$ with $i \neq k$, $j \neq l$. In this manner, all terms containing four different indices are handled.

ii) Increase of $P_{ij,kj,ij,kj}$, $P_{ij,\widetilde{kj},ij,\widetilde{kj}}$ $(i \neq k)$, etc. increases the right hand side of (4.15) more than the left hand side since $\mu > \nu$. Hence such an increase makes (4.15) strict.

iii) Decrease of $P_{ij,kj,ij,kj}$, $P_{ij,\widetilde{kj},ij,\widetilde{kj}}$ $(i \neq k)$, etc. makes (4.14) strict.

iv) $P_{ij,kj,\widetilde{ij},\widetilde{kj}}$ etc. does not occur in (4.15).

v) A decrease of $P_{ij,\widetilde{ij},ij,\widetilde{ij}}$ makes (4.13) strict.

vi) An increase of $P_{ij,\widetilde{ij},ij,\widetilde{ij}}$ makes (4.15) strict.

We now have handled all terms occuring in (4.15). Thus (4.15) is proved.

This is the desired estimate.

We now determine the implications of equality in (9) in case of rank 1. *i.e.* for p or q equal to 1.

The following result is a special case of [S] but we derive it here because a similar reasoning will lead to stronger results for quotients of quaternionic hyperbolic space or the hyperbolic Cayley plane.

Corollary 2. *Any harmonic map from a compact quotient of* $SU(p,q) /\ SU(p) \times SU(q)$ *into a Riemannian manifold* N *of non-*

positive curvature operator is pluriharmonic.

As explained in [C2], the curvature condition on N may be weakened to those occuring in [S] and [C2]. We recall that the preceding corollary allows an easy proof of Mostow's rigidity theorem for compact quotients of complex hyperbolic space. Namely, if the image is also a quotient of complex hyperbolic space then any harmonic map which has real rank at least 3 at some point is holomorphic or antiholomorphic. This is the easiest of the cases considered by Siu in [S]. If domain and image then both have the same dimension, such a map, if a homotopy equivalence, has to be a diffeomorphism and Royden's version of Yau's Schwarz lemma ([R], [Y]) implies that it is an isometry.

Proof of Corollary 2.

We use holomorphic normal coordinates $z^1 \ldots z^m$. We only need to consider the case $p = 1$, or $q = 1$ as otherwise the map is totally geodesic so that its Hessian vanishes. Pluriharmonicity is the weaker condition

$$f_{\alpha\bar{\beta}} = 0 \text{ for all } \alpha, \beta \tag{4.16}$$

(of course, $f_{\alpha\bar{\beta}} := \frac{\partial^2 f}{\partial z^\alpha \partial \bar{z}^\beta}$.) The argument, however, will work for any p and q. Now $\lambda_1 < 0$ in (9) whereas the curvature tensor of M is actually positive semifefinite when operating on tensors of the form $f_{\alpha\bar{\beta}}$. Therefore, equality in (9) implies (4.16).

The positive semidefiniteness can be seen as follows in our notation: We put

$$Z_{ij} = \frac{1}{2}\left(X_{ij} + Y_{ij}\right), Z_{\overline{ij}} = \frac{1}{2}\left(X_{ij} - Y_{ij}\right)$$

Then with indices $\underline{ij}$ corresponding to Z_{ij}, $\overline{ij}$ to $Z_{\overline{ij}}$. We then have to evaluate

$$
\begin{aligned}
&i^2\ R_{\underline{i}j,\overline{ij},\underline{k}l,\overline{kl}} < f_{\overline{ij},\underline{k}l}, \\
&\quad = -i^2 \sum_{\lambda} c^{\lambda}_{ij,\widetilde{\overline{ij}}}\, c^{\lambda}_{kl,\widetilde{\overline{kl}}} \langle f_{\overline{ij},\underline{k}l}, f_{ij,\overline{k\ell}} \rangle \\
&\quad = \sum_{\lambda} c^{\lambda}_{ij,\widetilde{\overline{ij}}}\, c^{\lambda}_{kl,\widetilde{\overline{kl}}} \langle f_{\overline{ij},\underline{k}l},\ f_{\underline{i}j,\overline{kl}} \rangle
\end{aligned}
$$

The above formulae for the structure constants show that this expression is always nonnegative. Since $R_{\alpha\beta\gamma\delta}$ is symmetric in the sense that it remains invariant if α, δ is exchanged against β, γ, this suffices to show positive semidefiniteness.

Remarks. 1) One may also see directly that $\lambda_1 < 0$ in (9) by evaluating the curvature tensor on pure derivatives. Namely

$$
i^2\ R_{\overline{ij},\underline{i}\underline{j},kl,\overline{kl}} \langle f_{\underline{i}j,\underline{k}l}, f_{\overline{ij},\overline{kl}} \rangle = -\sum_{\lambda} c^{\lambda}_{ij,\widetilde{\overline{ij}}}\, c^{\lambda}_{kl,\widetilde{\overline{kl}}} \langle f_{\underline{i}j,\underline{k}l}, f_{\overline{ij},\overline{kl}} \rangle
$$

and this is always nonpositive.

2) The results in the proof of Cor.2 and the previous remark on the curvature tensor of $SU(p,q)/SU(p) \times SU(q)$ are due to Calabi-Vesentini [CV] and Borel [B].

3.5. The symmetric space $Sp(p,q)/Sp(p) \times Sp(q)$ succumbs to a similar reasoning as the one for $SU(p,q)/S(U(p) \times U(q))$. The argument is computationally more conplicated, but conceptually simpler because $sp(p) \times sp(q)$ has trivial center. For this reason, we only sketch the argument, leaving out some computational details.

We shall adopt the following conventions:

If i is an index, we put

$$
i^{+} = i + p + q
$$

The indices a', b' will run from 1 to p, and so will i, k, m, r while j, l, m, s, a'', b'' will run from $p+1$ to q. The killing form of $sp(p,q)$ is given by

$$
B(A,B) = 2(p+q+1)\, tr\ AB
$$

We use the following basis for $sp\,(p,q)$: For a, $b = a'$, b' or a'', b''

$$
\begin{aligned}
X_a &= \frac{1}{2\sqrt{p+q+1}} i \left(E^{aa} - E^{a^+a^+}\right), \\
X_{a^+} &= \frac{1}{2\sqrt{p+q+1}} \left(E^{aa^+} - E^{a^+a}\right) \\
Y_{a^+} &= \frac{i}{2\sqrt{p+q+1}} \left(E^{aa^+} + E^{a^+a}\right) \\
X_{ab} &= \frac{1}{2\sqrt{2}\sqrt{p+q+1}} \left(\left(E^{ab} - E^{ba}\right) + \left(E^{a^+b^+} - E^{b^+a^+}\right)\right) \\
Y_{ab} &= \frac{1}{2\sqrt{2}\sqrt{p+q+1}} \left(i\left(E^{ab} + E^{ba}\right) - i\left(E^{a^+b^+} + E^{b^+a^+}\right)\right) \\
X_{ij} &= \frac{1}{2\sqrt{2}\sqrt{p+q+1}} \left(\left(E^{ij} + E^{ji}\right) - \left(E^{i^+j^+} + E^{j^+i^+}\right)\right) \\
Y_{ij} &= \frac{1}{2\sqrt{2}\sqrt{p+q+1}} \left(i\left(E^{ij} - E^{ji}\right) + i\left(E^{i^+j^+} - E^{j^+i^+}\right)\right) \\
X_{ab^+} &= \frac{1}{2\sqrt{2}\sqrt{p+q+1}} \left(\left(E^{ab^+} + E^{ba^+} - \left(E^{a^+b} + E^{b^+a}\right)\right)\right. \\
Y_{ab^+} &= \frac{1}{2\sqrt{2}\sqrt{p+q+1}} \left(i\left(E^{ab^+} + E^{-ba^+}\right) + i\left(E^{a^+b} + E^{b^+a}\right)\right) \\
X_{ij^+} &= \frac{1}{2\sqrt{2}\sqrt{p+q+1}} \left(\left(E^{ij^+} + E^{ji^+}\right) + \left(E^{i^+j} + E^{j^+i}\right)\right) \\
Y_{ij^+} &= \frac{1}{2\sqrt{2}\sqrt{p+q+1}} \left(i\left(E^{ij^+} + E^{ji^+}\right) - i\left(E^{i^+j} + E^{j^+i}\right)\right)
\end{aligned}
$$

In the sequel, an index with a will correspond to the corresponding matrix Y. For example

$$
c^{\tilde{a}\tilde{b}}_{ij^+,\widetilde{kl^+}} = 2\,(p+q+1)\,tr\,([X_{ij^+}, Y_{kl^+}]Y_{ab})
$$

The structure constants then are computed as

$$
\begin{aligned}
c^{a\prime}_{ij,\widetilde{kl}} = c^{a\prime}_{ij+,\widetilde{kl}+} &= -c^{a\prime+}_{ij,kl+} = c^{\widetilde{a'}\widetilde{b'}+}_{\widetilde{ij},kl+} \\
&= -c^{\tilde{a}\prime\tilde{+}}_{ij,\widetilde{kl}} = c^{\widetilde{a'}^{+}}_{\widetilde{ij},kl+} = \frac{1}{2\sqrt{p+q+1}}\delta_{jl}\delta_{ia\prime}\delta_{ka\prime} \\
c^{a\prime b\prime}_{ij,kl} = c^{\widetilde{a'}\widetilde{b'}}_{\widetilde{ij},\widetilde{kl}} = c^{a\prime b\prime}_{ij+,kl+} &= c^{\widetilde{a'}\widetilde{b'}}_{\widetilde{ij}+,\widetilde{kl}+} = \frac{\delta_{jl}}{2\sqrt{2}\sqrt{p+q+1}}\left(\delta_{ka\prime}\delta_{ib\prime} - \delta_{ia\prime}\delta_{kb\prime}\right) \\
c^{\tilde{a}\prime\tilde{b}\prime}_{ij,\widetilde{kl}} = c^{\tilde{a}\prime\tilde{b}\prime}_{ij+,\widetilde{kl}} &= -c^{a\prime b\prime+}_{ij,kl+} = c^{\widetilde{a'}\widetilde{b'}+}_{\widetilde{ij},\widetilde{kl}+} \\
&= c^{\tilde{a}\prime\tilde{b}\prime+}_{ij+,\widetilde{kl}} = -c^{\tilde{a}\prime\tilde{b}\prime+}_{ij,\widetilde{kl}+} = \frac{\delta_{jl}}{2\sqrt{2}\sqrt{p+q+1}}\left(\delta_{ka}\delta_{ib} + \delta_{ia}\delta_{kb}\right)
\end{aligned}
$$

The ones for a'', b'' are obtained from those for a' ,b' by interchanging j, l against i, k. The ones not listed here either equal those listed for symmetry reasons or vanish.

One has

$$
\begin{aligned}
&\sum_{\alpha,\beta} c^{a'b'}_{\alpha,\beta}\, c^{d'e'}_{\alpha,\beta} = \cdots = \sum_{\alpha\beta} c^{\tilde{a}^{\gamma}\tilde{b}^{\gamma}\tilde{+}}_{\alpha,\beta}\, c^{\tilde{d}^{\gamma}\tilde{e}^{\gamma}\tilde{+}}_{\alpha,\beta} = \frac{q}{2\,(p+q+1)}\delta_{a'd'}\delta_{b'e'} \\
&\sum_{\alpha,\beta} c^{a'}_{\alpha,\beta} c^{d'}_{\alpha,\beta} = \cdots = \frac{q}{2\,(p+q+1)}\delta_{a'd'}
\end{aligned}
$$

We thus have

$$a_1 = \frac{q}{2\,(p+q+1)}$$

and similarly

$$a_2 = \frac{p}{2\,(p+q+1)}$$

Likewise

$$b_1 = \frac{2p+1}{4\,(p+q+1)},\, b_2 = \frac{2q+1}{4\,(p+q+1)}$$

and

$$\mu = 2a_1b_1 + 2a_2b_2 = \frac{4pq+p+q}{4\,(p+q+1)^2}$$

One then computes

$$
\begin{aligned}
-\sum_{\alpha\beta\gamma\delta} & R_{\alpha\beta\gamma\delta} P_{\alpha\beta\gamma\delta} \\
&= \frac{1}{4(p+q+1)} \sum_{ijkl} \{ P_{ij,kj,il,kl} + \ldots + P_{ij+,kl+,il+,kl+} + \ldots \\
&\qquad + 2P_{ij+,kl+,\widetilde{il}+,\widetilde{kl}+} + \ldots \} \\
&\quad + \frac{1}{4(p+q+1)} \sum_{ijkl} \{ P_{ij,il,kj,kl} + \ldots + P_{ij+,il+,kj+,kl+} + \ldots \\
&\qquad + 2P_{ij+,il+,\widetilde{kj}+,\widetilde{kl}+} + \ldots \}
\end{aligned}
$$

with the same pattern as in the $SU(p,q)$ case.

Consequently, one may again trade off terms in the first bracket against those in the second one by means of the Bianchi identity and proceed as there. Since here, we do not have a center, the argument is conceptually simpler. We therefore omit the details as the pattern is clear.

For Theorem 3, one shows as in the proof of Cor.2 in 3.4 that second derivatives of f of the form *e.g*

$$
\begin{gathered}
(X_{ij} + iY_{ij})(X_{ij} - iY_{ij})f \\
(X_{ij} + iY_{ij+})(X_{ij} - iY_{ij+})f \\
(X_{ij+} + iY_{ij})(X_{ij+} - iY_{ij})f \\
(X_{ij+} + iY_{ij+})(X_{ij+} - iY_{ij+})f
\end{gathered}
$$

vanish. The representation theoretic argument of [C2; §1] can then be used to conclude the vanishing of the Hessian of f as in [C2; pf. of Theorem 3.3]

3.6. The hyperbolic Cayley plane is treated by the argument of 3.1 plus the analogue of the one at the end of 3.5. Again, one uses the result of [C2; §1] to conclude the vanishing of all second derivatives once the preceding arguments yield the vanishing of those of a certain type. Otherwise, the details are straightforward modifications of those presented above and hence omitted.

3.7. It remains to treat the spaces EII, EVI, EIX, FI, G (in the terminology of [He, p. 518] and [B, p. 316]). For all these spaces, k is the sum of two simple ideals

$$k = k_1 \oplus k_2 \quad \text{with} \quad k_1 = su\,(2)$$

Furthermore, k_1 acts on p as in the standard representation of su(2) on $\mathbf{C}^2$ (see the table in [B; p. 316]). Similar to the situation in **3.1**, we may then choose our basis X_λ so that

$$X_{m+1}, X_{m+2}, X_{m+3} \in z$$
$$X_{m+4}, \cdots, X_{m+k} \in k_2$$

and so that $ad_p X_{m+1}$, $ad_p X_{m+2}$, $ad_p X_{m+3}$ are represented by the matrices

$$\left(e^{m+1}_{\alpha\beta}\right)_{\alpha,\beta=1,\cdots m} = \frac{c_1}{\sqrt{m}}\begin{pmatrix} 0 & 1 & & & & 0 \\ -1 & 0 & & & & \\ & & \cdot & & & \\ & & & \cdot & & \\ & & & & 0 & 1 \\ 0 & & & & -1 & 0 \end{pmatrix}$$

$$\left(e^{m+2}_{\alpha\beta}\right)_{\alpha,\beta=1,\cdots m} = \frac{c_1}{\sqrt{m}}\begin{pmatrix} i & 0 & & & & 0 \\ 0 & -i & & & & \\ & & \cdot & & & \\ & & & \cdot & & \\ & & & & i & 0 \\ 0 & & & & 0 & -i \end{pmatrix}$$

$$\left(e^{m+3}_{\alpha\beta}\right)_{\alpha,\beta=1,\cdots m} = \frac{c_1}{\sqrt{m}}\begin{pmatrix} 0 & i & & & & 0 \\ i & 0 & & & & \\ & & \cdot & & & \\ & & & \cdot & & \\ & & & & 0 & i \\ 0 & & & & i & 0 \end{pmatrix}$$

for some $c_1 > 0$. We have

$$\sum_{\alpha,\beta} c^{\lambda}_{\alpha\beta} c^{\mu}_{\alpha\beta} = -tr\,(ad_p X_\lambda ad_p X_\mu)$$

Since

$$\begin{aligned} -S_{\lambda p} &= \beta\,(X_\lambda,\ X_\mu) = tr\,(ad_g X_\lambda ad_g X_\mu) \\ &= tr\,(ad_k X_\lambda ad_k X_\mu) + tr\,(ad_p X_\lambda ad_p X_\mu) \\ &= tr\,\left(ad_{su(z)} X_\lambda ad_{su(z)} X_\mu\right) + tr\,(ad_p X_\lambda ad_p X_\mu) \\ &\qquad \text{for } \lambda, \mu = m+1, m+2,\ m+3, \end{aligned}$$

the value of c_1 may easily be determined.

Furthermore, one easily checks that in all cases considered in the present subsection, we have

$$a_1 \geq a_2$$

since a_i is given by the difference between the Killing form of g restricted to k_i and the Killing form of k_i (of [M; p. 318]). As in **3.1**, we compute

$$\begin{aligned} &\sum_{\alpha,\eta} R_{\alpha\beta\gamma\delta} R_{\eta\beta\gamma\delta} \langle f_\alpha, f_\eta\rangle \\ &= \sum_{\alpha,\beta,\eta} \left(a_1 \sum_{i=1}^{3} c^{m+i}_{\alpha\beta} c^{m+i}_{\eta\beta} + a_2 \sum_{i=4}^{k} c^{m+i}_{\alpha\beta} c^{m+i}_{\eta\beta} \right) \langle f_\alpha,\ f_\eta\rangle \\ &= \sum_{\alpha,\beta,\eta} (a_1 - a_2) \sum_{i=1}^{3} c^{m+i}_{\alpha\beta} c^{m+i}_{\eta\beta} \langle f_\alpha,\ f_\eta\rangle - \sum_{\alpha,\eta} a_2\ R_{\alpha\eta} \langle f_\alpha,\ f_\eta\rangle \\ &= -\left(\frac{6}{m}(a_1 - a_2) + a_2\right) \sum_{\alpha,\eta} R_{\alpha\eta} \langle f_\alpha,\ f_\eta\rangle \\ &= \left(\frac{3}{m}(a_1 - a_2) + \frac{a_2}{2}\right) \sum_{\alpha} \langle f_\alpha,\ f_\alpha\rangle \end{aligned}$$

We thus have

$$\mu = \frac{6}{m}(a_1 - a_2) + a_2$$

Then as in **3.1**

$$-\sum_{\alpha,\cdots\delta} R_{\alpha\beta\gamma\delta}P_{\alpha\beta\gamma\delta} \leq \sum_{\alpha,\cdots\delta}(a_1-a_2)\sum_{i=1}^{3} c_{\alpha\beta}^{m+i}c_{\gamma\delta}^{m+i}P_{\alpha\beta\gamma\delta} + \sum_{\alpha,\beta} a_2\, P_{\alpha\beta\alpha\beta}$$

and for $i = 1, 2, 3$

$$\sum_{\alpha,\cdots\delta} c_{\alpha\beta}^{m+i}c_{\gamma\delta}^{m+i}P_{\alpha\beta\gamma\delta} \leq \frac{2}{m}\sum_{\alpha,\beta} P_{\alpha\beta\alpha\beta}$$

using the Bianchi identity.

Altogether

$$-\sum_{\alpha,\cdots\delta} R_{\alpha\beta\gamma\delta}P_{\alpha\beta\gamma\delta} \leq \left(\frac{6}{m}(a_1-a_2)+a_2\right)\sum_{\alpha,\beta} P_{\alpha\beta\alpha\beta} = \mu\sum_{\alpha,\beta} P_{\alpha\beta\alpha\beta}$$

which is the desired estimate.

It follows from these classification of irreducible symmetric spaces of noncompact type (see [He; p. 516 and p. 518])that the preceding cases **3.1** — **3.7** cover all such spaces.

References

[A1] S.I. Al'ber, *On n-dimensional problems in the calculus of variations in the large*, Sov. Math. Dokl., **5** (1964), 700-704.

[A2] S.I. Al'ber, *Spaces of mapping into a manifold with negative curvature*, Sov. Math. Dokl., **9** (1967), 6-9.

[B] A. Borel, *On the curvature tensor of the Hermitian symmetric manifolds*, Ann. Math., **71** (1960), 508-521.

[CV] E. Calabi and E. Vesentini, *On compact, locally symmetric Kähler manifolds*, Ann. Math., **71** (1960), 472-507.

[C1] K. Corlette, *Flat G-bundles with canonical metrics*, J. Diff. Geom., **28** (1988), 361-382.

[C2] K.Corlette, Archimedean superigidity and hyperbolic geometry, Ann. Math., **135** (1992), 165-182.

[DO] K. Diederich and T. Ohsawa, *Harmonic mappings and disk bundles over compact Kähler manifolds*, Publ. Res. Inst. Math. Sci. (Kyoto University), **21** (1985), 819-833.

[DZ] J. Dodziuk, *Vanishing theorems for square-integrable harmonic forms*, Proc. Indian Acad. Sc. (Math. Sc.), **90** (1981), 21-27.

[Dn] S. Donaldson, *Twisted harmonic maps and the self-duality equations*, Proc. London Math. Soc., **55** (1987), 127-131.

[ES] J. Eells and J. Sampson, *Harmonic mappings of Riemannian manifolds*, Amer. J. Math., **85** (1964), 109-160.

[G] M. Gromov, *The foliated Plateau problem* I, II, GAFA, **1** (1991).

[GS] M. Gromov and R. Schoen, *Harmonic maps into singular spaces and p-adic superrigidity for lattices in groups of rank one*, Publ. I.H.E.S., **76** (1992), 165-246.

[Ha] P. Hartman, *On homotopic harmonic maps*, Can. J. Math., **19** (1967), 673-687.

[He] S. Helgason, *Differential geometry, Lie groups and symmetric spaces*, Academic Press, 1978.

[JY1] J. Jost and S.T. Yau, *The strong rigidity of locally symmetric complex manifolds of rank one and finite volume*, Math. Ann., **271** (1985), 143-152.

[JY2] J.Jost and S.T. Yau, *On the rigidity of certain discrete groups and algebraic varieties*, Math. Ann., **278** (1987), 481-496.

[JY3] J. Jost and S.T. Yau, *Harmonic maps and group representations*, in: B.Lawson and K.Tenenblat (eds.), *Differential geometry and minimal submanifolds*, Longman Scientific, 1991.

[KN] S.Kaneyuki and T.Nagano, *On certain quadratic forms related to symmetric Riemannian spaces*, Osaka Math. J., **14** (1962), 241-252.

[L] F. Labourie, *Existence d' applications harmoniques tordues à valeurs dans les variétés à courbure négative.*

[Mg1] G.A. Margulis, *Discrete groups of motion of manifolds of nonpositive curvature*, AMS Transl., **190** (1977), 33-45.

[Mg2] G.A. Margulis, *Dircrete subgroups of semisimple Lie groups*, Springer, 1991.

[M] Y. Matsushima, *On the first Betti number of compact quotient spaces of higher-dimensional symmetric spaces*, Ann. Math., **75** (1962), 312-330.

[Mk1] N. Mok, *Aspects of Kähler geometry on arithmetic varieties*, Proc. Symp. Pure Math., Vol. 52, part 21, 335-396.

[Ms] G. Mostow, *Strong rigidity of locally symmetric spaces*, Ann. Math. Studies, **78**, Princeton, 1973.

[MSY] N. Mok, Y.T.Siu and S.K.Yeung, *Geometric supperrigidity*, Inv. Math., **113(1)** (1993), 57-83.

[R] M. Royden, *The Ahlfors-Schwarz lemma in several complex variables*, Comment. Math. Heler, **5.5** (1980), 547-558.

[Sa] J. Sampson, *Applications of harmonic maps toKähler geometry*, Contemp. Math., **49**, 125-134.

[S] Y.T. Siu, *The complex analyticity of harmonic maps and the strong rigidity of compact Kähler manifolds*, Ann. Math., **112** (1980), 73-111.

[SZ] R. Spatzier and R.Zimmer, *Fundamental groups of negatively curved manifolds and actions of semisimple groups*, Top., **30** (1991), 591-601.

[Y] S.T. Yau, *A general Schwarz lemma for Kähler manifolds*, Amer. J. Math., **100** (1978), 197-203.

[Z] R. Zimmer, *Ergodic theory and semisimple groups*, Birkhäuer, 1984.